Handbook of Nuclear Power Plants

Handbook of
Nuclear Power Plants

Edited by **Matt Fulcher**

*C*LANRYE
INTERNATIONAL

New Jersey

Published by Clanrye International,
55 Van Reypen Street,
Jersey City, NJ 07306, USA
www.clanryeinternational.com

Handbook of Nuclear Power Plants
Edited by Matt Fulcher

International Standard Book Number: 978-1-63240-282-0 (Hardback)

Printed in the United States of America.

Contents

Preface   VII

Chapter 1   **Strategic Environmental Considerations of Nuclear Power**   1
Branko Kontić

Chapter 2   **Investigation on Two-Phase Flow Characteristics in Nuclear Power Equipment**   25
Lu Guangyao, Ren Junsheng, Huang Wenyou,
Xiang Wenyuan, Zhang Chengang and Lv Yonghong

Chapter 3   **Analysis of Emergency Planning Zones in Relation to Probabilistic Risk Assessment and Economic Optimization for International Reactor Innovative and Secure**   43
Robertas Alzbutas, Egidijus Norvaisa and Andrea Maioli

Chapter 4   **Deterministic Analysis of Beyond Design Basis Accidents in RBMK Reactors**   61
Eugenijus Uspuras and Algirdas Kaliatka

Chapter 5   **Evolved Fuzzy Control System for a Steam Generator**   95
Daniela Hossu, Ioana Făgărăşan,
Andrei Hossu and Sergiu Stelian Iliescu

Chapter 6   **The Gap Measurement Technology and Advanced RVI Installation Method for Construction Period Reduction of a PWR**   113
Do-Young Ko

Chapter 7   Cross-Flow-Induced-Vibrations
in Heat Exchanger Tube Bundles: A Review   145
Shahab Khushnood, Zaffar Muhammad Khan,
Muhammad Afzaal Malik, Zafarullah Koreshi,
Muhammad Akram Javaid, Mahmood Anwer Khan,
Arshad Hussain Qureshi, Luqman Ahmad Nizam,
Khawaja Sajid Bashir, Syed Zahid Hussain

Chapter 8   Radiochemical Separation of Nickel for ^{59}Ni and ^{63}Ni
Activity Determination in Nuclear Waste Samples   195
Aluísio Sousa Reis, Júnior, Eliane S. C. Temba,
Geraldo F. Kastner and Roberto P. G. Monteiro

Chapter 9   Radiobiological Characterization Environment
Around Object "Shelter"   209
Rashydov Namik, Kliuchnikov Olexander, Seniuk Olga,
Gorovyy Leontiy, Zhidkov Alexander, Ribalka Valeriy,
Berezhna Valentyna, Bilko Nadiya, Sakada Volodimir, Bilko Denis,
Borbuliak Irina, Kovalev Vasiliy, Krul Mikola and Georgy Petelin

Chapter 10   Analysis of Primary/Containment Coupling
Phenomena Characterizing the MASLWR Design
During a SBLOCA Scenario   257
Fulvio Mascari, Giuseppe Vella, Brian G. Woods,
Kent Welter and Francesco D'Auria

Chapter 11   AREVA Fatigue Concept – A Three Stage Approach to
the Fatigue Assessment of Power Plant Components   285
Jürgen Rudolph, Steffen Bergholz,
Benedikt Heinz and Benoit Jouan

Chapter 12   Phase Composition Study of Corrosion Products at NPP   309
V. Slugen, J. Lipka, J. Dekan, J. Degmova and I. Toth

Permissions

List of Contributors

Preface

This book has been an outcome of determined endeavour from a group of educationists in the field. The primary objective was to involve a broad spectrum of professionals from diverse cultural background involved in the field for developing new researches. The book not only targets students but also scholars pursuing higher research for further enhancement of the theoretical and practical applications of the subject.

Nuclear power plant is described as a power plant in which nuclear energy is converted into heat for use in producing steam. This book caters to different issues, ranging from thermal-hydraulic analysis to the safety analysis of a nuclear power plant. The objective of the book is to discuss the challenging ideas that can be executed in and used for the development and advancement of future nuclear power plants. This book will expose readers to the world of innovative research and advancement of future plants.

It was an honour to edit such a profound book and also a challenging task to compile and examine all the relevant data for accuracy and originality. I wish to acknowledge the efforts of the contributors for submitting such brilliant and diverse chapters in the field and for endlessly working for the completion of the book. Last, but not the least; I thank my family for being a constant source of support in all my research endeavours.

Editor

Strategic Environmental Considerations of Nuclear Power

Branko Kontić
Jožef Stefan Institute
Slovenia

1. Introduction

The key topics of this chapter are i) comparative evaluation of various energy options, and ii) radioactive waste disposal. Both are treated from the strategic planning and assessment points of view and are supported by a discussion of multi-objective decision-making. Environmental considerations are foremost. The discussion is focused on the uppermost level of societal energy planning, and attempts to answer strategic questions concerned with the comparative evaluation of various energy options and waste disposal. It is guided by a number of questions as illustrated in Table 1. The Table also indicates in which sub-chapter a certain, more specific discussion can be found.

The author is a natural scientist, experienced in research and preparation of different types of environmental impact and risk assessments. At the present time – January 2012 - after more than 30 years of practice in the field he is astonished by the increasing inefficiency of formal guidance on evaluation of environmental impacts. He wonders why is this so and is especially disappointed when seeing that even the highest administrative level EU institutions, the DG Environment and DG Regional Policy, do not succeed in implementing the guides on performing strategic environmental assessments. For example, the DG Regional Policy and Cohesion provided a guide for the ex-ante evaluation of the environmental impact of regional development programmes in 1999 (EC, 1999) as complementary to the Handbook on Environmental Assessment of Regional Development Plans and EU Structural Fund Programmes (EC, 1998). These were a kind of predecessor of the EU Directive 2001/42/EC (usually referred to as the strategic environmental assessment - SEA Directive). Despite the fact that the guides clearly stress the importance of establishing an interactive relationship between evaluation and planning – the objective of the integration is to improve and strengthen the final quality of the plan or programme under preparation – more than 10 years afterwards Member States fail to follow them and report on a number of difficulties in SEA implementation (EC, 2009). The most important deficiency in the current practice of SEA in certain EU countries is still the approval/permitting context of the use of SEA instead of the planning context and optimisation of plans, and the mixed use (misuse) of project level environmental impact assessment - EIA and SEA. SEA is very often used for the evaluation of specific projects, while EIA is used at higher, i.e. strategic, levels, sometimes even for the evaluation of sustainability of plans and programmes (Kontić & Kontić, 2011). This situation stimulated the author to prepare the present condensed overview of research and consultancy results on strategic considerations of nuclear power. His aim is that this will contribute to the desired change of implementation of strategic evaluation in the area of energy production and elsewhere.

Comparative information about the environmental impacts of various energy systems can assist in the evaluation of energy options and consequent decision making. Over the last thirty years a number of studies have attempted to quantify such impacts for a wide range of energy sources. These estimations have taken different approaches, from impacts of fuel acquisition through to waste disposal (IAEA, 2000). Recent major studies have been completed and new studies begun in which nuclear power is either supported – justification through e.g. climate change issues or low-carbon society – or criticised – justification through e.g. accidents at Chernobyl and Fukushima, or waste related issues. The results of the studies provide useful insights and help to promote further studies of impacts for many technologies and sites. However, the strategic level of these considerations still remains less well covered and a number of questions are still unanswered. This chapter is aimed as a contribution to filling these gaps.

Related to the radioactive waste issue, the siting of a disposal facility or final repository is a task with unique traits that are clearly associated with changes in the surrounding world. A number of questions can be posed regarding how ongoing and future changes in technology, views, politics and practices in other parts of the world, concerning e.g. energy supply, nuclear power and nuclear waste, may affect national decisions regarding the approach and decisions involved in successful and safe disposal of the waste. National trends in politics, economy and opinion also influence events and views, locally and nationally (SKB, 2011). The decision-making process has to fulfil certain democratic expectations and criteria: openness, transparency, participation. So far, known and applied approaches have not been efficient or effective in solving the primary issue of participatory decision-making in this area, i.e. proper, fair and balanced consideration of specific priorities and interests. Neither weight assignment, as a representative method rooted in (expert) opinion and value judgements, nor methods based on statistics and probability theory (applicable for measurable attributes) have proved successful for this purpose. Maybe 'approval voting' (Laukkanen et al., 2002) is the closest to what is widely understood as participative/democratic decision making. It appears, on the other hand, that a continuous engagement process, sound and consistent, scientifically supported and respected by all involved parties, which deals adequately with uncertainties related to long-term predictions/evaluations – as applied in Finland and Sweden – can provide satisfactory results (SKB, 2011). The approach applied in Slovenia for identifying and approving a site for a low and intermediate level radioactive waste disposal facility could also be seen as being successful, and is presented in more detail in Section 3. In summary, it builds on social acceptance of predictive uncertainty based on so-called "local partnership" i.e. the community is actively involved in the siting process and has a right of *veto*, together with a comprehensive investigation of the perceptions of the types of consequences rather than the likelihood of their occurrence. The underlying basis of the approach is that it is more promising to investigate which consequences of a certain alternative are more likely to be accepted by society than how likely these consequences are to occur. Thus, as many feasible alternatives as possible should be evaluated, so that the parties involved can express their preferences rather than just "yes/no", or "accept/reject" responses. This is clearly in line with the basic philosophy of SEA and strategic considerations of nuclear power.

Questions/Issues	Comments/Specification
What are the energy needs? What are the energy issues? What are the strategic energy goals?	The questions are inter-connected. At the country level these questions need to be answered in a solid, transparent and inter-disciplinary way. It is the responsibility of politics to ensure full and proper involvement of societal* planners in answering these questions. In the process of answering the questions it is necessary to know where to get information/data and who to involve; the answers should be reliable, valid, and trustworthy. See subchapter 2.1.
Spatial planning and strategic environmental assessment; Territorial impact assessment	Energy policy should be integrated with spatial planning procedures at high planning levels. Planning and strategic environmental impact evaluations should be integrated. See subchapter 2.1.
What are the expected outcomes of strategic considerations? What forms of auditing have to be implemented to achieve trust in the answers about strategic policy? Who are the decision-makers?	Early involvement of interested parties, early input by decision-makers with their guiding elements, and clarification/agreement on representation issues associated with different social groups should be resolved and implemented in the process of creating a trustworthy energy policy. See subchapter 2.1.
Why choose nuclear technology? Is nuclear power a good choice?	Solid and transparent comparative assessment of the various options should first be made on the strategic level, i.e. without detailed information on environmental status at potential sites for different options. This requires proper comparative environmental indicators. For example, indicators on specific air emission from different technologies (e.g., radioactivity from NPPs, and CO_2 from coal fired power plants) should not be directly used for comparison. Rather, common consequences in the environment, which these emissions may cause, should be the subject of comparison. See subchapters 2.1 and 2.2.
Which uncertainties have to be considered when deciding about energy options? Is trustworthiness of planners and scientists just another imperative? How to distinguish between facts and values? What is the role and credibility of regulators in the process of approving long-term predictions of environmental and health impacts?	At least the sources and types of uncertainty should be clearly explained when quantification is not feasible (e.g., long-term future predictions cannot be checked/verified at the present time, so performance assessment results of a particular radioactive waste repository for the next million years cannot be quantified, either in terms of environmental or societal changes). Scientific truth related to siting of the repository should be tested in the communication process at international, regional and local levels. See subchapters 3.1, 3.2, and 3.3.

* By societal planning is here meant an integration of all sectoral planning, including environmental.

Table 1. Questions and issues in strategic considerations of nuclear power

2. Comparative evaluation of environmental impacts of various energy systems

2.1 Energy planning, assessment and decision-making

In a very general terms, when one gets involved in planning it is strongly recommended to consult the theoretical background to the topic and its integration with strategic evaluation. As an initial and philosophical reading one may choose Nigel Taylor's article Planning theory and the philosophy of planning (Taylor, 1980) where the author provides an overview and explanation of the relationship between values and facts and the logical distinction that can be made (and thus between ethics and knowledge). The sections on Ethics and Planning, and Knowledge and Planning, clearly explain the reasoning necessary when making strategic choices related to development plans.

2.1.1 Key parties involved

The evolution towards a more comprehensive approach to electricity system planning emerged from a broader recognition of the need to identify the broad social responsibility of the power sector. The concept of social responsibility covers a number of issues ranging from local employment to rational exploitation of national resources. It implies a comprehensive analysis of natural resource requirements and social, health and environmental impacts arising at all steps of the energy chains constituting the electricity generation system (IAEA, 2000).

Integration of the power system analysis and planning process within the social and economic context can be considered as a shift from minimising costs (i.e. direct cost of electricity production) to maximising effectiveness. The concept of maximising effectiveness should be understood, in a broad sense, as an attempt to find solutions optimised from the view point of society as a whole. In this context, the planning process is aimed at seeking the preferred supply and demand side options and the strategies for solving present problems in the power sector (e.g. supply shortages, high costs with unclear externalities, non-compliance with environmental policy goals and regulations). This, at the same time as addressing various objectives of the electricity utilities, integrates the various actors in the energy and other economic sectors and, more generally, all interested and affected parties (IAPs) (IAEA, 1999; IAEA, 2000).

This shift in emphasis requires a comprehensive consideration of the overall objectives underlying the development of the power sector and of the parameters (attributes), data and assumptions that have to be taken into account in analysing alternative technologies for electricity production, and the electricity system as a whole. In particular, the power sector has to be analysed as one part within the overall economic and social context (URS, 2010).

In recent years, the traditional utility oriented decision making process has changed to involve a larger number of actors. Figure 1 shows a schematic diagram of the respective roles and responsibilities of the three main groups of actors involved in the overall planning, assessment and decision-making process. Decision makers have the key responsibility for identifying the problems needing solution and for choosing from among the possible solutions derived by decision support studies, according to their own values and priorities, as well as the political and social context. Interested and affected parties have an important role to play in the overall process, and their viewpoints and concerns have to be recognised and taken into account, insofar as is feasible, at each step, starting at the very beginning. The role of electricity analysts/planners is to formulate the decision maker's problems in an analytical framework and to derive alternative possible solutions, taking into account relevant constraints (e.g. emission limits, public health goals, land-use interests) imposed by regulators and concerns expressed by IAPs (IAEA, 2000).

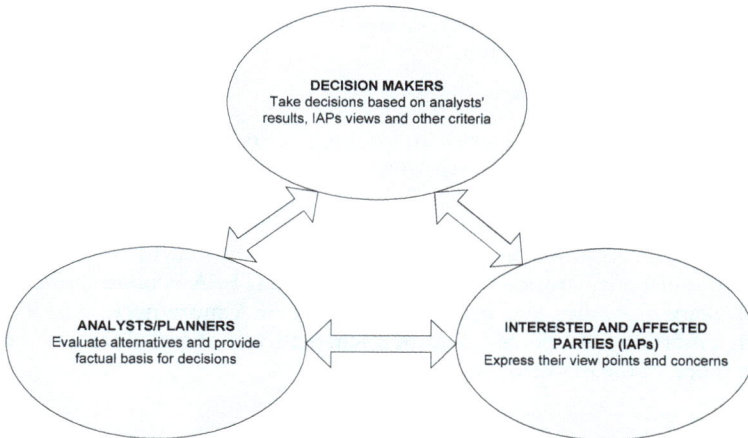

Fig. 1. Schematic diagram of interactions in the decision making process (IAEA, 2000)

2.1.2 Planning and strategic assessment

The production and consumption of electricity lead to environmental impacts which must be considered in making decisions on the way in which to develop energy systems and energy policy. The key to moving towards rational energy development lies in finding the 'balance' between the environmental, economic and social goals of society and integrating them at the earliest stages of project planning, programme development and policy making. The environmental consequences of energy production and use must be known in order to manage and choose energy products and services. The requirements for information in support of corporate and/or government planning and decision making are changing, there being a clear emergence of concerns for environmental accountability. Thus, there is a need to integrate the environment more effectively into all aspects of energy planning and decision making, in order to make current decisions environmentally prudent, economically efficient and socially equitable, both now and for the future. Assessing environmental impacts associated with different energy systems through the use of a framework which facilitates comparison will permit consistent and transparent evaluation of these energy alternatives.

Tiering of environmental evaluation

Appraising sustainability

Sustainability appraisal (SA) has recently emerged as a policy tool whose fundamental purpose is to direct planning and decision-making towards sustainability. Its foundations lie in well-established practices such as strategic environmental assessment (SEA), applied to policies, plans and programmes, and in project environmental impact assessment (EIA). The distinguishing feature of sustainability appraisal, when compared with others, e.g. SEA, is that the concept of sustainability, not just the environment, lies at its core. However, as explained below, comprehensive SEAs also deal with all three components – environment, economy and society - in a balanced way. No matter which type of assessment is applied at the highest planning level, either SA or SEA, its aim is to provide answers in a comparative manner and to assist in the process of identifying the most suitable alternative, e.g. energy option.

Strategic environmental assessment

SEA is a systematic process for evaluating the environmental consequences and for identifying the adverse effects of emerging environmental and/or health risks of a proposed policy, plan or programme. This is necessary in order to ensure that they are fully included and appropriately addressed at the earliest appropriate stage of decision making, on a par with economic and social considerations. As such, SEA may also include social and economic considerations. Due to these features SEA is often interchanged with SA, however, some countries and practitioners make SEA more narrow in its scope and almost purely environment oriented. Figure 2 schematically shows different combinations of depth and scope of the assessment.

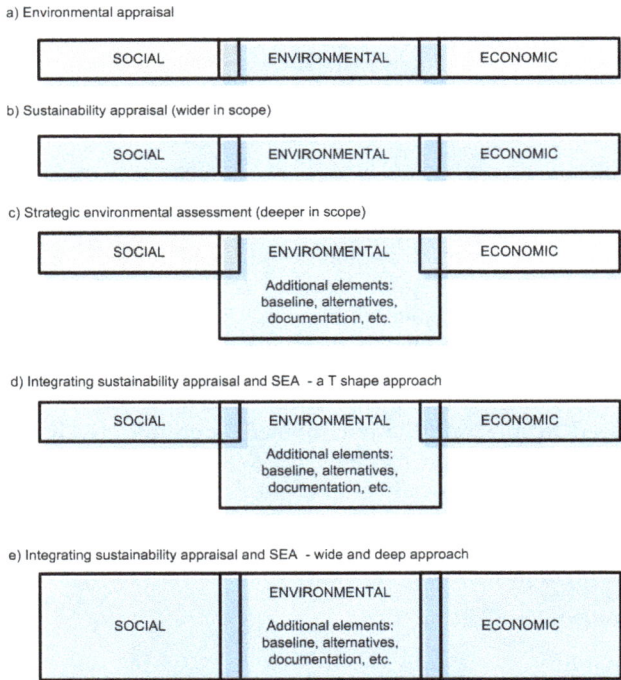

Fig. 2. Evolution from environmental appraisal to comprehensive/integrative SEA (Therivel 2005)

SEA deals with impacts that are difficult to consider at the project level. It deals with cumulative and synergistic impacts of technologies or multiple projects. This is very difficult to address by individual project oriented EIAs.

SEA promotes a better consideration of alternatives and affects the decision-making process at a stage where more alternatives are available for consideration. The following characteristics of SEA should be recognised (Therivel, 2005):

1. SEA is a tool for improving the strategic action, not a post-hoc snapshot;
2. In order to fit into the timescale and resources of the decision-making process, SEA should focus on key environmental/sustainability constraints, thresholds and limits at the appropriate plan-making level. It should not aim to be as detailed as a project oriented environmental impact assessment;
3. SEA aims to identify the best alternative for the development and implementation of policies, plans and programmes;
4. SEA aims to minimize negative impacts, optimize positive ones, and to compensate for the loss of valuable (environmental and other) features and benefits.

Project related environmental impact assessment (EIA)

EIA is the selected technology and location linked consideration. Environmental assessment is specific, concrete, and deep. The endpoint is to determine clearly the environmental changes in terms of their scope, intensity and tolerability. Risks are assessed quantitatively. Very specific indicators of environmental quality may be applied.

Integration of strategic planning and environmental evaluation

Figure 3 provides a synthesis of the desired integration between strategic planning and tiering environmental evaluations. A brief overview of present issues and their possible resolution at different planning stages is also given. One should not overlook the importance of a loop from the fifth planning step (Plan implementation; Licensing) back to the step 2a informing all planning steps between success and issues in the plan implementation. This loop actually acts as a special form of historical monitoring of the plan implementation.

Comparative evaluation approach and its indicators

Multi-objective analysis (MOA) is aimed at facilitating comprehensive and consistent consideration, comparison and trade-offs of economic (financial), supply security, social, health and environmental attributes of selected alternative energy options or systems (could also be technologies for electricity production). These technologies are usually classified as thermal and non-thermal, or renewable and non-renewable, and include nuclear, coal, natural gas, biomass, hydro, PV-Photo Voltaic, and wind systems. MOA is expected to assist in the systematic evaluation of options according to multiple objectives/criteria which are different and which may not be measured on an interval (or even ordinal) scale. It should be understood that MOA is not primarily a method that can be used to derive impacts, but rather a method that places different types of impact on a comparable basis and facilitates comparisons between impacts originally estimated and expressed in different units (IAEA, 2000).

Environmental Assessment - Sustainability and/or strategic level - Project level	Development/Spatial Planning process - Economic; societal development - Spatial (land-use) organisation and licensing, integration of economic development and environmental protection goals	Issues & Resolution

Environmental Assessment column (flowchart boxes):

Sustainability Assessment (SA)

Analysis of the development needs, technological options and environmental limitations

Strategic Environmental Assessment (SEA)

Existing knowledge about new emerging technology/risks

Identification and analysis of technological and spatial (site) alternatives

Detailed information about hazard/risk for a new technology

Preliminary allocation of new technology based on SEA including risk analysis

Environmental Impact Assessment (EIA)

Detailed environmental impact assessment and risk analysis on a project level

Analysis of the impact synergy of multiple project developments

Development/Spatial Planning process column (flowchart steps):

STEP 1 — Strategy, initiatives, goals, possibilities/options

STEP 2a — General definition and terms of new emerging technology/risk

STEP 2b — Plan development, definition of alternatives — Draft plan, description of alternatives;

Preliminary environmental analysis; need for additional information on technology, hazards, risks

STEP 3 — Specification of the plan — Plan proposal justified by SEA including risk analysis

STEP 4 — Negotiations, plan confirmation — Plan approval

STEP 5 — Project development proposal — Plan implementation; Licensing

Issues & Resolution column:

Present issue: Environmental issues/risks are not being considered at this highest planning level stage.
Resolution: Development proponents should present the development needs through strategies, goals and options. These needs should be checked by sustainability assessment and approved in the context of societal development. Planners should provide integration of sustainability and strategic environmental assessment.

Present issue: Plans do not deal with alternatives comprehensively (adequately) – lack of resources, time consuming, no Cost- Benefit Analysis (CBA).
Resolution: Reservation/assurance of resources for the analysis of alternatives should be a requirement in the planning process. Comparative assessment of alternatives is a key for decision-making and final justification.

Present issue: Lack of systematic and clear (transparent) justification of the plan proposals.
Resolution: SA & SEA, when integrated with planning, act as the key source of information for justifying plan proposal from the earliest planning stages

Present issue: EIA has no potential, role or power to justify the project.
Resolution: EIA should act as the final justification step in the tiering process of environmental assessment: SA – SEA – EIA. Consequently, the licensing process should be transparent and is expected to be widely accepted as the societal control of desired development with fewer conflicts, quicker implementation of economic benefits, etc.

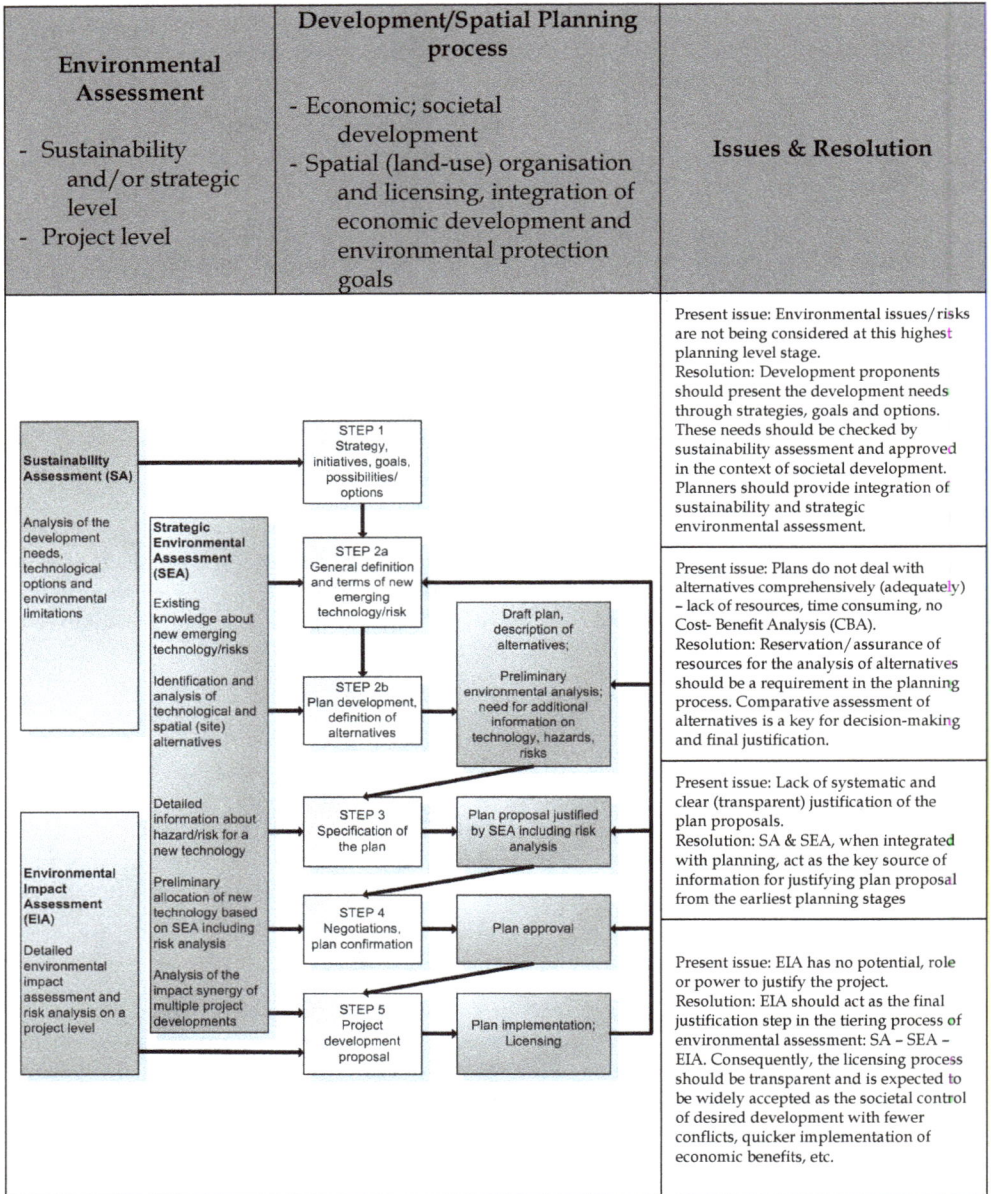

Note: The term "new emerging technology/risk" relates to any new development having a potential for causing environmental damage, e.g. a new generation of nuclear reactors, hydrogen based fuel technology, nano technology, etc.

Fig. 3. A schematic presentation of a possible integration of the development and spatial/land-use planning process with environmental evaluations (SA, SEA, EIA)

The main objectives of MOA are:

- to provide quantitative information where it is difficult to quantify the impacts directly;
- to display risk–benefit trade-offs that exist between different impact indicators;
- to facilitate comparisons and trade-offs;
- to facilitate understanding of the 'values' that need to be placed on different attributes.

The impact of each option under consideration should be represented using the units of measurement appropriate for each indicator or attribute. For example, impact indicators could be:

- The proportion of area utilized in the area (e.g. as a measure of land use impacts associated with each option referring to shares of existing and planned land-use);
- Health determinants affected/changed due to implementation of the alternative energy option.

Table 2 indicates a set of aggregated indicators; these need to be developed further into measurable (possibly quantifiable) sub-indicators, so as to enable clear, verifiable, reproducible, and transparent evaluation. How this could be done in a comprehensive and transparent manner shows the example of Eurelectric RESAP - Renewables Action Plan (Eurelectric, 2011); the WG Environmental Management and Economics of the Eurelectric RESAP was tasked with an evaluation, based on existing literature – 296 selected worldwide studies - of the sustainability of renewable energy sources (RES) and other technologies over their whole life cycle (IPCC, 2011). The quantitative indicators applied in comparative evaluation were, e.g., carbon footprint, health impacts, water use, land use, biodiversity, raw materials, energy payback, etc. No matter the approach of selecting the indicators, caution should be exercised to ensure that the sub-indicators are chosen on the basis of:

- Relevance: indicators should reflect the overall objectives of the study;
- Directionality: indicators must be defined in a manner that ensures that their magnitude can be assessed and interpreted. This can be accomplished by specifying indicator measurement in terms of maximizing or minimizing, increasing or maintaining, etc.;
- Measurability: it should be possible to measure quantitatively or estimate directional impacts of each alternative on each indicator, in the unit of measurement that is appropriate for the indicator. Directionality and measurability together determine interpretability, i.e. they permit an interpretation of impacts as being good/bad or better/worse on each indicator;
- Manageability: in order to make assessments comprehensible and to facilitate effective comparison, the number of sub-indicators should not be too large.

Once the impact analyses have been consolidated, all the data should be expressed in a common metric, or 'standardized', so that the indicators can be compared and assessed. For example, impact indicators can be presented on an interval scale (e.g. from 0 to 1). The scale would indicate the relative effect of each fuel chain option being considered, on the basis of the relative magnitude of the impact indicator.

The process can be standardized as follows (adapted from Canter & Hill, 1979 and combined with IAEA, 2000):

Main (aggregated) indicators	Goals/objectives as a basis for specification of sub-indicators and development of the evaluation criteria
Cost/Value	Development of competitive (least cost) electricity production
Supply Reliability	The energy payback ratio
Economic/Technological Advancement	Development of an electricity system expansion plan that minimises greenhouse gas emission
Risk/Uncertainty Management	Enhancement of the welfare of local communities; growth of social capital across region
Environmental and Health Impacts	Protection and improvement of the health of all residents and workers (good access to health care, reduced health inequalities, affordability of safe and quality nutrition, availability of recreation zones/infrastructure, nursing/work/social inclusion for elderly people, clean and healthy environment, safe urban areas, etc.)
Welfare of local and regional communities	Changes/improvements in regional and local employment Improvement of economic benefit to the community (to reduce disparities in income; access to jobs, housing, and services between areas within the region and between segments of the population; access to better and effective education; energy efficiency; etc.)
	Maintenance of high and stable levels of economic growth (good accessibility to business within the region, stronger linkages between firms and the development of specialism within area, local strengths and economic value locally, emergence of new and high technology sectors and innovations, etc.). Effective protection of the environment (maintenance and enhancement of the quality and distinctiveness of the landscape; making towns more attractive places to live in; maintenance and improvement of the quality of air, ground and river water; reduced contribution to climate change (greenhouse gases); moving up through the waste management hierarchy; prudent use of resources – to reduce consumption of undeveloped land, natural resources, greenfield sites; to reduce need to travel; to apply reasonable, long-term land-use planning considering open space; improvement of resource efficiency; etc.)

Note on sustainable development: Sustainable development *does not mean* having less economic growth. On the contrary, a healthy economy is better able to generate the resources for environmental improvement and protection, as well as social welfare. It also does not mean that every aspect of the present environment should be preserved at all cost (*extremism, fundamentalism*). What it requires is that decisions throughout society are taken with proper regard to their environmental impact and implications for wide social interests. Sustainable development *does mean* taking responsibility for policies and actions. Decisions by the government or the public must be based on the best possible scientific information and analysis of risk, and a responsible attitude towards community welfare. When there is uncertainty and the consequences of a decision are potentially serious, precautionary decisions are desirable (see Hansson, 2011 for further discussion on applying the precautionary principle). Particular care must be taken where effects may be irreversible. Cost implications should be communicated clearly to the people responsible.

Table 2. A list of main indicators to be applied in comparative multi-objective assessment

a. For each indicator, the analyst should identify the best value (e.g. highest contribution to employment) and the worst value (least contribution to employment) from the alternatives under consideration.
b. Then, the impact scale should be arranged on a horizontal axis from the best value (at the origin on the scale) to the worst value (at the extreme of the scale). The scale will depend on the units of measurement used in the impact assessment for each indicator.
c. Then, the standardized values of the impact indicators should be represented on the vertical axis, the same for all indicators and ranging from 0 to 1.

Finally, an indicator value of 1 should be assigned to the best option and 0 to the worst. The other options are then located according to their impact values on the line joining the best and worst.

Once the impact data are standardized, the following three methods could be used for the aggregation of results (IAEA, 2000; Kontić et al., 2006):

• Weighting; weight should be assigned to each indicator on the basis of its relative importance, for instance in a comparison of human health, global environmental impacts and land occupation (land-use impacts). Sensitivity analysis of the weighting should be performed in terms of investigating the difference in final comparative assessment results due to assignment of different weight values to a particular indicator (at least three justified variations should be considered); the final amalgamation method can be weight summation.
• Aggregation rules; based on standardization of the indicator's values, and a tree structure of the whole set of indicators where a root of the tree represents the ultimate aggregated value; pairs or sets of multiple indicators should be aggregated and evaluated by means of the »if-then« approach. In this way the aggregation rules should be developed as an alternative to weighting. A final score is derived by comparing aggregated values at the tree root for the treated alternatives. This approach is described in detail in (Bohanec, 2003) while an example of a decision tree specifying evaluation indicators is presented in Figure 4.
• Trade-offs; the final product of the analysis should be presented as a description of trade-offs in either tabular or graphical form. Goal programming can employ the amalgamation method which ranks the alternatives on the basis of the deviation from a goal or target that analysts (decision makers) would like to see achieved: the less the deviation, the closer to the goal, and thus the higher the alternative is ranked.

The analysts' view on the three methods and results achieved should be a part of the conclusions.

Presentation of the evaluation results

The analytical study should provide a systematic comparative assessment of the consequences (costs, benefits, impacts and risks) of alternative energy options (technologies). For decision-making purposes, these results need to be evaluated and presented in a coherent way. The evaluation and presentation of the results should focus on pointing out the main findings and conclusions that could support decision taking.

In order to assist decision makers effectively, analysts should present their results in a transparent manner (no "black boxes"), focusing on the verifiable results and their interpretations. In particular: input data and assumptions should be specified clearly and the boundaries and limits of the study should be indicated; comparison of alternatives should be based upon indicators that have been estimated quantitatively and qualitatively.

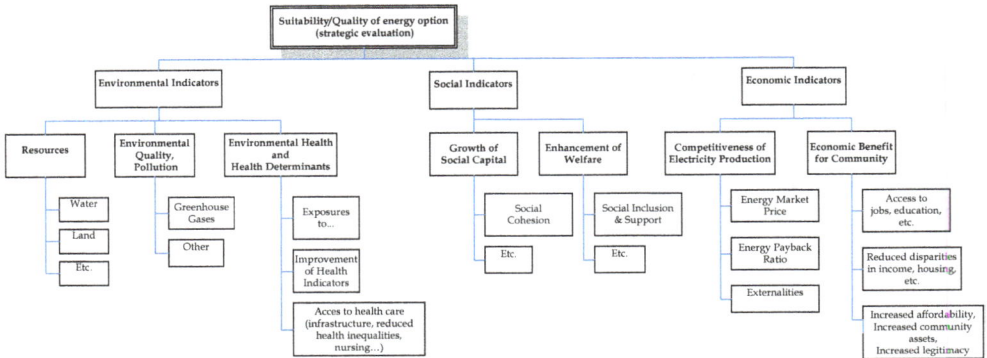

Fig. 4. A set of comparative assessment indicators for different energy options at the strategic evaluation level. The set is organised in a (decision) tree structure.

In general, the presentation of the results has to be adapted to the target audience of the study. The primary audience will be decision makers. However, in most cases, the study will also be disseminated to, and used by, interested and affected parties, e.g. local communities or NGOs. In both cases, the audience has not the same experience and knowledge on technical and economic issues as do the analysts. Therefore, results should be presented clearly and concisely, pointing out the main findings and outcomes.

2.2 Multi-objective decision making

Multi-objective decision making builds on previous multi-objective (sometimes called multi-attribute) valuation of the alternatives. Because the different ways to solve the problem tend to be mutually exclusive, the selection of the "best" option requires the formulation of trade-offs among the different attributes used to evaluate the performance of the several possible alternatives. Such trade-offs require a multi-objective analysis (see above) in order to assess and compare the relative merits of the alternatives. In practice, a multi-objective analysis usually does not yield a single optimal alternative. Therefore, the choice of the "best" solution requires that the decision maker's preferences and value trade-offs among conflicting objectives be clearly articulated and made explicit in the selection process. A vast number of publications on multi-attribute decision making is available from which one can extract useful information and guidance on how to perform such decision modelling. The following selection may serve as an introductory reading to the comprehensive overview of approaches, methods and tools for different multi-objective decision applications (Bohanec & Rajkovič, 1999; Bohanec , 2003; Munda et. al., 1995).

3. Radioactive waste disposal

3.1 Perception of radioactive waste disposal issues

The recent international perspective can be found in the report "Resource or waste? The politics surrounding the management of spent nuclear fuel in Finland, Germany, Russia and Japan" (SKB, 2011). A clear historical divide can be discerned between countries that decided to reprocess spent nuclear fuel and those that chose final disposal. Three of the countries mentioned – Japan and Russia and, in an earlier phase, Germany, have considered spent nuclear fuel as a resource rather than as waste, and for that reason invested in reprocessing. The report provides an account of how and why these countries chose different alternatives; why, despite a common basic approach, they gradually came to aim at completely different strategies and methods for spent nuclear fuel management. Today Germany has totally abandoned its previous reprocessing strategy, Russia has maintained its strategy, but also steered certain operations toward direct disposal, and Japan has recently completed a major industrial reprocessing facility. The issue of final disposal is, however, far from solved in Germany and Japan. In order to understand why different countries have chosen one alternative over another, and how a strategy changed over time, the authors chose to elaborate on eight key dimensions. Five of these relate to nuclear power issues, such as whether or not a country produces nuclear weapons, has an expanding or stagnating nuclear power sector, weak or strong competence in the field of nuclear energy, good or poor prerequisites for a final repository, and whether or not it has domestic uranium resources. Three other dimensions cover political characteristics, i.e. whether or not the country had or has a strong or weak anti-nuclear movement, whether it is a democracy or a dictatorship, and whether or not it is characterized by strong or weak local political power. The latter aspect is seen as essential to issues of local acceptance of a spent nuclear fuel repository. The reasons behind different choices appear to be the military use of spent nuclear fuel and the absence of democratic discussion (Russia), consensual political decision-making (Finland), and situations of strong political opposition and local disputes (Germany and Japan).

In the project "Nuclear waste: From an Energy Resource to a Disposal Problem" (SKB, 2011) Jonas Anshelm analyzed the nuclear waste debate since the 1950s, including issues of risk, responsibility, design of a final repository and safety of the technology. The author points to the importance of elucidating the different kinds of answers that have been given concerning these issues in different time periods. The challenge is to understand how changing technological, political, economic and scientific circumstances have influenced perceptions and debates. Such clarification can broaden the perspective and facilitate an understanding of the complexity of the issue. The project observes shifts in meaning and public opinion changes regarding central aspects on the nature of nuclear waste – as a resource or as a waste, and the characteristics of the waste – as well as of its associated risks. Likewise, issues of who has responsibility for the final repository, what should be considered scientific facts concerning bedrock characteristics, and the sustainability of the technological solutions, have been subjected to controversy throughout the period. It is striking, Anshelm notes, that central actors have been both utterly confident in their opinions and able to assume totally different points of view in new situations. This characterization applies to both proponents and opponents of nuclear power. In summary, this contribution illustrates that what is perceived to be true, valid, correct, morally right, and rational with respect to the debated issue has recurrently been subject to renegotiation and change during the past half-century. This has resulted in a number of serious conflicts since the 1970s. The issue has currently reached a level of stabilization and does not exemplify a strong national or local controversy. It is, however, reasonable to assume that current views on what is true and right regarding the nuclear waste issue – on which there is some consensus today – will, in the future, also be subjected to renegotiations in the light of scientific, technological, economic and political reorientations. This already appears to have been triggered in a number of countries, e.g. Germany, Japan, Slovenia, by the consequences from the damaged NPP Fukushima I after the quake and tsunami in March 2011. It could be viewed that this accident encouraged the German government to announce that it will bring forward the closure of its nuclear power stations to 2022, 14 years earlier than originally planned, while Japan considers a review of plans for construction of new NPPs, just like Slovenia in its new National Energy Programme currently under debate.

3.2 Waste disposal siting

Radioactive waste should be disposed of in a way that guarantees its isolation from the biosphere. The release of potentially harmful substances - radionuclides - must be prevented or limited to levels that do not harm human health or the environment (IAEA, 1994). In this context, the issue of a proper siting process gains importance, especially in terms of selecting a site that has geological, hydrological, seismic, morphological and other characteristics that would not contribute to the release of radioactivity from a repository and subsequent exposure of the population. The site selection process is therefore a critical step in the overall site acquisition process. Countries are seeking their own ways on how to achieve these goals. As regard Slovenia it may be seen as a successful example concerning low and intermediate level waste (LILW) disposal. However, a strategy for the management of spent nuclear fuel and high level waste (HLW) is still under consideration.

The benefits of strategic environmental considerations in the process of siting a repository for LILW are clearly presented in Dermol & Kontić, 2011. The benefits have been explored by analyzing differences between the two site selection processes. One is a so-called official site selection process, which was implemented by the Agency for radwaste management (ARAO); the other is an optimization process suggested by experts working in the area of environmental impact assessment (EIA) and land use (spatial) planning. The criteria on which comparison of the results of the two site selection processes has been based are spatial organization, environmental impact, safety in terms of potential exposure of the population to radioactivity released from the repository, and feasibility of the repository from the technical, financial/economic and social points of view (the latter relates to consent by the local community for siting the repository). The site selection processes have been compared with the support of the multi-objective decision expert system named DEX –Decision EXpert (Bohanec & Rajkovič, 1999). The results of the comparison indicate that the sites selected by ARAO meet fewer suitability criteria than those identified by applying strategic environmental considerations in the framework of the optimization process. This result stands when taking into account spatial, environmental, safety and technical feasibility points of view. Acceptability of a site by a local community could not have been tested, since the formal site selection process had not yet been concluded at that time. Now the consent has been granted and ARAO is about to start construction of the repository in 2012. This approach to siting and comparison of the two site selection processes may serve as an example of transparent and inclusive - the local partnership has been established – way of dealing with radioactive waste disposal.

3.3 Uncertainties

3.3.1 The time perspective of nuclear waste

The most discussed aspect of nuclear waste is its longevity. Previously nuclear waste was the only issue for social decision-making that was widely discussed in very long time perspectives. Today climate change is discussed in such long time perspectives, and we also have a general discussion on sustainable development that does not have any time limits (Hansson, 2011). Hansson further says that discussions on decisions related to very long time perspectives include the issue of *how* to evaluate outcomes in the future. For example, is the value of a human life similar or dissimilar if it relates to assessing a final repository in e.g. 10,000 years hence, or in our time? And how should uncertain outcomes be evaluated? We seldom know about the consequences, in a hundred year perspective, of a decision taken today. This uncertainty has often resulted in not caring for the long-term consequences of the actions. The nuclear waste issue has become a pioneering case in the sense that uncertainties have not hindered us from considering long-term consequences seriously. Hansson's concluding observations are that it is not the uncertainty *per se* that has resulted in the high attention and controversy regarding future effects of a nuclear waste repository, but rather the combination of certainty in specific areas (e.g. radioactive decay over time, etc.) and uncertainty in other areas (e.g. future generations' knowledge, intentions, etc.). Finally Hansson notes that the International Climate Panel (IPCC) focuses on a time perspective of around 100 years and utilizes a kind of "trimmed discounting" in the work. He concludes that this is rather unprincipled reasoning, and suggests that much would be achieved by approaching the climate change issues in a way similar to that of nuclear waste.

3.3.2 Approval context of waste disposal

Regulators all over the world formally base their decisions about the acceptability of a particular radioactive waste disposal system upon the related performance (safety) assessment. The key element of this assessment is dose evaluation. However, the requirements for the certainty/accuracy/validity of such evaluation are not clearly defined in advance and are a subject of development in the dose evaluation process itself. Therefore, dose evaluation, as well as the associated licensing procedure that builds on compliance assessment, seems to be a less appropriate approach due to the uncertainty involved. An alternative method for assessing human exposure in the framework of long-term safety assessment should be developed. Such a method, integrated with the concept of reasonable assurance (IAEA, 1997), should build on indicators of future exploitation of the environment – therefore a clear link to spatial planning in site selection process where human activities remain the basis for future exposure assessment (Kontic et al., 1999).

Since this approach is more fundamental, direct and transparent than dose or risk assessment, it is expected that it will be more powerful in confidence building among different social groups, i.e. scientists, regulators, the public and politicians. Eventually, it is also expected to be effectively applied in comparative evaluations of various energy options. In addition, certain ethical dilemmas in the licensing process connected to regulatory decision making in the presence of uncertainty and in the context of the disposal of long lived radioactive wastes, could also be reduced if dose or risk are avoided as individual numerical safety indicators.

3.3.3 Scientific treatment of specific uncertainty and predictions related to spent fuel

In this sub-section an analysis of the impact of uncertainty (associated with the quantity of radioactive waste produced by Krsko NPP in its anticipated operational time) on the waste-disposal strategy, particularly the selection of the disposal option, is presented. The dilemma is whether to build a shallow land repository for the LILW and to treat all high-level and long-lived waste separately; or to adopt deep geological disposal as the option for all waste types produced in the country. Tightly connected to these questions is the credibility of the evaluation of health consequences due to radioactive waste disposal. Indicators can, for example, be the dose and risk in the presence of uncertainty associated with the waste characteristics on the one hand, and societal characteristics and human habits in the distant future on the other. The approach and methods applied in the analyses were as follows:

- First, information about the present status of the waste was gathered. Attention was paid to the variability and accuracy of data on quantities, the radionuclide inventory and the activity of different types of waste.
- Then, based on this information, what may most likely be expected (with regard with these waste characteristics) by the end of the anticipated operational period of Krsko NPP, i.e. 2023, was estimated. The ORIGEN2 computer code was used for calculating isotope generation, activity build-up and depletion, and the decay heat of spent fuel (Croff, 1983), while a specific code was developed for calculations associated with LILW.
- The total activity, its time dependent change and the identification of radionuclides that mainly contribute to the activity in long timeframes, were applied as key information for discussing waste-disposal options for spent fuel (HLW). Changes (variations, uncertainty) in these characteristics were evaluated based on the technical specifications that are in place after the replacement of steam generators at the plant in the 17th fuel cycle in 2002. The variations considered were 3-5 % of U-235 in the fuel, and an operational period of the plant of five years more or five years less than that envisaged. The basic estimate was that Krsko NPP uses fuel with 4% U-235 in all future cycles and that it operates for 35 years.

The key input data for calculating burn-up and fuel characteristics in future cycles is not available at the moment. Consequently, certain assumptions had to be made. These were:

- 35 fuel cycles are assumed for the operational period of Krsko NPP.
- The average cycle burn-up is 12,000 MWd/tU. This value was adopted based on the following: The average number of effective days of full power operation per cycle is

324. Using 1,876 MW as the nominal power of the plant, and 48.7 t of uranium per cycle, one obtains 11,857 MWd/tU. When this is rounded off, 12,000 MWd/tU for burn-up and 320 effective days of operation at full power are obtained.

- A twelve-month cycle was assumed (i.e. the cycle lasts 365 days); the operational period is 320 days and the cooling (decay) period between cycles is 45 days (actually used for refuelling and maintenance).

- One batch of fuel consists of 40 elements, containing 16.24 tonnes of uranium, and on average represents one-third of the total amount of fuel in the cycle (there are three different batches in the reactor during operation). Each batch remains in the reactor for the three following cycles – except the first, second, penultimate and last batches. The real situation is more complicated but corresponds roughly to these assumptions.

- Being aware of the differences between these assumptions and the real operational data for Krsko NPP, a screening calculation of the activity of spent fuel for the first 13 cycles was made, for the purpose of further calibrating the model. The results for the model and those based on operational data differ very little – see Table 3 for details.

- The content (mass) of uranium isotopes per fuel batch is given in Table 4.

- The mass of zircaloy (Zr-40) per batch is 4012.5 kg; the mass of oxygen (O-16) is 2183.5 kg.

- The average power per tonne of uranium is 37.5 MW; the average power of the batch is 609 MW.

	Activity after 7th cycle (Bq)	Cooling period (days)	Activity after 13thcycle (Bq)	Cooling period (days)
Source of operational data: NPP Krsko, and Ravnik and Železnik, 1990	7,5E+18	45	N.A.	N.A.
Source of operational data: NPP Krsko, and Božič (1998)	9,2E+18	45	2,5E+19	32
Model prediction	9,9E+18	45	1,2E+19	45
Difference (%), rounded	9-33		208	

Note: One should note that these differences are very low, taking into account that the order of magnitude of the values is E+19, and that the cooling periods differ, after the 13-th cycle, between the model and the real data. The latter is important if the activity of spent fuel decreases rapidly during the cooling period; this would mean that the activity drops by a factor of two over a two week period, i.e. in the period between 32 and 45 cooling days, which is the difference between the real data and the model.

Table 3. Comparison of the calculated (model) results and operational data

The calculated changes of activity of activation products (AP), actinides (ACT), fission products (FP) and total activity per fuel batch with time are presented in Figure 4. The illustration is for model Batch 6; however, the figures are similar for other model batches. Batch 6 goes into the reactor in the fourth cycle (year). At the moment of irradiation, the total activity immediately increases by about eight orders of magnitude. Before that, the activity is constant at a level of $1.9*10^6$ MBq (the activity of approximately 16 tonnes of non-irradiated fuel). During irradiation, this activity rises slightly from 1.23 to $1.28*10^{14}$ MBq, while during the cooling period of 45 days it drops by approximately two orders of magnitude. Each batch stays in the reactor for three successive cycles (except the first, the second, the penultimate and the last), whereupon the batch goes into the spent fuel pit for ultimate cooling and decay. It should be noted that the scale of both axes is logarithmic, so that the origins of axes are avoided in the illustrations.

Batch-enrichment (%)	Isotope (kg)			
	U-234	U-235	U-236	U-238
2.1	2.44	341.04	2.11	15894.25
2.6	3.25	422.24	2.59	15811.91
3.1	3.89	503.44	3.09	15729.58
3.4	4.22	551.67	0.81	15683.29
3.6	4.55	584.64	0.65	15650.49
3.9	5.36	633.36	1.30	15599.82
4.0	5.85	649.60	2.03	15581.96

Table 4. Mass of uranium isotopes in the fuel (per batch)

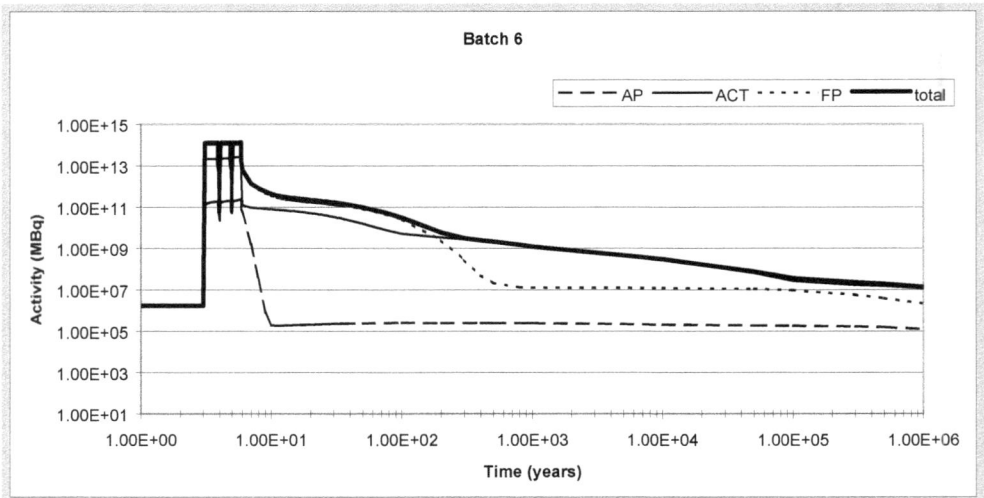

Fig. 4. Activity of model batch 6 over a million years

Model results for all the fuel are presented in Figure 5, which shows time changes in total activity. With regard to activity during the first 34 cycles, an almost linear increase can be identified due to the collection of spent fuel in the spent fuel pit - one batch per cycle/year. After the 35th cycle, i.e. at the end of the assumed operation of the plant, all three batches from the reactor will be placed in the spent fuel pit at the same time, which is seen as an noncontinuous increase in activity. Activity then decreases, depending on the radionuclides contained in the spent fuel. Note again that the scale of the axes is logarithmic. The values of total activity and decay heat for all spent fuel at selected time-points are summarised in Table 5.

The model adequately represents the overall operation of the plant. This was proved in the process of calibrating the model, where data for the past thirteen cycles were used for comparison. However, fuel enrichment, as well as other key operational elements in future cycles, may not remain constant, since an upgrade of the plant's power in parallel with the replacement of the steam generators has been achieved. Extension of the fuel cycles was also adopted/made. This was the reason for the analysis of the changes in the activity and radionuclide inventory of spent fuel, due to different fuel enrichment and the prolonged operation of the plant. The adopted variation in fuel enrichment was 1% above and below the value applied in previous calculations, i.e. 4% of U-235.

Complete spent fuel

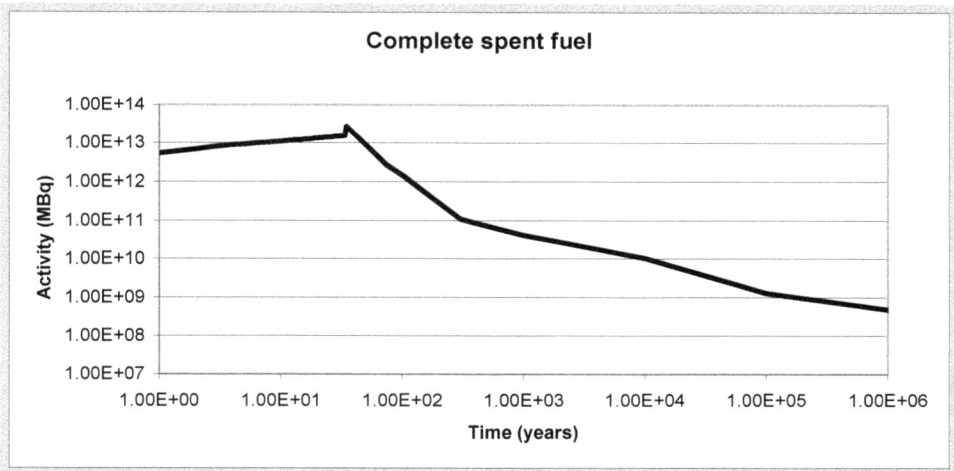

Fig. 5. Total activity of all the spent fuel as a function of time

With regard to the prolonged operation of the plant, a five-year variation was applied. All the variations were simulated for the period following the replacement of the steam generators, i.e. after the 17th cycle. The differences are presented in Figures 6 and 7, respectively. It is clear that the differences are so small that they can be neglected, since they are not relevant to the overall waste management strategy. Moreover, the conclusion which can be drawn from this result is that no benefit can be expected in terms of improved safety connected with radioactive waste disposal whether Krsko NPP were closed down immediately or operated for almost a further 12, or even 40 years as the new National Energy Programme suggests.

The results of the modelling show that the main contributors to fuel activity during the period approximately 200 years after irradiation are the fission products; after that, actinides will prevail. The total expected activity of the spent fuel after one million years is $4{,}8*10^{14}$ Bq. The main contributors to this activity are the radionuclides of the U- and Np-chains. Residual thermal power is about $1.0*10^5$ W approximately 200 years after irradiation, about $1.0*10^4$ W after 10,000 years, and about 250 W after one million years.

The problem of uncertainty, which can be treated scientifically, is manageable. It was recognised that the basic characteristics of this waste can be accurately predicted, since all the sources of uncertainty are well defined, understandable and therefore controllable. Residual uncertainty does not change the overall picture of the waste, which would mean that the predictions could clearly be used as a basis for policy-making, i.e. creating a strategy for radioactive waste management, decision-making and also for communication with the public.

Time (years)	Total activity (MBq)	Decay heat (W)
1	$5.30*10^{12}$	$5.55*10^5$
2	$6.72*10^{12}$	$7.19*10^5$
3	$7.92*10^{12}$	$8.71*10^5$
4	$8.70*10^{12}$	$9.54*10^5$
5	$9.23*10^{12}$	$1.01*10^6$
10	$1.09*10^{13}$	$1.13*10^6$
15	$1.21*10^{13}$	$1.22*10^6$
20	$1.31*10^{13}$	$1.31*10^6$
25	$1.40*10^{13}$	$1.38*10^6$
30	$1.48*10^{13}$	$1.45*10^6$
35	$2.69*10^{13}$	$2.72*10^6$
75	$2.63*10^{12}$	$3.08*10^5$
100	$1.46*10^{12}$	$2.23*10^5$
300	$1.08*10^{11}$	$8.44*10^4$
1,000	$4.11*10^{10}$	$3.52*10^4$
10,000	$1.04*10^{10}$	$8.23*10^3$
100,000	$1.28*10^9$	$6.82*10^2$
300,000	$8.12*10^8$	$3.80*10^2$
1,000,000	$4.81*10^8$	$2.53*10^2$

Table 5. Total activity and decay heat of all the fuel from the Krsko NPP at selected time-points over a million years

Influence of fuel enrichment (% of U-235) on activity of actinides;
ACT3 stands for 3%U-235, ACT4 for 4%, and ACT5 for 5%

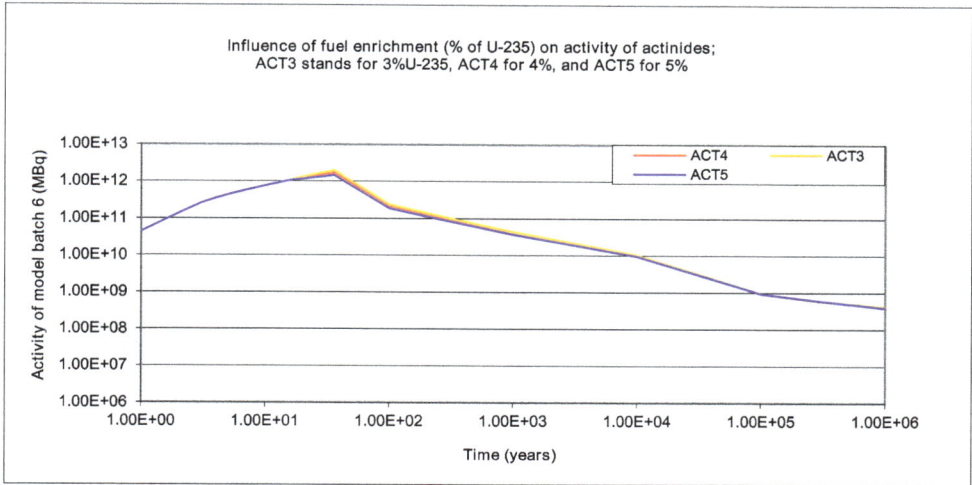

Fig. 6. Influence of fuel enrichment on activity of actinides (it is assumed that the change in fuel enrichment starts with the 17th cycle)

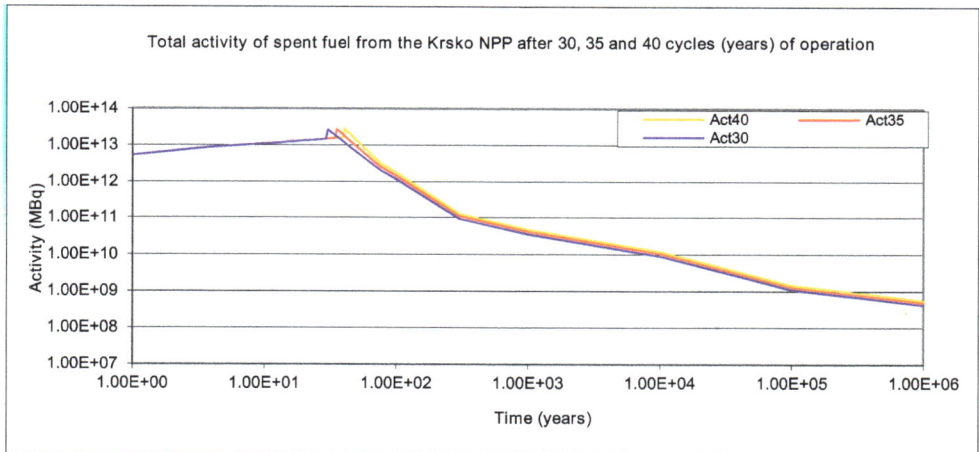

Total activity of spent fuel from the Krsko NPP after 30, 35 and 40 cycles (years) of operation

Fig. 7. Influence of extended or shortened operation of the Krsko plant on the activity of actinides in the complete spent fuel (the basic estimate is that the plant will operate 35 cycles)

With regard to confidence building connected to radioactive waste disposal, the recommendation is that prompt, clear and complete informing of all interested parties and the general public should take place. It should be clearly stated that the spent fuel from Krsko NPP, and a part of the decommissioned waste, will remain radioactive above today's prescribed levels for hundreds, thousands or even a million years from now. Consequently, a strategy built upon waiting for the activity to "disappear" cannot be effective. Doubts and uncertainties regarding safety assessments in a timeframe of a million years should also be revealed. At the same time, efforts should be made to present the concept of reasonable assurance (IAEA, 1997) as the most reliable method, and as the basis upon which a waste management strategy can rely.

4. Concluding remarks

Strategic environmental considerations of nuclear power should inevitably cover the following (however, not restricted to):

- Comparative evaluation of electricity generation technologies; the evaluation should, in addition to topical consideration, thoroughly deal with the ways on how to overcome specifics and details of individual analysis of a certain technology which is usually influenced by the specific characteristics of the site compared to others in its category, the manufacturing and design characteristics, the power, lifetime and the operating conditions. Results are therefore difficult to transfer from a country to another or one generation unit to another, as most major environmental, economic and social impacts, with the exception of e.g., climate change, are heavily site-dependent. Application of proper indicators in such a comprehensive comparative evaluation may be of practical help and guidance;
- The energy system as a whole; the electricity grid and market issues are rarely taken into consideration when making comparative evaluation. Similarly, the issue of increased share of intermittent RES, its impact on the energy system, and consequent need for the adaptation of environmental impact appraoches by taking into account actual share of each generation technology as provided in the energy system;
- Uncertainties; uncertainties and limitations of various methodologies may be acknowledged by the authors of the studies but those are rarely taken into account when results (or only some of them) are used by policymakers. Strategic considerations should provide guidance/recommendations on how to deal with the uncertainties in the decision-making process associated with comparative evaluation of different electricity generation technologies. This is especially relevant when deciding about long-term impacts, e.g. nuclear waste disposal or societal and spatial consequences of climate change.

5. References

Bohanec, M. & Rajkovič, V. (1999). Multi-Attribute Decision Modeling: Industrial Applications of DEX, *Informatica*, Vol. (23), 487–491.

Bohanec, M. (2003). Decision Support *in* D. Mladenic, N. Lavrac, M. Bohanec, S. Moyle (eds.), *Data Mining and Decision Support: Integration and Collaboration*, Kluwer Academic Publishers, 23–35

Božič, M. (1998). Izračun aktivnosti in zakasnele toplote goriva NE Krško, Seminar-podiplomski študij jedrske tehnike, Ljubljana (A Seminar: Activity and residual heat calculation at the NPP Krško; available in Slovene only)

Canter, L. W. & Hill, L.G. (1979). *Handbook of Variables for Environmental Impact Assessment*, Ann Arbor Science Publishers Inc., Michigan

Croff, A.C. (1983). Origen2: A versatile computer code for calculating the nuclide compositions and characteristics of nuclear materials, *Nuclear Technology*, No. 62, 335-352

Dermol, U. & Kontić, B. (2011). Use of strategic environmental assessment in the site selection process for a radioactive waste disposal facility in Slovenia, *Journal of Environmental Management*, Vol. (92), 43-52

EC - European Comission (1998). Handbook on Environmental Assessment of Regional Devlopment Plans and EU Structural Fund Programmes, EC-DGXI: Environment, Nuclear Safety and Civil Protection, Bruxelles

EC - European Comission (1999). Environment and Sustainable Development. A Guide for the ex-ante evaluation of the environmental impact of regional development programmes, *Series: Evaluation and Documents* No. 6, Bruxelles

EC - European Comission (2009). Study concerning the report on the application and effectiveness of the SEA Directive (2001/42/EC): Final report, European Commission, DG ENV, COWI, 12.07.2011, Availlable from http://ec.europa.eu/environment/eia/pdf/study0309.pdf

Eurelectric (2011). 20% Renewables by 2020: a eurelectric action plan. Summary Report, Available at http://www.eurelectric.org/RESAP/About.asp (9 January 2012)

Hansson, S.O. (2011). Ethical and philosophical perspectives on nuclear waste, *in* B. Berner, B.-M. Drottz Sjöberg, E. Holm (eds.), *Social Science Research 2004-2010,Themes, results and reflections*, ISBN 978-91-978702-2-1, CM Gruppen AB, SKB, Stockholm, 137-149

IAEA (1994). Siting of geological disposal facilities, *Safety Guide*, Safety Series, Vienna, 4-19

IAEA (1997). Regulatory decision making in the presence of uncertainty in the context of the disposal of long lived radioactive wastes, IAEA-TECDOC-975, 22-23

IAEA (1999). Health and environmental impacts of electricity generation systems: Procedures for comparative assessment, TRS No. 394, IAEA, Vienna

IAEA (2000). Enhanced electricity system analysis for decision making – A Reference Book, DECADES Document No. 4, IAEA, Vienna

IPCC (2011). Special Report on Renewable Energy Sources and Climate Change Mitigation (SRREN), Available at http://srren.ipcc-wg3.de/ (9 January 2012)

Kontić, B., Kross, B. C. & Stegnar, P. (1999). EIA and long-term evaluation in the licensing process for radioactive waste disposal in Slovenia, *Journal of Environmental Assessment Policy and Management*, Vol. (1), No. 3, 349-367

Kontić, B., Bohanec, M. & Urbančič, T. (2006). An experiment in participative environmental decision making, *The Environmentalist*, Vol. (26), 5-15

Kontić, B. & Kontić, D. (2011). A viewpoint on the approval context of strategic environmental assessments, *Environmental Impact Assessment Review*, doi: 10.1016/j.eiar.2011.07.003

Laukkanen, S., Kangas A. & Kangas J. (2002). Applying voting theory in natural resource management: A case of multi-criteria group decision support, *Journal of Environmental Management*, Vol. (64), 127-137

Munda, G., Nijkamp, P. & Rietveld, P. (1995). Monetary and non-monetary evaluation methods in sustainable development planning, *in* K. G. Willis, K. Button, P. Nijkamp (eds.), *Environmental Valuation Volume II: Multi-Attribute Programmes, Validity, Allocation Issues in Case Studies*, Series Environmental Analysis and Economic Policy, Edward Edgar Publishing, Inc., Northampton

Ravnik, M., Železnik N. (1990). Izračun izotopske sestave in aktivnosti goriva JE Krško, IJS-DP-5851, Ljubljana (Technical Report on isotopic content and activity of spent fuel at Krško NPP; available in Slovene only)

SKB – Swedish Nuclear Fuel Management Company (2011). Distinctive international strategies, *in* B. Berner, B.-M. Drottz Sjöberg, E. Holm (eds.), *Social Science Research 2004-2010,Themes, results and reflections*, ISBN 978-91-978702-2-1, CM Gruppen AB, SKB, Stockholm, 42-44

Taylor, N. (1980). Planning theory and the philosophy of planning, *Urban Studies*, Vol. (17), 159-172

Therivel, R. (2005). *Strategic Environmental Assessment In Action*, Earthscan, London, pp. 7-19.

URS (2010). Preliminary report on environmental impacts of different energy technology options for Slovenia, WSMS-OPS-09-0001, Washington

2

Investigation on Two-Phase Flow Characteristics in Nuclear Power Equipment

Lu Guangyao, Ren Junsheng, Huang Wenyou, Xiang Wenyuan,
Zhang Chengang and Lv Yonghong
China Guangdong Nuclear Power Holding Co. Ltd.
China

1. Introduction

Two-phase flow exits in many nuclear power equipments, such as containment sump strainer, steam generator, steam turbine, control rod drive mechanism and so on. Experimental investigations are carried out to study the two-phase flow patterns and their transitions in these nuclear power equipments. And the results show that the two-phase flow patterns and their transitions are quite different from those in normal circular tubes.

For tube-bundle channel heat transfer enhancement technique, it has great advantages in high heat transfer efficiency and compact configuration without complex machining or additional surface processing, which has been successfully used in steam generator in nuclear power station and other industrial equipments. The characteristics of boiling flow and heat transfer have an important impact on these industrial equipments. It was found that heat transfer characteristics of fluid flowing in tube-bundle channels were different from those in circular tubes on account of the special geometric frame and different flow patterns in tube-bundle channels (Petigrew & Taylor, 1994). Boiling flow and heat transfer is a complex issue of two-phase flow, and many studies have been conducted in this research area.

Experiments of boiling flow in tube-bundle channels were carried out in order to simulate boiling flow and heat transfer in fuel module of nuclear reactor (Bergles, 1981). The experimental results showed that there were many different aspects of flow patterns and their transitions in tube-bundle channels. Grant & Chisolm (1979) and Ma (1992) made studies of air-water two-phase flow in tube-bundle heat exchangers, in which the vertical flow and horizontal flow experiments were conducted respectively. Grant & Chisolm (1979) found that there were mist flow, bubbly flow, intermittent flow, stratified-mist flow and stratified flow in tub-bundle channels. But Ma (1992) detected stratified flow and wave flow in horizontal channels, and bubbly flow and intermittent flow in vertical channels, which was different from the results gained by Grant & Chisolm (1979). Chan & Shoukri (1987) conducted visualization experiments of refrigerant-113 flowing in tube-bundle channels, of which the tubes were arranged 1×1, 3×1, 3×3, and 9×3 respectively. Sadatomi & Kawahara (2004) carried out experiments in tube-bundle channels, of which the tubes were arranged 2×3. On the basis of experiments, Sadatomi protracted flow pattern maps.

On the basis of the results obtained before, studies are carried out to investigate the characteristics of refrigerant 113 flowing in a tube-bundle channel, of which the tubes are arranged 2×2. Furthermore, flow visualization experiments are carried out with high-speed camera. Through the comparison between the experimental results and other works, it is found flow regimes and their transitions in the tube-bundle channel are different from other normal circular tubes.

In the event of LOCA (Loss-of-Coolant Accident) or HELB (High Energy Line Break) within the containment of a light-water reactor (LWR), the primary safety concern regarding the long-term recirculation is that accident-generated debris and resident debris may be transported to the recirculation sump screens, which would result in adverse blockage effects and loss of the pumps net positive suction head (NPSH) of the emergency core cooling system (ECCS) and the containment spray system (CSS). The debris may starve the sump and the head loss of water flow through the containment sump strainers may be so large that it would exceed the available NPSH margin of pumps of ECCS and CSS systems. The pressure loss due to the accident-generated debris accumulated on the sump screens would be computed through the debris types and contents which are determined to be destroyed and transported. So the computational researches were carried out to investigate the characteristics of containment sump strainers by Lu et al. (2011a, 2011b, 2011c).

For control rod drive mechanism (CRDM) in nuclear power station, it works in high pressure and high temperature condition and single-phase water is adopted as the working liquid. But two-phase flow would come into being in the cold-state test and hot-state test. The high speed camera is also used in the CRDM visualization tests in cold-state to investigate the characteristics of two-phase flow.

2. Experiments

2.1 Experiments of tube-bundle channels

A schematic diagram of the experimental apparatus is given in Fig.1, which is adopted to study the boiling two-phase flow in tube-bundle channels. In the experiments, all measured

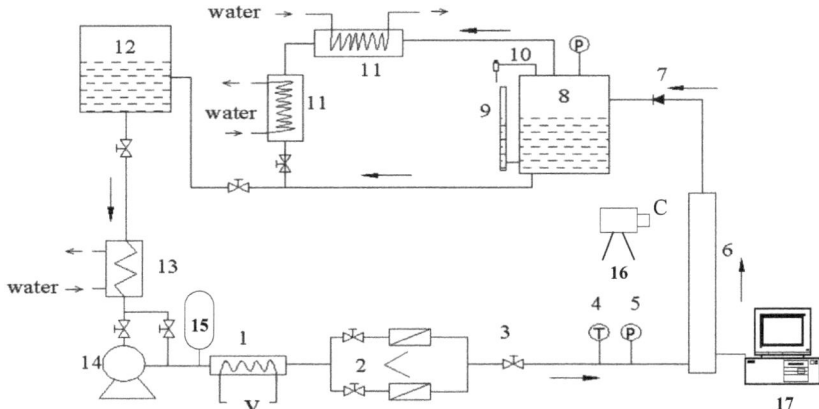

1. Preheater 2. Flowmeter 3. Control valve 4. Thermocouple 5. Pressure meter 6. Test section 7. Check valve 8. Seperator 9. Liquid-height meter 10. Relief valve 11. Condenser 12. Liquid tank 13. Cooler 14. Magnetopump 15. Manostat 16. High-speed camera 17. Data collector

Fig. 1. Schematic diagram of experimental apparatus (tube-bundle channel)

data are recorded at the intervals of 10ms. The error in the pressure drop measurement is less than ±2%. The error in flow rate measurement is less than ±3%, and for temperature measurement is less than ±1%.

The test section of tube-bundle channel is composed of four straight electrical heaters and a thimble, which is shown in Fig. 2. The thimble is made of PMMA pipe for the purpose of flow visualization, of which the inner diameter is 35mm and the length is 2100mm. The dimension of the electrical heater is Φ8mm and 1500mm long; the gap between electrical heaters is 5mm; the width between electrical heater and thimble is 4.5mm. There are, along flow direction, three pressure sensors and five thermocouples. The length between pressure sensors is 200mm and that between thermocouples is 100mm.

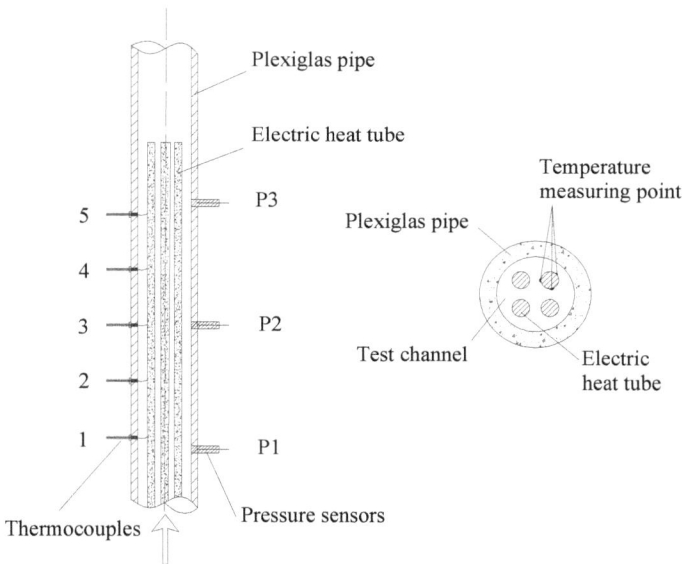

Fig. 2. Structure of test section

A schematic diagram of sub-channels partition is shown in Fig. 3. There are 9 sub-channels in the present test section.

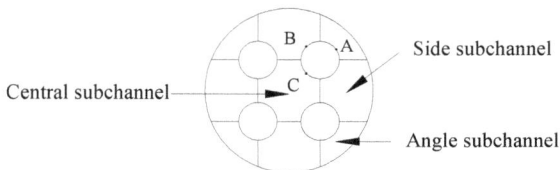

Fig. 3. Schematic diagram of sub-channels partition

2.2 Experiments of containment sump strainers

For the tests of containment sump strainers, the schematic diagram of the experimental apparatus is given in Fig.4.

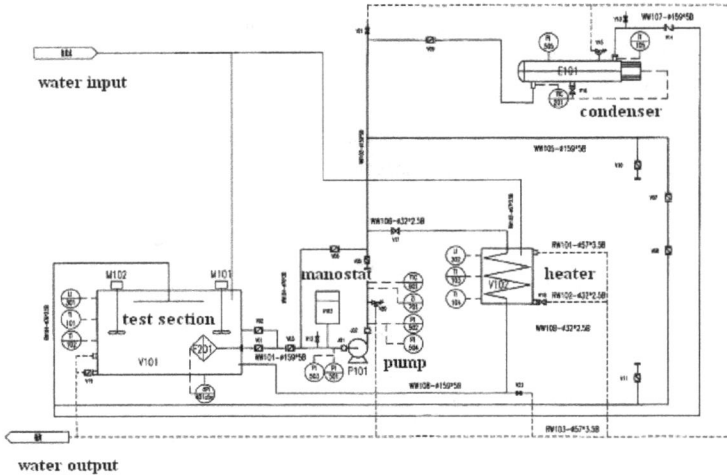

Fig. 4. Schematic diagram of experimental apparatus for containment sump strainers

In order to monitor the tests, the test apparatus includes flow rate measurements, temperature measurements, and differential pressure measurement. High speed camera is adopted to perform the visual observation of the two-phase flow and the debris bed during the test and after the test. The photos of the test system are shown in Figs 5~7.

Fig. 5. Photograph 1 of the test system (containment sump strainers)

Fig. 6. Photograph 2 of the test system (containment sump strainers)

Fig. 7. Photograph 3 of the test system (containment sump strainers)

2.3 Experiments of control rod drive mechanism

For the tests of control rod drive mechanism (CRDM) in nuclear power station, the schematic diagram of the experimental apparatus is given in Fig.8. And the photos of the test system are shown in Figs 9~11.

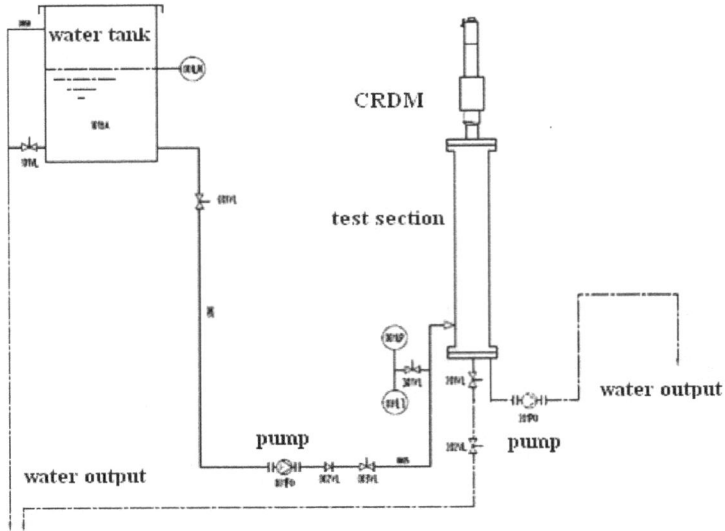

Fig. 8. Schematic Diagram of Experimental Apparatus (CRDM)

Fig. 9. Photograph 1 of the test system (CRDM)

There are four floors in the test bench, whose site area is 5m×5m and height is 25m. The photo of the test bench is shown in Fig. 9. The first floor is used to provide ground support for the test section and to settle the pump and pipeline. The second floor is adopted to upper support for the test section. The test section is joined with CRDM by J seal weld on the second floor. The photo of partial test section is shown in Fig. 10 and Fig. 11. CRDM is settled on the third floor. And the fourth floor is used to settle the crane and provide moving space for the driving rod.

Fig. 10. Photograph 2 of the test system (CRDM)

Fig. 11. Photograph 3 of the test system (CRDM)

3. Results and discussion

3.1 Test of tube-bundle channel

High-speed camera is adopted to carry out experiments of the two-phase boiling flow in the tube-bundle channel for different heat flux and different flow rate. Several representative pictures obtained are displayed in Fig.12.

	$q=1.5\times10^4 w/m^2$	$q=2.7\times10^4 w/m^2$	$q=4.2\times10^4 w/m^2$	$q=5.1\times10^4 w/m^2$
$m=63.3$ (kg/m²s)				
	Bubbly flow	Churn flow	Churn-annular flow	Annular flow
$m=133.9$ (kg/m²s)				
	Bubbly flow	Bubbly flow	Bubbly- churn flow	Churn flow
$m=281.8$ (kg/m²s)				
	Bubbly flow	Bubbly flow	Bubbly flow	Churn flow

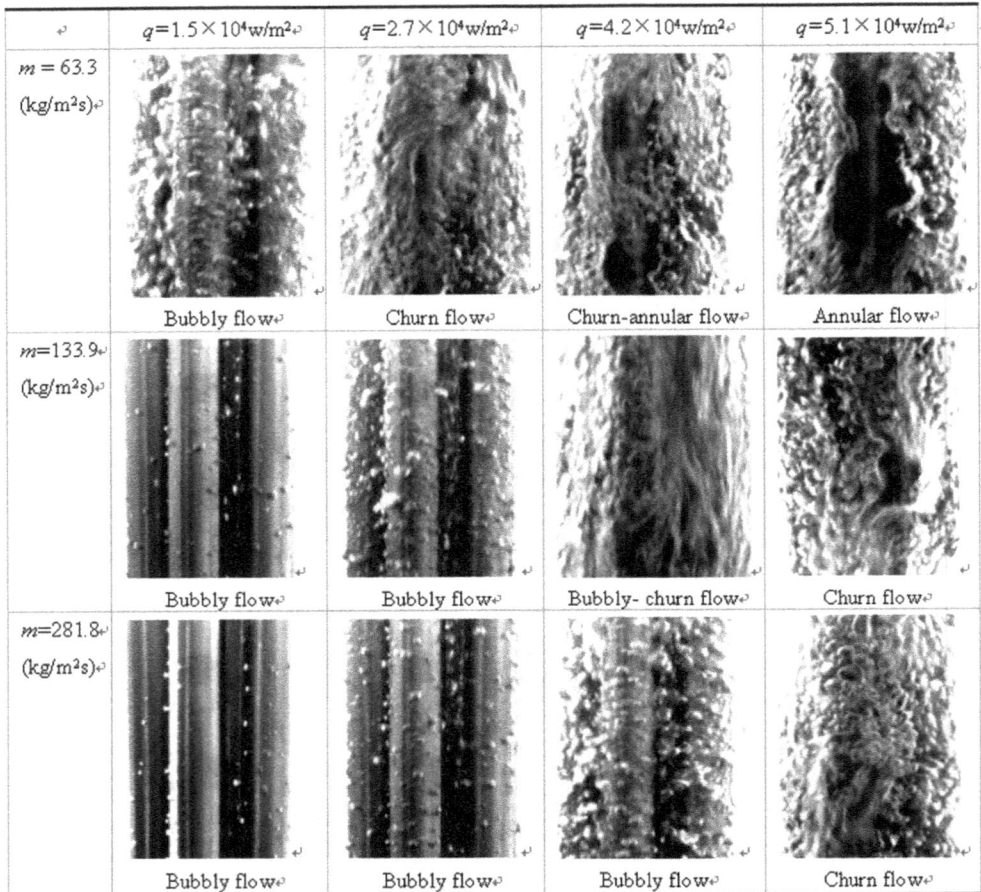

Fig. 12. Flow patterns in a vertical tube-bundle channel

Fig. 12 shows that there are four main flow patterns, bubbly flow, bubbly-churn flow, churn flow and annular flow, which is different from flow patterns of two-phase boiling flow in a circular tube. Through the analyses, it is shown that there may be two reasons for these differences. Firstly, the geometric dimensions cause the different flow patterns. The tube-bundle channel is divided into several sub-channels by the tubes, as shown in Fig. 12. And the inner tubes divide the large bubble and make disturbance on two-phase flow. Furthermore, flows in the sub-channels interact and enhance the complexity of two-phase flow in the tube-bundle channel. Secondly, the heating mode of boiling flow in the tube-bundle channel and that in the circular tubes are different. When flowing in a circular tube, the fluid is surrounded and heated by the wall of the tube. On the other hand, in a tube-bundle channel, the fluid surrounds the tube bundle, which acts as the heating source. Then, all of these might cause differences between flow patterns in vertical tube-bundle channels and that in vertical circular tubes (Hewitt & Roberts, 1969).

Two-phase boiling flow in tube-bundle channels exhibits several main flow patterns, bubbly flow, bubbly-churn flow, churn flow and annular flow, as shown in Fig. 13. There are differences from the results gained by Grant & Chisolm (1979) and Ma (1992). These differences might be caused by the different test conditions. The characteristics of flow patterns and transitions in the present experiments are analyzed as follows.

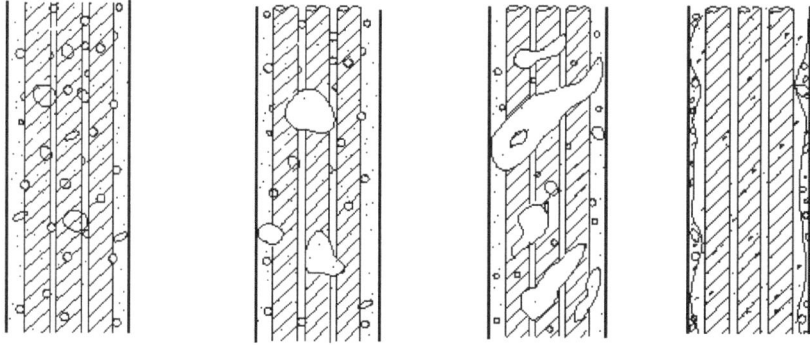

(a) Bubbly flow   (b) Bubbly- churn flow   (c) Churn flow   (d) Annular flow

Fig. 13. Schematic diagram of flow pattern in a tube-bundle channel of upward flow

(1) Bubbly flow: In the experiments, bubbles begin to come into being in the liquid when the heat flux is small. The mainstream is liquid and discrete bubbles are dispersed in the mainstream, which indicates that bubbly flow in circular tubes is similar to that in the tube-bundle channel, as shown in Fig. 13(a). With the sustaining heating, the dimension and the quantity of bubbles increase gradually along the flow direction, and discrete bubbles begin to aggregate to combine to be large bubbles. Confined by the narrow space in the tube-bundle channel, large bubbles transfigure to be oval or crescent.

(2) Bubbly-churn flow: When the little discrete bubbles aggregate and combine to be large bubbles, the tubes agitate and divide these large bubbles. This makes there be no slug flow in the tube-bundle channel. And the flow pattern begins to be transformed from bubbly flow to churn flow.

(3) Churn flow: With the augmentation of heat flux, discrete bubbles continue to aggregate and combine and the aggregation bubbles become bigger in size. These aggregation bubbles present unstable state, on account of the agitation and division of the tubes. And the aggregation bubbles begin to burst into many little discrete bubbles with unequal geometric dimensions. Then, the flow pattern is transformed to churn flow.

When churn flow occurs in the tube-bundle channel, there might be many bubbles with unequal geometric dimensions. The liquid moves up and down in the channel and the two-phase flow exhibits surge state.

(4) Annular flow: Heat flux continues increasing, and the dryness fraction in the channel increases too. When the vapor content is higher than that of churn flow, the liquid block is smashed and the vapor unites to be a continuous axle center in the core of the tube-bundle channel. The liquid film goes upward along the wall of PMMA pipe. Thus, annular flow occurs. The liquid film might be broken due to the effect of the vapor wave, as shown in Fig. 4 and Fig. 13 (d).

In the experiments, there is no mist flow due to the limit of the experiment condition, which is detected by Grant & Chisolm (1979).

There may be several differences between flow patterns and their transition in circular tubes and that in tube-bundle channels. Firstly, slug flow is one of the main flow patterns for two-phase flow in a circular tube. But the tube-bundle channels, the inner tubes divide the large bubbles and flows in the sub-channels interact, which makes bubbles can not aggregate and combine to be a slug. Thus there is a lack of slug flow in the tube-bundle channel and the flow pattern is transform directly from bubbly flow to churn flow. Secondly, churn flow, as the transition from slug flow to annular flow, exists transitory in circular tubes. Under some circumstances, there might be a lack of churn flow in circular tubes. But in the tube-bundle channel, churn flow presents itself as one of the main flow patterns. And churn flow exists in many experimental conditions and present a long time in the tube-bundle channel, as shown in Fig. 12.

With the same flow rate, there may be different flow patterns in the tube-bundle channel due to the different heat flux. For example, when the flow rate is 281.8 (kg/m^2s), bubbly flow occurs in the tube-bundle channel where the heat flux is comparatively small. But the amount of bubbles begin to increase with the enhanced heat flux. And the flow pattern changes from bubbly flow to churn flow when the heat flux is equal to 5.1×10^4 (w/m^2). This difference in flow patterns will be more obvious along with the decrease of the flow rate, which is shown in Fig. 12. When the flow rate is 133.9 (kg/m^2s), there will appear three flow patterns, which are bubbly flow, bubbly-churn flow and churn flow, under different heat flux in the tube-bundle channel. And when the flow rate is getting smaller, the flow patterns with differnet heat flux will be differentiated more distinctly. As shown in Fig. 12, there appear four different flow patterns, which are bubbly flow, churn flow, churn-annular flow and annular flow, when the flow rate is equal to 63.3 (kg/m^2s). From above all, it is shown that heat flux will affect the flow pattern more remarkably under the small flow rate condition.

Furthermore, it is found that there may be two different flow patterns in the same cross-section when the experiments run. The flow pattern transitions exhibit unsynchronized in different sub-channels. This unsynchronized phenomenon is caused by the different geometric dimensions, different heat flux and different quantity of discrete bubbles generated in different sub-channels. It was found the same phenomenon in a tube-bundle channel by adopting microprobe to detect the flow pattern transitions (Bergles 1981).

The flow pattern map obtained in the experiments is shown in Fig. 14. Comparisons are made with the results gained by Hewitt & Roberts (1969), which is figured by dashed line. Two-phase flow patterns and their transition in vertical circular tubes were experimentally studied by Hewitt & Roberts. The mass flow rate in the present experiments is less than that of Hewitt & Roberts owing to the experimental condition limits. In Fig. 6, the abscissa $\rho' j_f^2$ and the ordinate $\rho'' j_g^2$ are calculated by (Hewitt & Roberts, 1969),

$$\rho' j_f^2 = \frac{G^2 (1-x)^2}{\rho'} \tag{1}$$

$$\rho'' j_g^2 = \frac{G^2 x^2}{\rho''} \tag{2}$$

It is shown in Fig. 14 that there are great differences between flow pattern transitions in a tube-bundle channel and that in a circular tube. The generation regions of bubbly flow and churn flow in a tube-bundle channel move left in Fig. 14, compared with the region in a circular tube. And it is shown that the regions of bubbly flow and churn flow are larger in a tube-bundle channel, which is caused by that the inner tubes have effects of disturbance and division on the bubbles and these effects make it impossible for discrete bubbles to converge and unite to be a slug. In addition, on account of being heated by the inner tubes, the fluid generates bubbles in the core of the channel, which makes the probability increase for bubbles to converge and unit to be a continuous axle center in the core of the tube-bundle channel, and then the flow pattern transform to annular flow in a tube-bundle channel is earlier than that in a circular tube.

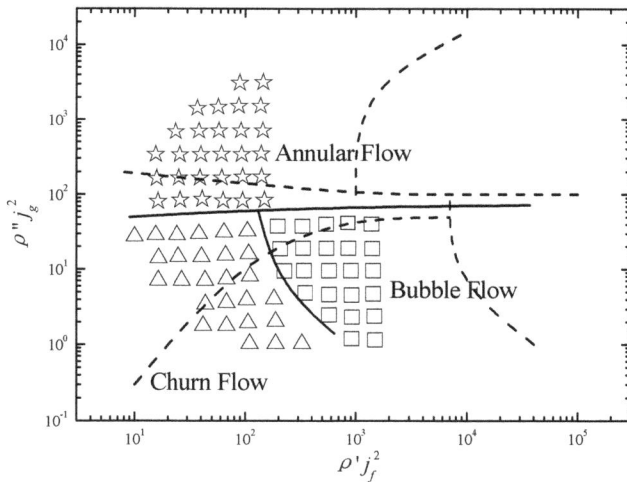

Fig. 14. Flow pattern map of two-phase boiling flow in a tube-bundle channel of upward flow

3.2 Test of containment sump strainer

In the tests of containment sump strainer, the solid debris such as fiber and particulate is put into the test system. And then the liquid-solid two-phase flow comes into being. Three main tests are carried out to measure pressure drop of the debris bed, which are clean screen test, thin bed test and full fiber/particulate load test. If there is vortex on the water surface, air may be inhaled into the pump, and then cavitation erosion would appear. So the vortex formation is carefully observed in these tests.

Clean screen test is performed without debris, which is shown in Fig. 15.

The thin bed test is conducted for the purpose of determining the amount of fiber fines which are necessary to completely cover the strainer. When the full coverage of the strainer screen is visually observed, the strainer screen should be photographed. The post-test photograph of thin bed test is shown in Fig. 16. It can be seen that the strainer is completely covered by fibrous debris and the thin bed is formed.

Fig. 15. Photograph of clean screen test

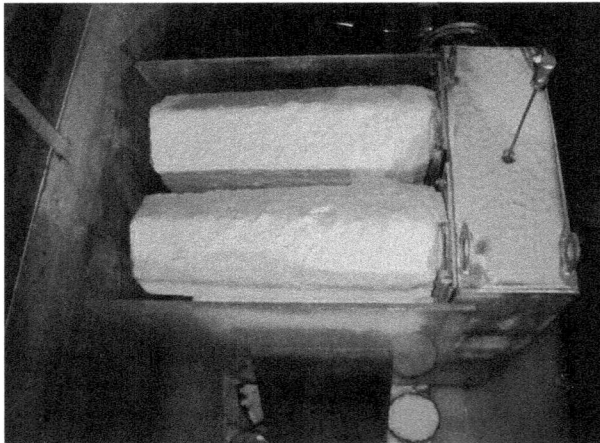

Fig. 16. Post-test photograph of thin bed test

Full fiber/particulate load test is performed to determine the head loss associated with the maximum fibrous and particulate debris load. In this test, full debris load is put into the test tank and the strainer is covered by more debris than that in the thin bed test. And the post-test photograph of full fiber/particulate load test is shown in Fig. 17.

As the safety nuclear apparatus, the containment sump strainers filter the debris out of the recycling water and provide the filtered water for the emergency core cooling system (ECCS) and the containment spray system (CSS). In order to keep the normal operation of ECSS pump and CSS pump, the containment sump strainers must guarantee sufficient NPSH (net positive suction head). Then the pressure loss due to the accident-generated debris accumulated on the sump screens should be one of the most important parameters of the containment sump strainers. And liquid-solid two-phase flow will appear when the accident-generated debris is flushed into the recycling water. NPSH of ECCS pump and CSS pump will be directly affected by this two-phase flow, of which the characteristics are important and significant for researchers to investigate.

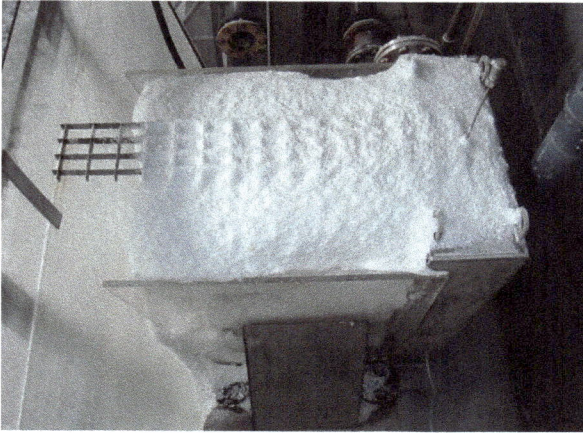

Fig. 17. Post-test photograph of full fiber/particulate load test

In the tests of containment sump strainer, the solid debris such as fiber and particulate is put into the liquid. And the liquid-solid two-phase flow comes into being, which is different from the flow patterns in tube-bundle channels. The liquid-solid two-phase flow and the debris bed covering on the containment sump strainer are carefully observed and recorded in the tests.

In the recycling course, the ECCS and CSS systems would adopt the water in the containment as pump source when the water in PTR001BA was used up. The debris generated by LOCA or HELB would be transported to the containment ground floor with the elevation of -3.5m. And a fraction of debris would accumulate on the sump screen which could induce pressure loss and might lead to the pump failure of ECCS and CSS systems. The debris transportation fraction to the sump strainers is analyzed by numerical simulation.

The authors take Daya Bay PWR for example to establish a 3-D computational model, with the purpose of studying the debris types and contents transported to the sump strainers. According to the actual dimension of the containment sump strainer in Daya Bay nuclear power station, a 3-D CAD model is established, as shown in Fig. 18. The altitude of the CAD model is ranging from -3.5m to 0m.

Then this CAD model is imported into Gambit, and the computational grid can be plotted. The cooper mode is adopted in defining the computational grid. In the complex locality and the key position, the computational grid is refined to resolve the important features. For the main part of the model, 5×5×5cm mesh spacing is induced in x-y-z directions. And the total cell amount in the model is 7,166,332, which is shown in Fig. 19.

In CFD model, the water temperature in the containment is set as 120°C and the pressure is 1.99bar. In this circumstance, the water is sub-cooled and the water density is 943kg/m^3, and the viscosity is 2.32×10^{-4} Pa·s.

According to the mass conservation principle and momentum conservation principle, continuity equation and momentum equation are established.

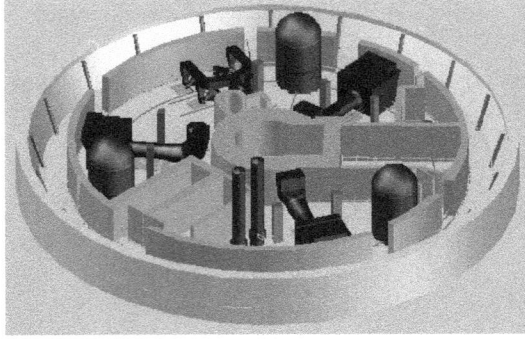

Fig. 18. 3-D CAD model of containment

Fig. 19. Computational grid of containment

$$\frac{\partial}{\partial t}\rho + \frac{\partial}{\partial x_i}(\rho u_i) = 0 \tag{3}$$

$$\frac{\partial}{\partial t}(\rho u_i) + \frac{\partial}{\partial x_j}(\rho u_i u_j) = -\frac{\partial p}{\partial x_i} + \frac{\partial}{\partial x_j}\left(\mu\frac{\partial u_i}{\partial x_j}\right) + S_i \tag{4}$$

Where, ρ is the liquid density, kg/m³. u is the flow velocity, m/s. t is time, s. μ is the dynamic viscosity Pa·s. S is the source term.

In the CFD calculation, the general transportation equations are established with the use of k-ε technique.

$$\frac{\partial}{\partial t}(\rho k) + \frac{\partial}{\partial x_i}(\rho k u_i) = \frac{\partial}{\partial x_j}\left[\left(\mu + \frac{\mu_t}{\sigma_k}\right)\frac{\partial k}{\partial x_j}\right] + G_k - \rho\varepsilon + S_k \tag{5}$$

$$\frac{\partial}{\partial t}(\rho\varepsilon) + \frac{\partial}{\partial x_i}(\rho\varepsilon u_i) = \frac{\partial}{\partial x_j}\left[\left(\mu + \frac{\mu_t}{\sigma_\varepsilon}\right)\frac{\partial\varepsilon}{\partial x_j}\right] + C_{1\varepsilon}\frac{\varepsilon}{k}G_k - C_{2\varepsilon}\rho\frac{\varepsilon^2}{k} + S_\varepsilon \tag{6}$$

$$\mu_t = \rho C_\mu \frac{k^2}{\varepsilon} \qquad (7)$$

$$G_k = \mu_t \left(\frac{\partial u_i}{\partial x_j} + \frac{\partial u_j}{\partial x_i} \right) \frac{\partial u_i}{\partial x_j} \qquad (8)$$

Where, μ_t is the turbulent viscosity coefficient. G_k is the turbulent energy generated by time-average velocity gradient. σ_k and σ_ε are turbulent Prandtl number of k equation and ε equation.

The velocity field of water flow in the containment is shown in Fig. 20. The water would tumble the sunken debris along the ground floor or lift debris over a curb in the area where the water velocity is high enough.

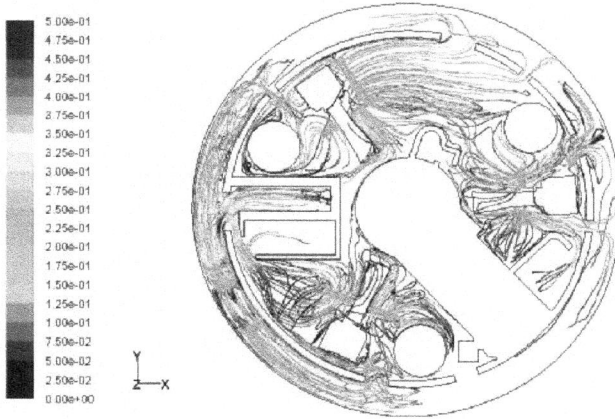

Fig. 20. Velocity field in containment

The experimental and numerical results gained above can provide necessary basis for the analysis of the properties of containment sump strainer and for the design of new-type containment sump strainers.

3.3 Test of control rod drive mechanism

In the beginning tests of control rod drive mechanism (CRDM), two phase flow would come into being because there is air dissolved in the liquid and hidden in the groove of driving rod. The air would be decomposed from the water and be extruded out of the groove when there is disturbance. For example, electrifying of coil component, up-down movement of driving rod, swing in and out of gripper component would give rise to disturbance and make the bubbles appear in the liquid. There is small amount of air dissolved or hidden in the water and two-phase flow would not occur in the test after CRDM moves several days.

In the tests, there is no bubble observed and the air would not be decomposed from the water when there is no disturbance, which is shown in Fig. 21(a). There are three coil components in CRDM, which are lifting coil, moving coil and stationary coil. And bubbles begin to come into being and there appear several separate bubbles when the coil component is electrifying and the gripper swings into the driving rod, which is shown in

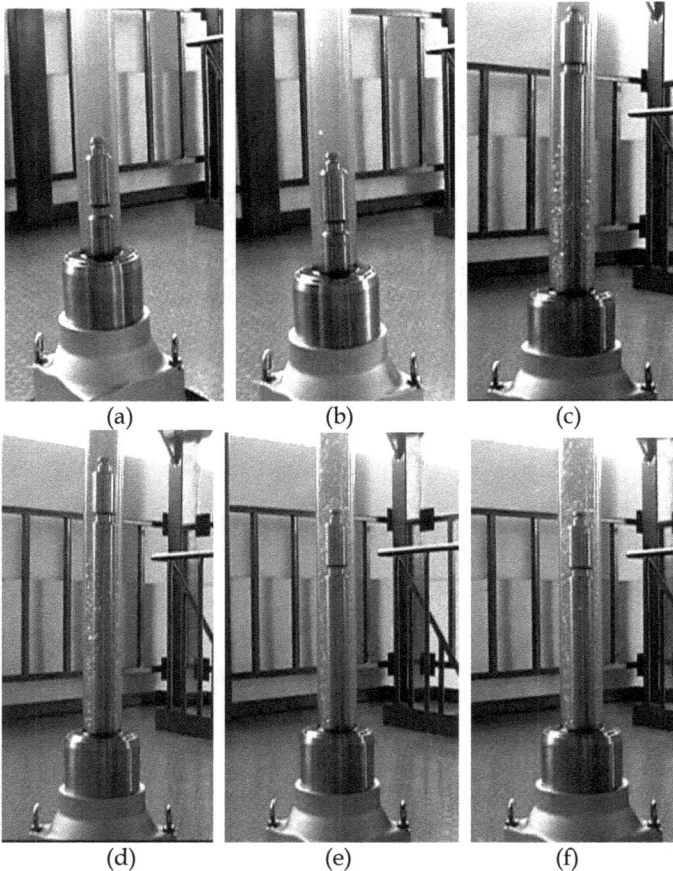

Fig. 21. Rod lifting tests

Fig. 21(b). In the tests of rod lifting, the driving rod goes upwards, the electromagnet is engaged and grippers swing into and out of the driving rod one by one. At this time, large numbers of bubbles appear in the liquid and bubbly flow come into being, as shown in figures 21(c) ~(f). In the tests of rod inserting, the phenomenon of downwards movement of the driving rod is similar to that of upwards movement.

In the tests of rod dropping, the power of three coils is cut off, the grippers swing out of the driving rod, and the driving rod free falls in the rod travelling house. In this process, the disturbance is transitory and bubbles appear in the liquid, of which the amount is less than that of rod lifting. When the bubbles go up to the bottom of the test section and there is no disturbance again, no bubbles will generate in the liquid, as shown in Fig. 22.

In the cold tests of rod lifting, rod inserting and rod dropping, only bubbly flow comes into being due to that the amount of gas dissolved in the liquid is small. After CRDM moves several days, the gas dissolved are all driven out of the liquid and there will not appear two-phase flow.

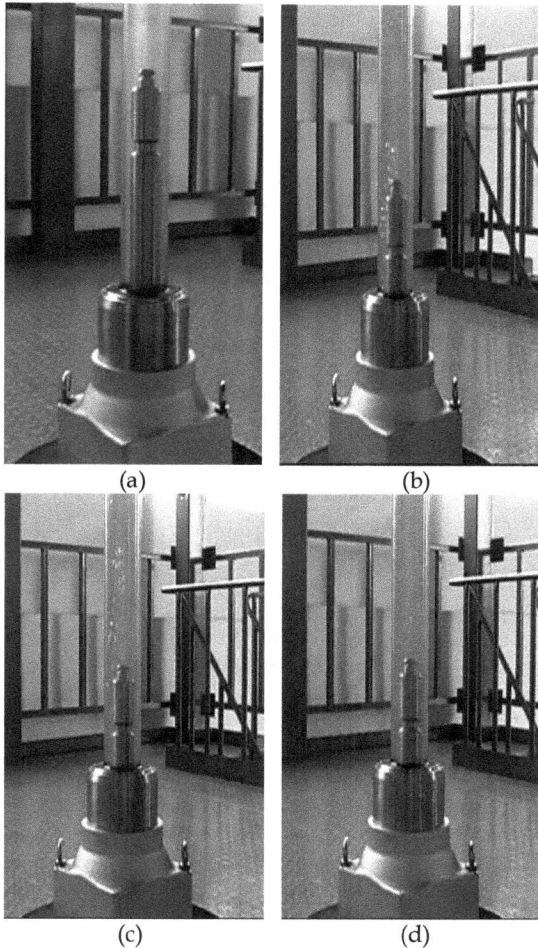

(a)   (b)

(c)   (d)

Fig. 22. Rod dropping tests

4. Conclusion

On the basis of the experimental results, the conclusions are obtained,

(1) It is found that there are several main flow patterns, bubbly flow, bubbly-churn flow, churn flow and annular flow in the tube-bundle channel. And there are great differences between flow patterns and their transitions in a tube-bundle channel and that in a circular tube.

(2) Experiments show that there may be two different flow patterns in the same cross-section of the tube-bundle channel. And the flow pattern transitions exhibit unsynchronized in different sub-channels. This unsynchronized phenomenon is caused by the different geometric dimensions, different heat flux and different quantity of discrete bubbles generated in different sub-channels.

(3) The flow pattern map is drawn on the basis of experiments. Comparisons are conducted between flow pattern transition in a tube-bundle channel and that gained by Hewitt & Roberts. The results show that the regions of bubbly flow and churn flow in a tube-bundle channel are larger than that in a circular tube. In addition, the flow pattern transforming to annular flow is earlier in a tube-bundle channel than that in a circular tube.

(4) Three main tests are carried out to measure pressure drop of the containment sump strainers. The photographs are taken and the vortex is not observed in these tests.

(5) In the tests of control rod drive mechanism, two-phase flow is observed. The results of upwards movement and drop movement of the driving rod are compared and analyzed.

5. Acknowledgments

The authors would like to acknowledge the financial support from the National Science and Technology Program of China (Grant No. 2011BAA06B00). The authors would also like to thank Messrs. Yu Jiang., Zhou Jianming, Wu Wei, Bai Bing, Lu Zhaohui for the helpful discussions.

6. References

Bergles, A. E. (1981). *Two-phase flow and heat transfer in the power and process industries*. ISBN: 978-0070049024 , McGraw-Hill Inc., Washington, US

Chan A. M. C. & Shroukri M. (1987). *Boiling characteristics of small multi-tube bundles. Journal of Heat Transfer*. Vol.109, pp.753-760. ISSN 0022-1481, New York, USA

Grant I. D. R. & Chisolm D. (1979). Two-phase flow on the shell-side of a segmentally baffled shell and tube heat exchanger. *Journal of Heat Transfer*. Vol.101, pp.38-42, ISSN: 0022-1481

Hewitt, G. F. & Roberts, D. N. (1969). *Studies of two-phase flow patterns by simultaneous X-ray and flash photography*, Atomic Energy Research Establishment, ISSN: 0029-5450, Harwell, England

Lu, G. Y.; Ren, J. S.; Zhang, C. G. & Lin, P. (2011). Debris transport calculation of nuclear power plant containment, *19th International Conference On Nuclear Engineering*, ISBN: 978-0-7918-4351-2, Osaka, Japan, October 24 - 25, 2011

Lu, G. Y.; Ren, J. S.; Zhang, C. G. & Lin, P. (2011). Investigation on pressure drop characteristics and disposal optimization of conflux channels of containment sump strainers, *19th International Conference On Nuclear Engineering*, ISBN: 978-0-7918-4351-2, Osaka, Japan, October 24 - 25, 2011

Lu, G. Y.; Ren, J. S.; Zhang, C. G. & Xiang, W. Y. (2011). Application and development of containment sump strainers in PWR power stations in China, *19th International Conference On Nuclear Engineering*, ISBN: 978-0-7918-4351-2, Osaka, Japan, October 24 - 25, 2011

Ma Weimin. (1992). Experimental investigations on two-phase flow in heat exchangers. *Xi'an Jiaotong University*, ISSN: 1671-8267

Petigrew, M. J. & Taylor C. E. (1994).Two-phase flow-induced vibration. *Journal of Pressure Vessel Technology*, Vol.166, pp. 233-253, ISSN: 0094-9930

Sadatomi, M. & Kawahara, A. (2004). Flow characteristics in hydraulically equilibrium two-phase flows in a vertical 2×3 rod bundle channel. *International Journal of Multiphase Flow*, Vol.30, pp.1093–1119, ISSN: 0301-9322

Analysis of Emergency Planning Zones in Relation to Probabilistic Risk Assessment and Economic Optimization for International Reactor Innovative and Secure

Robertas Alzbutas[1,2], Egidijus Norvaisa[1] and Andrea Maioli[3]
[1]Lithuanian Energy Institute
[2]Kaunas University of Technology
[3]Westinghouse Electric Company
[1,2]Lithuania
[3]USA

1. Introduction

Probabilistic Risk Assessment (PRA) tehniques applied to the definition of Emergency Planning Zone (EPZ) have not reached the same level of maturity when dealing with external events as PRA methodologies related only to internal events (Alzbutas et al., 2005). This is even of greater importance and relevance when PRA is used in the design phase of new reactors (IAEA-TECDOC-1511, 2006; IAEA-SSG-3, 2010; IAEA-SSG-4, 2010).

The design of the layout of a Nuclear Power Plant (NPP) within its identified site, with the arrangement of its structures, as well as the definition of the EPZ around the site can be used to maximise the plant safety related functions, thus further protecting nearby population and environment. In this regard, the design basis for NPP and site is deeply related to the effects of any postulated internal and external hazardous event and the possibilities of the reactor to cope with related accidents (i.e., to perform the plant safety related functions).

Among the objectives for advanced reactors there is the aim to establish such a higher safety level with improved design characteristics that would justify and enable revised emergency planning requirements. While providing at least the same level of protection to the public as the current regulations, ideally, but still not realistically, the total elimination of hazards' consequences would result in the EPZ coincidinge with the site boundary, thus, there would be no need for off-site evacuation planning, and the NPP would be perceived as any other industrial enterprise.

In this chapter, the International Reactor Innovative and Secure (IRIS) is adopted as a prime example of an advanced reactor with enhanced safety. The IRIS plant (Carelli, 2003, 2004, 2005) used a Safety-by-Design™ philosophy and such that its design features significantly reduced the probability and consequences of major hazardous events. In the Safety-by-Design™ approach, the PRA played a key role; therefore a Preliminary IRIS PRA had been developed along with the design, in an iterative fashion (Kling et al., 2005). This unprecedented application of the PRA techniques in the initial design phase of a reactor was also extended to the external event with the aim of reviewing the EPZ definition. To achieve this particular focus was dedicated to PRA and Balance Of Plant (BOP).

For the design and pre-licensing process of IRIS, the external events analysis included both qualitative evaluation and quantitative assessment. As a result of preliminary qualitative evaluation, the external events that had been chosen for more detailed quantitative assessment were as follows: high winds and tornadoes, aircraft crash and seismic activity (Alzbutas et al., 2005, Alzbutas & Maioli, 2008).

In general, the analysis of external events with related bounding site characteristics can also be used in order to optimize the potential future restrictions on plant siting and risk zoning. Due to this and Safety-by-Design™ approach, IRIS, apart from being a representative of innovative and advanced reactors, had the necessary prerequisite, (i.e., excellent safety), for attempting a redefinition of EPZ specification criteria, IRIS was therefore used as a test-bed.

The work presented in this chapter was performed within the scope of activities defined by the International Atomic Energy Agency (IAEA) Co-ordinated Research Project (CRP) on Small Reactors with no or infrequent on-site refuelling. Specifically, it was relevant to "Definition of the scope of requirements and broader specifications" with respect to its ultimate objective (revised evacuation requirements), and to "Identification of requirements and broader specifications for NPPs for selected representative regions" considering specific impact on countries with colder climate and increased interest for district heating co-generation.

The economic modelling and optimization presented in the second part of the chapter was concentrating on the evaluation of possibilities to construct a new energy source for Lithuania. The MESSAGE modelling tool was used for modelling and optimization of the future energy system development (IAEA MESSAGE, 2003). In this study, the introduced approach was applied focusing on Small and Medium nuclear Reactor (SMR), which was considered as one of the future options in Lithuania. As an example of SMR, the IRIS nuclear reactor was chosen in this analysis.

If IRIS with reduced EPZ could be built near the cities with a big heat demand is, it could be used not only for electricity generation, but also for heat supply for residential and industrial consumers. This would allow not only to reduce energy prices but also to decrease fossil fuel consumption and greenhouse gas emissions.

Finally, the analysis of uncertainty and sensitivity enabled to investigate how uncertain were results of this modelling and how they were sensitive to the uncertainty of model parameters (Alzbutas et al, 2001).

In summary, the study presented in this chapter consists of two main parts: the analysis of EPZ in relation to PRA with focus on external events, and the economic optimization of future energy system development scenarios with focus on sensitivity and uncertainty analysis in relation to initial model parameters. The study explicitly uses features of IRIS technology and a potentially reduced EPZ.

2. Approach used for IRIS

2.1 Safety-by-Design™ concept

The IRIS designers used the Safety-by-Design™ philosophy from its inception in 1999. Such a designing approach had been outlined in detail in previous works (Carelli, 2005), (Carelli, 2004); here it is suffice to remember that the key idea of the Safety-by-Design™ concept is to physically eliminate the possibility of occurrence or to reduce consequences of accidents, rather than focusing only on the mitigation phase.

The most evident implication of this design approach is the choice of an integral reactor configuration, where the integral reactor vessel (containing eight internal steam generators and reactor coolant pumps) and the internal control rod drive mechanism were introduced causing the consequential absence of large primary pipes. Such a configuration enabled to have either eliminated major design basis events such as Large Break LOCA (loss-of-coolant accident) or rod ejection and also to have significantly reduced the consequences of them.

This Safety-by-Design™ approach was used by the designers of IRIS to eliminate the possibility of occurrence of certain severe accidents caused by internal events and have been extended to the external events. The focus was on the balance of plant that had not been analyzed as extensively or explicitly as NPP accidents caused by internal events. However, since extreme external events, in general, have one of the largest contributions to the degradation of the defence in depth barriers, the external events, especially for new NPPs, represent a major challenge to the designer in order to determine siting parameters and to reduce the total risk.

2.2 External event analysis

The preliminary qualitative analysis and screening of external events considered for the IRIS PRA was, in general, based on the external events PRA methodology developed by the American Nuclear Society (ANSI/ANS-58.21, 2003) and on the PRA's of other NPPs (CESSAR-DC, 1997).

For the quantitative analyses, bounding site characteristics were used in order to minimize potential future restrictions on plant siting. The following four separate steps were performed in order to identify external events to be considered:

1. Initial identification of external events to be analysed in detail.
2. Grouping of events with similar plant effects and consequences.
3. Screening criteria establishment to determine which events are risk insignificant and can therefore be excluded from detailed quantitative analysis.
4. Each event evaluation against the screening criteria to determine if the event is risk-significant and thus requires further quantitative analysis.

PRA Guides and PRAs of existing plants were used as the sources for list of external events development in order to ensure that all external events already recognized as possible threats for IRIS were taken into consideration. The resultant set of external events represented a consensus listing of external events. Then, the list was reviewed in order to group all the external events that are likely to have the same impact on the plant. During this grouping the specific screening criteria were also applied to determine, which events are risk-insignificant and could be excluded from quantitative analysis.

The criteria used for excluding external events from detailed quantitative analysis are:

1. The plant design encompasses events of greater severity than the event under consideration. Therefore, the potential for significant plant damage from the event is negligible.
2. The event cannot occur close enough to the plant to have an effect on the plant's operation.
3. The event has a significantly lower mean frequency of occurrence than other events with similar uncertainties and could not result in worse consequences than those events.
4. The event is included, explicitly or implicitly, in the occurrence frequency data for another event (internal or external).
5. The event is slow in developing, and it can be demonstrated that there is sufficient time to eliminate the source of the threat or to provide an adequate response.

As it is evident form screening criteria, some external events may not pose a significant threat of a severe accident, if they have a sufficiently low contribution to core damage frequency or plant risk. So, the final step in the qualitative analysis process was the evaluation of each external event against the screening criteria to determine if the event was risk-insignificant and could be excluded from further analysis. Thus, the external events identified as described above were screened out in order to select only the significant events for detailed risk quantification.

As a result of the qualitative analysis or screening criteria application , the identified external events that had beed needed further quantitative scoping evaluation to determine their impact on the core damage were as follows: aircraft crash, high winds or tornadoes and seismic activity.

This list of external events that require an additional analysis was consistent with previous PRAs and with what had been suggested for analysis and the individual plant examination of external events (NUREG-1407, 1991). In addition, a few so called area events such as internal flooding and internal fires were also considered for IRIS. Also an impact of aircraft crash, that had been modelled and quantitatively analysed previously (Alzbutas et al., 2003) was included in the IRIS PRA and presented as an example related to risk zoning (Alzbutas & Maioli, 2008).

2.3 IRIS designing features

The Safety-by-Design™ approach, used by the designers of IRIS to eliminate the possibility of occurrence of certain severe accidents caused by internal events, had been extended to the external events.

The normally operating IRIS systems and their non-safety, active backup systems were typically located within substantial structures that can withstand some degree of external event challenges. This equipment included the backup diesel generators. IRIS had non-safety related backup diesels for normally available active equipment that could bring the plant to cold shutdown conditions.

IRIS plant safety features, once actuated, relied on natural driving forces such as gravity and natural circulation flow for their continued function. These safety systems did not need diesel generators as they are designed to function without safety-grade support systems (e.g. AC power, component cooling water, or service water) for a period of 7 days.

All the IRIS safety related equipment, including the batteries that provide emergency power, and the passive habitability system, were also located within concrete structures. The reactor, containment, passive safety systems, fuel storage, power source, control room and backup control were all located within the reinforced concrete auxiliary building and were protected from on-site explosions.

Actually, IRIS had a very low profile, which was very important when considering aircraft crash, especially by terrorists. The IRIS containment was completely within the reinforced concrete auxiliary building and one-half of it (13 m) was actually underground. The external, surrounding building was only about 25 m high, thus offering a minimal target. The integral vessel configuration eliminated loop piping and external components, thus enabling compact containment (see Figure 1) and plant size.

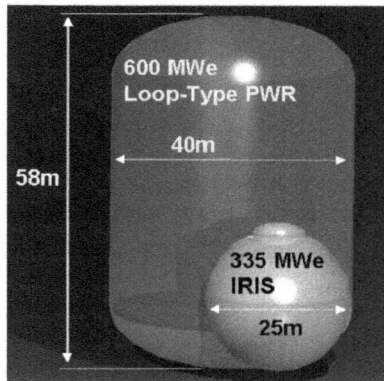

Fig. 1. IRIS Containment

The Refuelling Water Storage Tank (RWST) which is the plant's ultimate heat sink would be also protected from some external events by locating it inside the reinforced concrete auxiliary building structure. In addition, the IRIS RWST was designed to be replenished by alternative water sources such as fire trucks, therefore being completely independent by the plant power resources.

Because of these and other reasons, it was expected that the impact of external events at the site would be lower than that for current plants. In addition, typical design approaches, that could contribute to achieve such robustness in advanced NPPs design are:

- Capability to limit reactor power through inherent neutronic characteristics in the event of any failure of normal shut-down systems, and/or provision of a passive shut-down system not requiring any trip signal, power source, or operator action.
- Availability of a sufficiently large heat sink within the containment to indefinitely (or for a long grace period) remove core heat corresponding to above-mentioned events.
- Availability of very reliable passive heat transfer mechanisms for transfer of core heat to this heat sink.

It was observed that the implementation of innovative design measures needs to be supported (and encouraged) by a rational, technical and non-prescriptive basis to exclude any severe accident (core melt need not be presupposed to occur). The rational technical basis should be derived from realistic scenarios applicable for the plant design. Most of the innovative reactor designs aimed to eliminate the need for relocation or evacuation measures outside the plant site, through the use of enhanced safety features in design. Many of these designs also aimed to take advantage of these advanced safety characteristics to seek exemption from maintaining a large exclusion distance around the nuclear power plants.

3. EPZ in relation to PRA

3.1 PRA application for IRIS design

In the Safety-by-Design™ approach, the Probabilistic Risk Assessment plays obviously a key role, therefore a Preliminary IRIS PRA was initiated, and developed with the design, in an iterative way. This unprecedented application of the PRA techniques in the initial design phase of the reactor and the deep impact that this had in the development of the project was described in already published papers (Carelli, 2004, 2005).

Summarizing this, it is possible to note, that the success of the IRIS Safety-by-Design™ and PRA-guided design in the internal and external events assessments (Carelli, 2004) is due to the effective interactions between the IRIS Design team and the IRIS PRA team (see Figure 2). The main task of the PRA team was to identify high risk events and sequences.

The IRIS Design team provided information concerning the IRIS plant and site design. It updated IRIS component/system description and design data. PRA team identified assumptions concerning IRIS plant and site design requirements. The design team then reviewed assumptions concerning IRIS plant and site design requirements.

A preliminary evaluation of internal and external events was performed in the Preliminary IRIS PRA, to determine if there were any unforeseen vulnerabilities in the IRIS design that could be eliminated by design during the still evolving design phase of the reactor. The preliminary analysis of external events included both quantitative and qualitative analyses. For the quantitative analyses, bounding site characteristics were used in order to minimize potential future restrictions on plant siting.

Referring to Figure 3, it can be seen that the initial PRA for internal events resulted in a Core Damage Frequency (CDF) of $2.0 \cdot 10^{-6}$. The PRA team then worked with the IRIS design team in order to implement design changes that improved plant reliability and to identify additional transient analyses that showed no core damage for various beyond design basis transients. The resulting CDF around $1.2 \cdot 10^{-8}$ was therefore obtained thanks to a combination of the Safety-by-Design™ features of the IRIS design, coupled with the insights provided by the PRA team regarding success criteria definition, common cause failures, system layout, support systems dependencies and human reliability assessment.

Being still in a design development/refinement phase, the PRA was kept constantly updated with the evolution of the design; moreover, all the assumptions required to have a reasonable complete PRA model capable of providing quantitative insights as well as qualitative considerations, were accurately tracked down and the uncertainties connected with such assumptions were assessed. These refinements of the Preliminary IRIS PRA yielded a predicted CDF from internal events around $2.0 \cdot 10^{-8}$.

Analysis of Emergency Planning Zones in Relation to Probabilistic Risk Assessment and Economic Optimization for
International Reactor Innovative and Secure

49

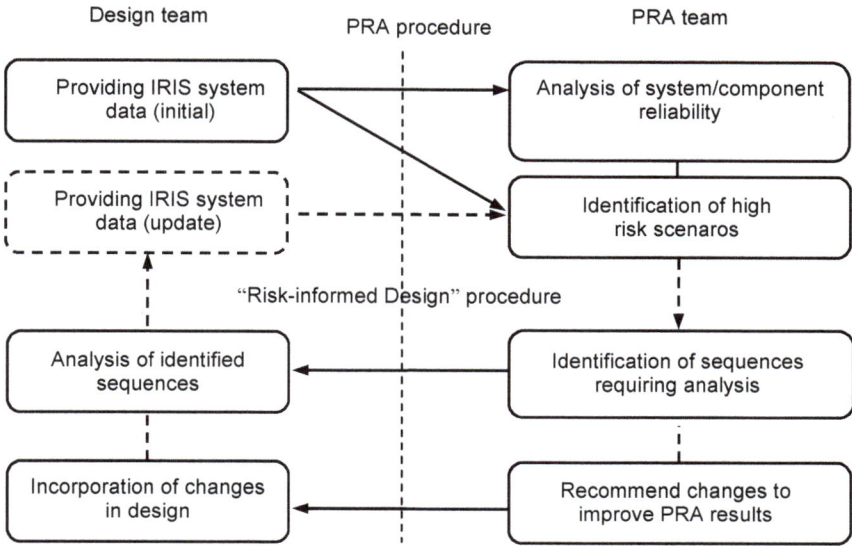

Fig. 2. IRIS Design and PRA Team Interactions

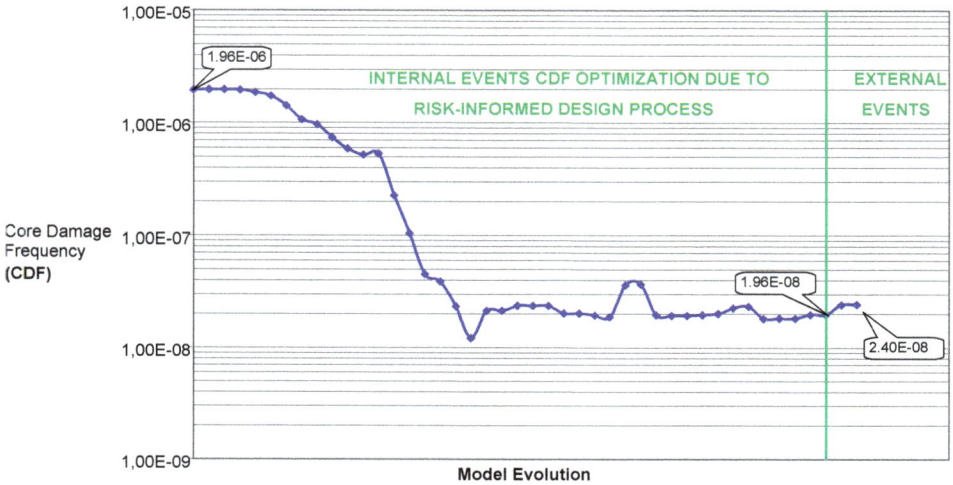

Fig. 3. IRIS Design CDF History

The same method was extended also to the external events. In comparison to events dominant in other plant PRA, the IRIS plant was expected to be significantly less vulnerable to some external events. In external events PRA, the focus was set on the plant BOP, that has not been analyzed as extensively or explicitly as accidents caused by internal events. In general, the IRIS plant arrangement structures were designed to minimize the potential for natural and manmade hazards external to the plant from affecting the plant safety related functions. The external events PRA insights were expected to help taking full advantage of the potential safety oriented features of the IRIS design and this implied probabilistic consideration of extreme winds, fires, flooding, aircraft crash, seismic activity, etc. In addition, it was shown that estimation of risk measures could be related to the site size and could be the input for the emergency zone planning.

3.2 Risk zoning practices

In fact, in the context of some severe external events, the assumption of continued availability of infrastructure required to administer emergency measures (for example roads and bridges) may not be valid. Under such situation, it is more effective to enhance the quality of the other levels of defence in depth. There is therefore, a need to define the scope of off-site emergency planning activities for advanced reactors, consistent with the ability of these reactor designs to meet enhanced safety objectives.

In some cases, such as the presence of a nearby airport, the consideration of the hazards may change risk zoning or eliminate a site from further consideration for an NPP, but most external hazards are either screened out from the necessity of being considered further or are taken into account in plant designing and siting. Risk zoning and siting is a matter for:

- The uncertainties of risk measures and influence to the public perception;
- Economic consideration (where power is needed, the availability of existing grid);
- Social and political factors;
- Topography affecting the dispersion of radio-nuclides through the atmosphere, rivers and ground-water;
- Political and safety consideration;
- Demographic characteristics;
- Hazards (natural and manmade).

Some IAEA Member States only address the risk to an individual member of the public, others have requirements to consider the potential aggregated effects to the population as a whole – societal risk.

Usually, off-site emergency measures are still seen as part of the Defence in Depth approach, which is mainly understood in deterministic sense, but to take full advantage of new reactor designs it should be moved towards a more probabilistic approach based on risk assessment with sensitivity and uncertainty analysis. The full benefit of innovative and evolutionary NPP requires the ability to licence without the need of an off-site Emergency Planning Zone.

In general, the desirability or possibility of reducing emergency response plans for accidents depends not only on the reactor type but also on a number of complex and intertwined factors including technical, societal, economical and cultural. The subject cannot be coupled directly and solely to the requirements for the external events but requires a separate consideration. Under the same subject also the risk-informed decision making related to the design basis accidents and severe accidents may be considered with the intent of moving away from somehow postulated risk zones and towards clearly calculated risk zones. Without such a change, related procedures and criteria, the issue of the emergency response plans cannot be resolved. In particular, in order to deal with external events and apply the risk-informed approach for plant design and siting, it is desirable to couple the PRA with analysis techniques of civil engineering.

3.3 Enhanced licensing framework

The ultimate objective for advanced NPPs is to establish an enhanced approach to licensing, reflecting improved safety characteristics of advanced reactors, that is expected to justify and enable revised (reduced or eliminated) emergency planning requirements, while providing at least the same level to protection to the public as the current regulations. Ideally, the emergency planning zone would coincide with (or be contained within) the site boundary, thus, there would be no need for off-site evacuation planning, and the NPP would become, relative to the general population, the same type of facility as any other industrial enterprise.

In order to contribute toward achieving this ultimate objective by addressing some of the relevant issues there is a need to consider the following research tasks:

- Critically evaluate current regulations to identify what changes are necessary to enable advanced licensing.
- Identify criteria based on technical, quantifiable parameters that may be used in support of the objective.
- Identify approach, based on a combination of deterministic modelling, probabilistic analysis, and risk management, which will enable assessment of advanced plants based on their key design operational and safety characteristics with respect to adequate emergency planning requirements.
- Prepare site-specific representative data (e.g., meteorological).
- Perform probabilistic analyses needed to support the proposed approach.
- Perform deterministic / dose evaluation analyses needed to support the proposed approach.
- Perform a detailed evaluation of the representative reactor utilizing the combined proposed approach.
- Identify, discuss and quantify the benefits attainable through the implementation of this objective, i.e., licensing with reduced emergency planning requirements.

In order to perform these tasks with the ultimate goal of developing a technology-independent approach, the design of IRIS was used as a testbed. IRIS was representative of innovative reactors, but because it was a LWR, its possible sequences and its behaviour under accident conditions was much better understood and predicted than that of some more distant new technologies. Moreover, it had the necessary prerequisite, excellent safety, due to its Safety-by-Design™ approach.

The related work was within the scope of activities defined within the International Atomic Energy Agency (IAEA) Co-ordinated Research Project (CRP) on Small Reactors with no or infrequent on-site refuelling. Specifically, it was relevant to "Definition of the scope of requirements and broader specifications" with respect to its ultimate objective (revised evacuation requirements) and to "Identification of requirements and broader specifications for NPPs for selected representative regions" considering specific impact on countries with colder climate and high interest for district heating co-generation.

It was expected that these results would contribute to ultimately defining a generic, country-independent approach and would support development of justification for reduced emergency planning through PRA analyses.

In addition, a study of the economic impact of revised licensing requirements on district heating was initiated. Thus the task was to perform economic study to evaluate positive economic effect on the nuclear district heating co-generation option, due to revised siting requirements with reduced emergency planning, which would allow placement of NPPs closer to population centres and allow them to be attractive option in heat energy supply market.

Finally, as part of this IAEA CRP a general methodology for revising the need for relocation and evacuation measures unique for NPPs for Innovative SMRs was developed and issued as IAEA publication (IAEA-TECDOC-1487, 2006).

Regarding further elaboration of the methodology it was suggested that external events and reasonable combinations of the external and internal events need to be included in the initial step of the methodology (accident sequence re-categorization), as for advanced reactors with the enforced inherent and safety by design features it might be that the impacts of external events would dominate the risk of severe accidents with possible radioactivity release. Work in this direction had already been started and was continued further, see (Alzbutas & Maioli, 2008) and (IAEA-TECDOC-1652, 2010).

3.4 Reduction of emergency planning zones

The developed approach for reduction of emergency planning zones (EPZ) was summarized in related IAEA document (IAEA-TECDOC-1652, 2010). It is applicable and recommended for all types of SMRs without on-site refuelling. The spatial extents of regulatory-mandated EPZ have historically been set according to conservative approaches for calculating bounding individual dose rates subsequent to a postulated accident sequence. The zones are not small – ranging up to 10 kilometers or even miles in radius. Moreover, regulations often require the reactor owner to provide for emplacement of infrastructure such as roads and bridges throughout the EPZ to facilitate public evacuation – as well as to periodic training and equipment supply to first responders. Current practice has been developed over many years specifically for the historical and current situation of large water-cooled reactor installations generating electricity for a regional grid.

Alternately, SMRs without on-site refuelling are being designed for local grids and some are even designed for cogeneration missions wherein the reactor must of necessity be placed very near the cogeneration application due to short heat transport distances. EPZ defined for large reactors on a one-size-fits-all basis can place a severe economic disadvantage on SMRs without on-site refuelling. For this reason, the IAEA CRP has conducted a review of the basis for the current regulations and has proposed a risk-informed methodology which could justify a reduced emergency planning zone extent on the basis of a smaller source term and a reduced probability of release for advanced SMRs, accounting for their passive safety and other risk reduction features. The methodology is not limited to small reactors without on-site refuelling, but is unique to many NPPs with innovative SMRs and larger reactors.

Within this methodology the information gathered from the PRA (both internal and external events) may be used to provide a basis for the redefinition of the EPZ defining criteria. The proposed approach consists of coupling the PRA results with deterministic dose evaluations associated to each relevant PRA sequence considered, and thus achieving a technically sound bases for the definition of a plant specific EPZ. In this approach the two basic components of risk (i.e. probability of occurrence and consequences of a given accident) are therefore explicitly combined. The EPZ radius then is defined as the distance from the plant such that the probability of exceeding the dose limit triggering the actuation of emergency procedure is equal to a specified threshold value. To identify this threshold value, detailed analysis of existing installations should be performed to infer the risk associated with the current EPZ definition.

The study conducted in the CRP included a sample application of the developed methodology for the IRIS-like SMR design under conditions of a particular site. This application indicated a potential for remarkable reduction of EPZ radius without increase in the public risk. However, to achieve this practically the proposed methodology first needs to be embraced by regulatory authorities. More details of the methodology and its trial application are provided in related IAEA document (IAEA-TECDOC-1652, 2010).

It must be noticed that the use of existing regulations and installations as the basis for this redefinition will not in any way impact the high degree of conservativism inherent in current regulations. Moreover, the remapping process makes this methodology partially independent from the uncertainties still affecting probabilistic techniques. Notwithstanding these considerations, it is still expected that applying this methodology to advanced plant designs with improved safety features will allow significant reductions in the emergency planning requirements, and specifically the size of the EPZ. In particular, in the case of IRIS it was expected that taking full credit of the Safety-by-Design™ approach of the IRIS reactor will allow a dramatic reduction in the EPZ requirement, while still maintaining a level of protection to the public fully consistent with existing regulations.

4. Case study for EPZ and economic optimization

The series of various studies were carried out in order to answer the question what energy sources should replace the lost nuclear electricity capacities (IAEA-TECDOC-1408, 2004; IAEA-TECDOC-1541, 2007). Currently all Baltic region countries cooperating and seeking to solve energy supply and energy security problems and planning to construct new nuclear reactors in Lithuania at existing NPP site. The last Ignalina Unit 2 RBMK-1500 was closed in the end of 2009 and Lithuania is considering both nuclear and fossil options for its replacement. In order to expand the research of the Lithuanian energy future another option to the analysis is also added: small and medium type nuclear reactor in the new site close to the cities with large heat demand. In general, it could be considered for small countries as alternative for the big nuclear units due to limitation imposed by the grid size and available financial resources.

The results of various studies concerning the future structure of power plants in Lithuanian energy system have showed that looking from the economical point of view the best options to replace Ignalina NPP are new nuclear unit or new combined cycle condensing units together with the existing and new units of Combined Heat and Power plants (CHP) (IAEA-TECDOC-1408, 2004; Norvaisa, 2005). Due to climate conditions in Lithuania, district heating presents a notable fraction of energy consumption in winter months, and infrastructure for its use is already in place in population centers in Lithuania. District heating is widely used in Lithuania (46% of total heat consumption), and the cities of Vilnius and Kaunas comprise the two largest consumers of district heat supply (see Table 1). So, in the future new cogeneration units are likely to be the best alternative for electricity and heat generation in Lithuania.

District heat (DH) supply:			Fuel structure:		
Total DH supply	GWh	9300	Natural gas	%	74
DH supply in Vilnius	GWh	3000	Renewables	%	19
DH supply in Kaunas	GWh	1600	Oil	%	5.5
Losses in DH	%	16	other	%	1.5

Table 1. Lithuanian district heating (DH) sector in 2009

Among the nuclear alternatives the 330 MW(e) IRIS-like reactor was used for the conceptual investigation, as it could be operated in either the electricity only or the co-generation mode. Thus, a case study was conducted to determine the best way to provide the electricity and district heat in Lithuania up to year 2025 and to assess the tactical implications that a reduced-radius emergency planning zone might have on least cost planning with the IRIS-like reactor operating in the electricity only versus the electricity/district heat (co-generation) mode (Norvaiša & Alzbutas, 2009).

The length of any newly required hot water/steam pipelines into the cities of Vilnius and Kaunas will depend on the radius of the emergency planning zone emplaced around the IRIS-like reactor site; these pipelines represent a cost due to construction and a cost due to heat losses. Both costs increase with the pipeline length and, thereby, affect the viability of the co-generation mode. The case study was conducted parametrically for pipeline lengths of 0.5, 5, 15, and 30 km.

4.1 Modelling of Lithuanian energy system

The IAEA's energy planning tool, MESSAGE, was used to model several alternate scenarios for Lithuania for a time horizon until 2025. The MESSAGE is an optimization model which from the set of existing and possible new technologies selects the optimal, in terms of selection criterion, mix of technologies capable to cover given country demand for various energy forms during the whole study period (IAEA MESSAGE, 2003). Table 2 lists the scenario options that were considered.

Figure 4 shows the base case, where the Ignalina NPP comes off line in 2009, and electricity and heat production is provided by the new fossil plants – some of which operate in electricity mode and some in co-generation mode. No nuclear plant is deployed in the base case. Figure 5 compares the total cost of this base case to the costs of the several options where IRIS-like NPP is deployed; clearly IRIS is a preferred option, no matter what configuration of its deployment.

At Lithuania's largest cities of Vilnius and Kaunas, the heat distribution pipelines already run through the neighbourhoods emanating from a massive heating plant sited at the outer edge of the city. Figures 6 and 7 show the evolution of electricity and district heat delivery by IRIS-like reactors operating in the cogeneration mode under the condition that the IRIS-like reactors can be sited at the city's edge within 0.5 km of the distribution header of the district piping network. This option is labelled 'IRIS cogeneration' in Figure 5 and is the overall lowest cost option. The optimization shows that by 2025, three IRIS-like reactors would have been deployed and would be supplying 44% of Lithuania's electricity (Figure 6) and 31% of Lithuania's district heat (Figure 7) centred in the cities of Vilnius and Kaunas.

No.	Scenario name	Description
1	No IRIS-like NPP	Base scenario: construction of IRIS-like NPP is not allowed
2	Co-generation with IRIS-like reactor	Construction of IRIS-like NPP (with co-generation option) is allowed in Vilnius and Kaunas cities. No additional heat supply network must be constructed. (0.5 km pipeline)
3	'IRIS EPZ' - IRIS-like NPP with larger emergency planning zone	Construction of IRIS-like NPP (with co-generation option) is allowed in Vilnius and Kaunas cities. The EPZ is parametrically assumed to be 5-30 km. Construction of IRIS-like units only for electricity production is also allowed in other locations.
4	IRIS-like NPP for electricity only	Construction of IRIS-like units used only for electricity generation is allowed (no co-generation option).

Table 2. Description of scenarios

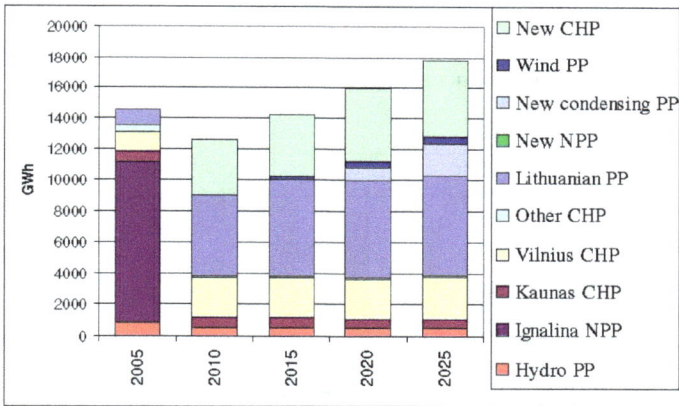

Fig. 4. Dynamics of electricity production in the case of 'No IRIS-like NPP' scenario (CHP – combined heat and power plant, PP – power plant)

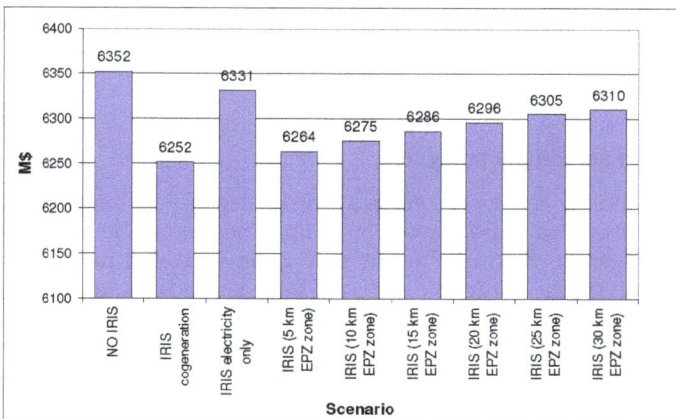

Fig. 5. Discounted total cost of energy system operation and development until 2025

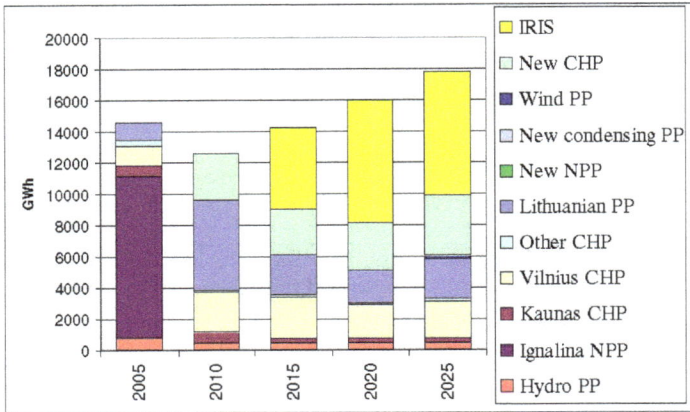

Fig. 6. Dynamics of electricity production in the case of 'IRIS cogeneration' scenario

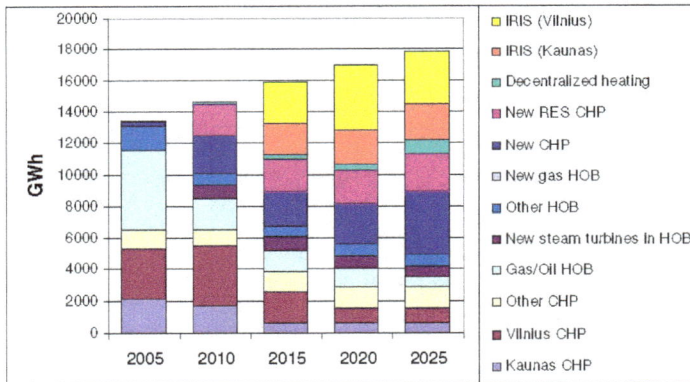

Fig. 7. Heat production by technologies in Lithuania for 'IRIS cogeneration' scenario (HOB – heat only boilers, RES – renewable energy sources)

4.2 Uncertainty and sensitivity analysis

Usually, the calculation results of such analysis depend on initial parameters and modelling techniques. In any case, having quite uncertain parameters, the influence of the main initial parameters to the main results may be investigated performing uncertainty and sensitivity analysis. In particular case, the analysis focuses on how the calculation results could change, when the initial parameters describing IRIS-like technology in MESSAGE model are changed. In addition, the most important parameters for the precision of calculation results were also indentified.

The main parameters and their possible values (describing the IRIS-like technology in the model) for different scenarios are presented in Table 3.

Par. No.	Parameter	Distribution type	Reference value	Min	Max
1	IRIS Investments, $/kW	Uniform	1410	1410	2000
2	IRIS fixed O&M costs, $/kW	Uniform	44.8	44.8	67.2
3	Discount rate, %	Uniform	5	5	10
4	IRIS starting year	Discrete	2015	2010	2025
5	Heat pipeline length, km	Discrete	0	0	30
6	Nuclear fuel cost, $/kWyr	Uniform	11.3	11.3	15

Table 3. Uncertain parameters and data for scenario generation

The reference values of some parameters presented in this table are taken from (Alzbutas & Maioli, 2008). The possible variations of these parameters are based on calculation assumptions. For instance, in the calculations it was assumed that the EPZ could change from 0 to 30 kilometers. In the model this is represented by the length of additional heat supply pipe in order to connect IRIS-like NPP with cogeneration option to the existing district heating network. In addition, the number of IRIS units in the MESSAGE model is adapting depending on specific conditions in the modelled energy system.

The distribution of total discounted costs of the energy system operation and development in the time period analyzed (the main result) are presented in the Figure 8.

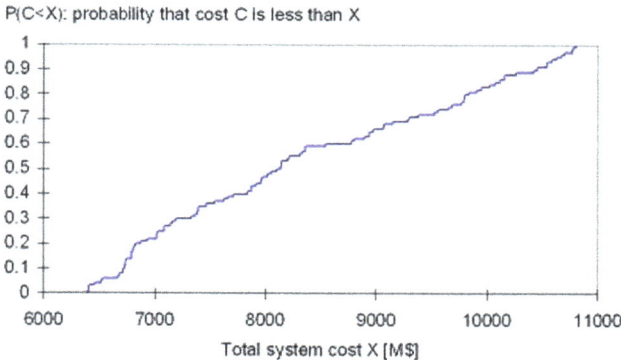

Fig. 8. Uncertainty of modelling result: empirical distribution function of total system cost

Following the uncertainty analysis the sensitivity measure PCC (see Fig. 9) describes how the initial conditions and model parameters (see Table 3) influence the result. From the sensitivity analysis we can see that the 3rd parameter (discount rate) has the largest (negative) influence on the total system costs (main modelling result). When this parameter increases, the considered model result decreases most significantly. Alternatively, the increase of nuclear fuel price (the increase of 6th parameter) in the considered range has the lowest influence.

In general, a high discount rate gives more weight or importance to present expenditures than to future ones, while a low discount rate reduces these differences and thus favours technologies that have high investment cost but low operation costs (for example NPPs).

PCC with respect to ordinary correlation, R**2=0.99

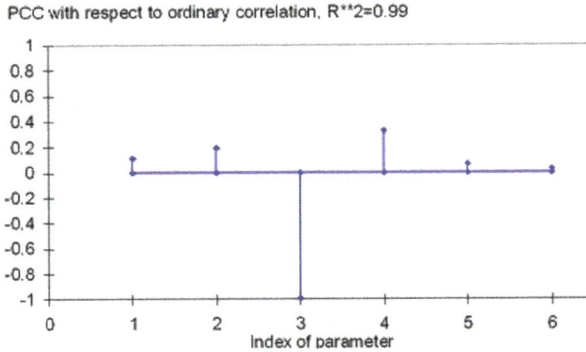

Fig. 9. Sensitivity measure and determination coefficient (R**2) for total system cost

In this case study, the sensitivity measure, which is a product of the statistical analysis, shows which sources of uncertainty are contributing most to the uncertainty in the predicted energy system performance (see Fig. 9). But it is possible that sensitivity measure, in this case a Partial Correlation Coefficient (PCC), explains too small a fraction of the variability of the model output values, for instance, if coefficient of determination is less than 0.5. However, for analyzed case the coefficient of determination is 0.99. Thus, in this case the sensitivity measure PCC in a very good way express the relation in variability and analyst can easily determine which model parameters should be controlled better in order to decrease unfavourable changes of results. Alternatively, the analyst can determine which parameters could be less precise without substantially affecting results.

5. Conclusions

1. While innovative design solutions are possible in an early design stage to cope with extreme internal events, the need for integrating external events considerations on a probabilistic basis at a relatively early design stage is going to be another challenge for effective and balanced use of PSA as a support of the design phase.
2. Further progress of PSA application and EPZ definition could be achieved via discussion with national regulatory authorities in those IAEA Member States that are considering performance-based and risk-informed licensing approaches for future NPPs.
3. Construction of SMR units is very attractive option (looking from economical point of view) for the future electricity and heat generation. The option with SMR cogeneration mode may cause the lowest total discounted cost among the scenarios analyzed.
4. In the case, when IRIS cogeneration unit should be installed away from existing district heating networks (due to EPZ), the attractiveness of this unit is decreasing gradually with distance, because of investment cost and heat losses in addition district heating pipelines.
5. The sensitivity analysis may be essential as it shows how particular parameter is important to the modelling results and where the accuracy of primary data could be increased (in order to decrease the uncertainty of the results). Alternatively, the analyst can determine which parameters could be less precise without substantially affecting results.
6. In our case, the discount rate has the highest influence on the total system costs, while the increase of nuclear fuel price in the considered range has the lowest influence to the total system costs.

6. Acknowledgment

Due to the international nature of the IRIS project, the cooperation can be treated as a trade mark of the IRIS project. This was even truer for the IRIS PRA. The authors wish to acknowledge the large support and valuable assistance of the IRIS heads M. D. Carelli and B. Petrovic as well as of other members of the IRIS PRA team, especially D. J. Finnicum and C. L. Kling. We also want to acknowledge the advises and useful discussions with L. E. Conway and L. Oriani from the IRIS design team. And last but not least we would like to extend thanks to J. Augutis and M. Ricotti for the great personal support provided during the initial stage of PRA related research. The publication of this chapter was funded by Westinghouse Electric Company, LLC.

7. References

Alzbutas R., Maioli A. (2008). Risk zoning in relation to risk of external events (application to IRIS design). *International journal of risk assessment and management.* ISSN 1466-8297. Vol. 8, No. 1/2, p. 104-122.

Alzbutas, R., Augutis, J., Krikštolaitis, R., Ušpuras, E. (2003). Uncertainty and Sensitivity Analysis in Aircraft Crash Modelling, ISSN 1642-9311, *Proceedings of The 3-rd Safety and Reliability International Conference (KONBiN'03)*, V2, pp. 267–274, Gdynia, Poland.

Alzbutas, R., Augutis, J., Maioli, A., Finnicum, D.J., Carelli, M.D., Petrovic, B., Kling, C.L., Kumagai, Y. (2005). External Events Analysis and Probabilistic Risk Assessment Application for IRIS Plant Design, *Proceedings of the 13th International Conference on Nuclear Engineering (ICONE-13)*, Beijing, China, Atomic Energy Press, CD: 8 p.

Alzbutas, R., Augutis, J., Urbonas, R. (2001). Risk and sensitivity analysis in relation to external events, ISBN 961-6207-17-2, *Proceedings of International Conference on Nuclear Energy in Central Europe 2001*, Portorož, CD308: 14 p.

ANSI/ANS-58.21 (2003). *External-Events PRA Methodology*, American Nuclear Society, March 3.

Carelli, M.D. (2003). IRIS: A global approach to nuclear power renaissance. *Nuclear News*, 46, No. 10, pp. 32-42.

Carelli, M.D., et al. (2004). The Design and Safety Features of the IRIS Reactor. *Nuclear Engineering and Design*, 230, pp. 151-167.

Carelli, M.D., et al. (2005). IRIS reactor design overview and status update. Proceedings of the American Nuclear Society-international congress on advances in Nuclear Power Plants (ICAPP'05). Vol. 5, pp. 451-459.

CESSAR-DC (1997). Combustion Engineering Standard Safety Report – Design Certification for System 80+ Design, *Combustion Engineering*, Inc. Volume 20.

IAEA MESSAGE. (2003). *Model for Energy Supply Strategy Alternatives and their General Environmental Impact.* Vienna, Austria: International Atomic Energy Agency.

IAEA-SSG-3. (2010). *Development and application of level 1 probabilistic safety assessment for Nuclear Power Plants.* IAEA safety standards series No. SSG-3. Austria: International Atomic Energy Agency. 195 p. ISBN 978-92-0-114509-3.

IAEA-SSG-4. (2010). *Development and application of level 2 probabilistic safety assessment for Nuclear Power Plants.* IAEA safety standards series No. SSG-4. Vienna, Austria: International Energy Atomic Agency. 82 p. ISBN 978-92-0-102210-3.

IAEA-TECDOC-1408. (2004). *Energy supply options for Lithuania: a detailed multi-sector integrated energy demand, supply and environmental analysis.* Vienna, Austria: International Atomic Energy Agency, 171 p. ISBN 92-0-110004-3, ISSN 1011-4289.

IAEA-TECDOC-1487. (2006). *Advanced nuclear power plant design options to cope with external events.* Vienna, Austria: International Atomic Energy Agency. 221 p. ISBN 92-0-100506-7.

IAEA-TECDOC-1511. (2006). *Determining the quality of probabilistic safety assessment (PSA) for applications in nuclear power plants.* Vienna, Austria: International Atomic Energy Agency. 172 p. ISBN 92-0-108706-3, ISSN 1011-4289.

IAEA-TECDOC-1541. (2007). *Analyses of Energy Supply Options and Security of Energy Supply in the Baltic States,* Vienna, Austria: International Atomic Energy Agency, 323 p. ISBN 92-0-101107-5, ISSN 1011-4289.

IAEA-TECDOC-1652. (2010). *Small reactors without on-site refuelling: neutronic characteristics, emergency planning and development scenarios,* Vienna: International Atomic Energy Agency. 94 p. ISBN 978-92-0-106810-1.

Kling, C.L., Carelli, M.D., Finnicum, D., Alzbutas, R., Maioli, A., Barra, M., Ghisu, M., Leva, C., Kumagai, Y. (2005). PRA improves IRIS plant safety-by-design. *Proceedings of the American Nuclear Society-international congress on advances in Nuclear Power Plants (ICAPP'05).* Vol. 5, pp. 3011-3019.

Norvaiša, E. (2005). *Modeling and analysis of sustainable development of Lithuanian power and heat supply sectors* [summary of dissertation]. Kaunas University of Technology. Lithuanian Energy Institute.

Norvaiša, E., Alzbutas, R. (2009). Economic and sensitivity analysis of non-large nuclear reactors with cogeneration option in Lithuania, *Energy, policies and technologies for sustainable economies: 10th IAEE European conference,* Vienna, Austria, International Association for Energy Economics, Cleveland OH. ISSN 1559-792X, p. 1-19.

NUREG-1407 (1991). *Procedural and Submittal Guidance for the Individual Plant Examination of the Externals Events (IPEEE) for Severe Accidents Vulnerabilities,* NRC.

4

Deterministic Analysis of Beyond Design Basis Accidents in RBMK Reactors

Eugenijus Uspuras and Algirdas Kaliatka
Lithuanian Energy Institute
Lithuania

1. Introduction

RBMK reactor belongs to the class of graphite-moderated nuclear power reactors that were designed in the Soviet Union in the 1950s. The usage of materials with low neutron absorption in RBMK design allows improving the fuel cycle by using cheap low-enriched nuclear fuel. In total 17 RBMK reactors have been built in Russia, Ukraine and Lithuania. One reactor is still under construction at Kursk Nuclear Power Plant (NPP). All three surviving reactors at Chernobyl NPP (Ukraine) were shutdown (the fourth was destroyed in the accident). Units 5 and 6 at Chernobyl NPP were under construction at the time of the accident; however, further construction was stopped due to the high contamination level at the site and political pressure. In Lithuania two reactors at Ignalina NPP were shutdown in 2004 and 2009. At present time no plans are made to build new RBMK type reactors, but in 2011, 11 RBMK reactors are still operating in Russia (4 reactors in Saint Petersburg, 3 – in Smolensk and 4 – in Kursk).

The RBMK reactor is a channel-type boiling water reactor. It has a huge graphite block structure, which functions as a moderator that slows down the neutrons produced by fission. The feature of RBMK type reactor is that each fuel assembly is positioned in its own vertical fuel channel, which is individually cooled by boiling water that is intended to remove the heat produced in it. The fuel channels are made of Zirconium and Niobium alloy similar to that used for fuel claddings. Reactor cooling system of RBMK has two loops, which are interconnected via the steamlines and do not have a connection on the water part. This is a difference from the vessel-type reactors.

The RBMK type reactors do not have full containment, preventing the environment from the radioactive material release. The absence of an overall containment suggests that in case of severe accident, the mitigation of fission products release to environment has to be based primarily on decreasing the extent of core damage, which is a key factor for the radiological consequences of accidents in RBMK. The degree of core damage is determined by the RBMK characteristics, such as the ability of the circulation loop to disintegrate and the multichannel nature of the core. Thus, depending on the type of accident, the damage of fuel assemblies can remain localized within a single fuel channel, a group of channels connected to the same group distribution header, or channels of a single loop (half of the core) or it can propagate to the entire core if complete loss of cooling occurs. Consequently, the severity of RBMK core damage depends on the degree and number of damaged fuel assemblies.

Another characteristic feature of RBMK is the graphite moderator. A positive property of such moderator is high heat capacity, which increases voided core heating time. This gives the operators more time to control the accident and to restore the failed equipment. At the same time, the existence of the graphite requires additional estimation of the graphite behavior at high temperature.

The mentioned specifics of RBMK reactors are affected on the design basis and beyond design basis accident sequences and necessary accident management measures, which are completely different from those in vessel type boiling water reactors. To understand the specifics of accidents in RBMK reactors the consequences of different accident groups were modeled by employing system thermal-hydraulic computer codes. This chapter presents the specifics of RBMK reactors, categorization of the Beyond Design Basis Accidents (BDBA) and specifics of the deterministic accident analyses in BDBA in RBMK. The results of the analysis were used for the development of Symptom-Based Emergency Operating Procedures and reactor cooldown strategies in case of beyond design basis accidents.

2. Specifics of RBMK reactors

A simplified heat flow diagram of RBMK reactor is provided in Figure 1 [1]. The reactor cooling water, as it passes through the core, is subjected to boiling in the fuel channels (2) and is partially evaporated. The steam-water mixture then continues to the large drum separator (3), the elevation of which is greater than that of the reactor. The water settles there, while the steam proceeds to the turbines (5). The remaining steam beyond the turbines is condensed in the condenser (7), and the condensate is supplied by the condensate pumps (8) into the deaerator (9). Deaerated water is returned by the feed pump (10) to the drum separator (4). The coolant mixture is returned by the main circulation pumps (11) to the core, where a part of it is again converted to steam. The reactor power is controlled using control rods (3).

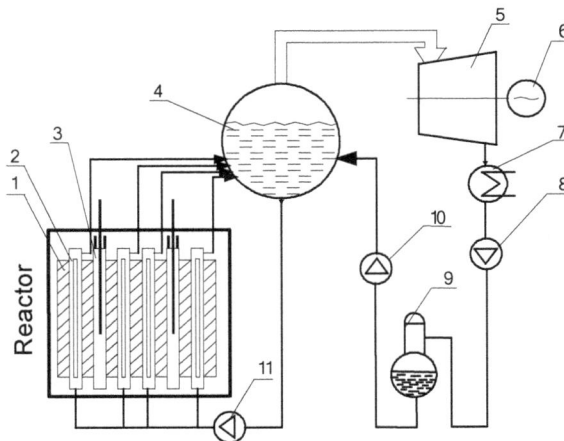

Fig. 1. Simplified RBMK-1500 heat flow diagram: 1 – graphite moderator; 2 - fuel channel; 3 – control rod; 4 - Drum Separator (DS); 5 - turbine; 6 - generator; 7 - condenser; 8 - condensate pump; 9 - deaerator; 10 - feedwater pump; 11 - main circulation pump

This fundamental heat cycle is identical to the Boiling Water Reactor cycle, extensively used throughout the world, and is analogous to the cycle of thermal generating stations. However, compared to BWRs used in Western power plants, the RBMK-1500 and RBMK-1000 have a number of unique features. The comparison of most important parameters of the reactor is presented in Table 1. As it is seen from the presented table, the values of specific power per fuel quantity are very similar for all reactors. The value of power per fuel rod length is the highest for RBMK-1500 reactor. To reach such high value, additional specifically designed spacers, which operate like turbulence enhancers to improve the heat transfer characteristics, are mounted in the fuel assemblies of RBMK-1500. Specific power per core volume in RBMK-1500 is higher than in RBMK-1000 reactor, but in BWR-type reactors this characteristic is approximately 10 times higher.

No.	Parameter	BWR*	RBMK-1000	RBMK-1500
1.	Thermal power, MW	3800	3840	4800
2.	Core diameter m	5.01	11.80	11.80
3.	Core height, m	3.81	7.0	7.0
4.	Core volume, m³	75	765	765
5.	Mean specific power per core volume, MW/m³	51	5.02	6.27
6.	Mean specific power per fuel quantity, MW/t	24.6	20.8	26.0
7.	Mean power per fuel rod length, kW/m	19.0	18.3	22.9

* General Electric design

Table 1. Comparison of BWR and RBMK reactor parameters

In RBMK-type reactors a part of Reactor Cooling System (RCS) above the reactor core is located outside the leaktight compartments. In the RBMK reactors design, these compartments are called Accident Localization System (ALS). This is different from the typical PWR or BWR plants, which have full containment [1]. The Drum Separators (DS) and a part of downcomers are contained in the DS compartments, which are connected to the reactor hall. Such compartments are not as strong as the leaktight compartments of ALS.

2.1 Specifics of RBMK reactor core

Nuclear fuel used in the RBMK-1500 (the reactor of Ignalina NPP in Lithuania) is slightly enriched with uranium in the form of uranium dioxide. According to RBMK-1500 design, low-enriched (2%) uranium fuel was used since the begging of Ignalina NPP exploitation. Later this fuel was mostly fully replaced by a little higher-enrichment (2.4% and 2.6%; 2.8%) uranium fuel with a burnable erbium absorber. The change of fuel allows improving safety and economic parameters of the plant.

Fuel pellets have a 11.5 mm outer diameter and are 15 mm long. The fuel pellets have hemispherical indentations in order to reduce the fuel column thermal expansion and thermo-mechanical interaction with the cladding. The 2 mm diameter hole through the axis of the pellet reduces the temperature at the center of the pellet, and helps to release the gases formed during the operation. The pellets placed into a tube with an outside diameter of 13 mm compose a fuel rod. The active length of RBMK-1500 fuel rod is approximately 3.4 m.

The tube (fuel cladding) material of the fuel rod is an alloy of zirconium with one percent niobium. The fuel rods are pressurized with helium and sealed. The fuel pellets are held in place by a spring. 18 fuel rods, arranged within two concentric rings in a central carrier rod, contain the fuel bundle with an inside diameter of 8 cm [1].

Active core height is 7 m in RBMK type reactors. Thus, the complete fuel assembly is made up of two bundles, which are joined by means of a sleeve at the central plane. The lower bundle of the fuel assembly is provided with an end grid and ten spacing grids. The central tube and the end spacer are also made from the zirconium-niobium alloy. The remaining spacers are made from stainless steel and are rigidly fixed (welded) to the central tube. Apart from the spacers, the top bundle also has intensifying grids, which act as turbulence enhancers to improve the heat transfer characteristics. The fuel tubes are mounted so that axial expansion of the upper or lower bundles takes place in the direction towards the center of the core. The total mass of uranium in one fuel assembly is approximately 110 kg [1].

The fuel channels, where the fuel assemblies are placed, consist of three segments: top, center and bottom. The center segment is an 8 cm inside diameter (4 mm thick wall) tube, made from zirconium-niobium alloy. The top and bottom segments are made from stainless steel tube. The center segment of fuel channel, set in the active core region, and zirconium-niobium alloy warrant the low thermal neutron absorption cross-section.

The fuel channel tubes are set into the circular passages which consist of aligned central openings of the graphite blocks and stainless steel guide tubes of the top and bottom core plate structures to maintain the core region hermetically sealed. The reactor core is constructed of closely packed graphite blocks stacked into approximately 2500 columns with an axial opening. Most of the openings contain fuel channels. A number of them also serve other purposes (e.g. instrumentation, reactivity regulation). The total mass of graphite is about 1700 tons. The fuel channels together with graphite stack are placed inside the leaktight reactor cavity.

The fuel channel tubes also provide cooling for the energy deposited in the graphite moderator of the core region. In order to improve heat transfer from the graphite stack, the graphite rings surround the central segment of the fuel channel. These rings are arranged next to one another in such a manner that one is in contact with the channel, and the other with the graphite stack block. The minimum clearance between the fuel channel and the graphite ring is 1.15 mm, and between ring and graphite stack – 1.38 mm. These clearances prevent compression of the fuel channel tube due to the radiation and/or thermal expansion of the graphite stack [1].

2.2 Flow paths of radioactivity release to the environment

The consequences of any accident at NPP (radioactivity release to environment) depend on what safety barriers were violated. As for any Light Water Reactor (LWR), 4 safety barriers (Table 2) could be distinguished for RBMK-1500. The provided comparison between the safety barriers of vessel-type reactors and RBMK-1500 indicates that each fuel channel corresponds to the reactor vessel and reactor cavity together with ALS and reactor building perform a function of containment.

RBMK-1500	Vessel-type reactors
Fuel pellet	Fuel pellet[1]
Fuel cladding	Fuel cladding
Fuel channel and reactor cooling system	Reactor vessel and reactor cooling system
Reactor cavity, ALS and reactor buildings	Containment

Table 2. Safety barriers of RBMK-1500 and vessel-type reactors

The fuel pellet contains most of the radioactive material. Some gaseous (e.g. Xenon, Krypton) and volatile (e.g. iodine, caesium) fission products is released from the fuel matrix to the gap between the fuel pellet and fuel cladding, but until the cladding remains intact, the radioactive materials are confined and do not enter into the coolant.

The fuel cladding can fail due to:

- thermal-mechanical interactions between the fuel and the cladding,
- or thermal-mechanical deformations of the cladding under positive or negative pressure differentials.

The first type of failure is typical for rapid and large power excursions (e.g., reactivity initiated power excursions) where hot and possibly molten UO_2 could come into contact with the cladding material. The RBMK fuel is similar to the fuel of any LWR; however, the probability of fuel damage in RBMK type reactors due to the reactivity initiated accidents is lower, because typical time of reactivity insertion in RBMKs is measured in seconds rather than in microseconds as for other LWR [2]. The second type of cladding failure is associated with cladding temperature excursions, either when the pressure in RCS is higher than the internal pressure (i.e. positive pressure differential), or when the internal pressure is higher than the coolant pressure in RCS (i.e. negative pressure differential). The positive pressure differential is possible in case when the pressure in RCS is maintained high without providing cooling to fuel. Under positive pressure gradients, hot cladding collapses onto the fuel pellet stack and deforms into gaps between the fuel pellets, which causes a failure of cladding. If the gap between fuel pellets is 2 mm or larger, then such fuel failure would appear at fuel cladding temperature of 1200 – 1300 ºC. The fuel cladding failure temperature decreases if the axial gap between the fuel pellets increases. Normally, the maximum gap between the fuel pellets in any fuel rod is 1.02 mm with the probability of 0.997. Thus, the fuel collapse probability for RBMK-1500 is very low at temperature level below 1200 ºC. The ballooning of fuel cladding is relevant to the accidents when the internal pressure is higher than the external one (i.e. negative pressure differential). The example of such accident is a large Loss of Coolant Accident (LOCA), when the fuel cladding temperature increases during a rapid pressure drop in the reactor cooling system. The internal pressure in fuel rods of RBMK-1500 is approximately 1.2 MPa during normal operation. If due to a large LOCA, the pressure in RCS decreased down to atmospheric, then the fuel cladding failure would appear due to ballooning at temperature 850 – 1000 ºC [2, 3]. Another potential for fuel cladding failure is the fuel cladding oxidation. The cladding oxidation is related to an embrittlement of fuel cladding that could potentially lead to a formation of fuel debris that can also obstruct the coolant flow path. The very rapid oxidation (reaction between steam and Zirconium) of fuel

---

[1] Some countries (e.g. France) does not consider the fuel pellet as a safety barrier

cladding starts at temperature level higher than 1200 °C. This chemical reaction is exothermic and if it occurred, a large amount of chemical heat would be generated and could lead to a melting of cladding, a liquefaction of fuel and possibly a blockage of coolant flow paths by relocated fuel materials. Summing-up all possible mechanisms, affecting integrity of fuel cladding, the acceptance criterion 700 °C was used for the safety analysis [2, 3]. It means that below such temperature fuel cladding integrity will be warranted.

If the fuel cladding loses its integrity (i.e., if it fails), a key barrier to the release of fission products is breached, and the coolant in the RCS becomes contaminated with radioactive fission products released from the fuel. However, until the RCS remains intact, fission products are confined inside piping and do not enter the compartments. If the RCS piping ruptured, then the contaminated coolant would be released to the compartments (see Figure 2).

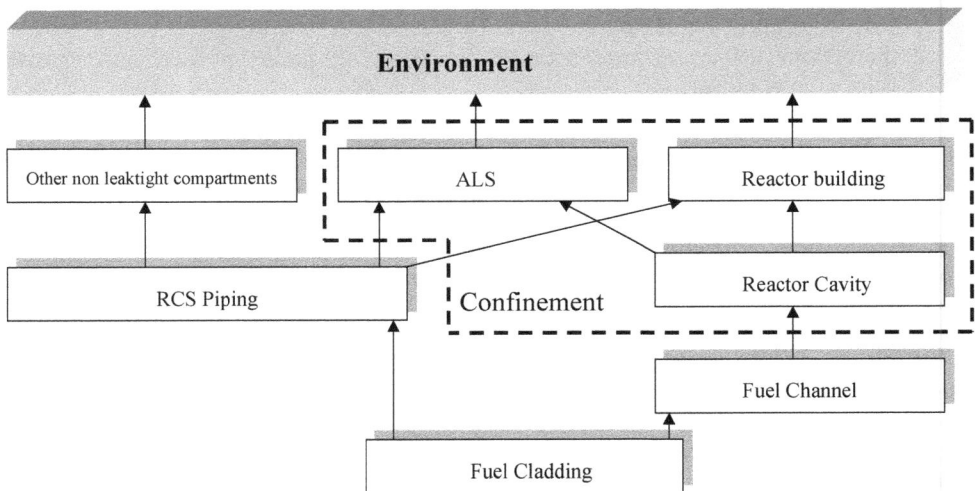

Fig. 2. Flow paths of radioactivity release in RBMK-1500

According to its function and location, the fuel channel of RBMK reactor corresponds to the pressure vessel of vessel-type reactors. Therefore, it is the most important part of RCS. If the Fuel Channel (FC) wall heats up while the internal pressure is elevated, it may expand until it contacts the surrounding graphite blocks [4]. In the RBMK reactor, the deformation of fuel channels is arrested at rather modest uniform strain values due to the contact of the deformed FC with surrounding graphite block. Experiments show that the contacted channel fails only if and when the graphite block is disrupted by the pressure load transmitted to it by the deformed channel. At nominal pressure in FC (7 – 8 MPa) the temperature of fuel channel failure is not less than 650 °C and it depends on the heat-up rate. Experiments showed that in case of a higher heat-up rate, when the FC rupture occurs, the temperature values are higher compared to the lower heat-up rate. It was also discovered that in order to obtain the corresponding deformations at lower pressures higher temperatures or higher heat-up rates are required [4]. The acceptance criterion of 650 °C for fuel channel walls was assumed for the safety analysis [2, 3].

The fuel channels together with graphite stack are placed inside the leaktight reactor cavity, which is formed by a cylindrical metal structure together with bottom and top metal plates (Figure 3). If FC ruptured, the steam- water mixture would be released to this cavity and come into contact with hot surfaces of the graphite stack (Figure 4). The Reactor Cavity (RC) performs the function of containment; therefore, the integrity of the cavity is of high importance. RC consists of the structures shown schematically on the left side of Figure 3, which summarizes the design pressures based on the most conservative assessments. The figure indicates that the minimum of permissible excess pressures is 214 kPa [3] i.e. the pressure, which corresponds to the weight of upper metal plate (2). According to the reports [5, 6, 7], the more realistic values are: 1) for the upper plate 300 kPa; 2) for the casing (5) 330 kPa and lower plate (6) 380 kPa. Thus, in any case the top metal plate is the weakest point in the structure of reactor cavity, but the excess pressure that could be withstood is at least 300 kPa. The failure of the bottom plate could be expected only in the case of low-pressure accident scenario if the molten fuel would accumulate on it. In such accident scenario the fuel would relocate downwards in the fuel channel boundaries by candling (melting, forming eutectics with the clad and structure, flow downwards, freezing, and then remelting) until it reaches the pipes below RC. Since these pipes become unrestrained if they melt, the molten material would flow out onto the surrounding floor. Thus, the fuel is not expected to accumulate on the bottom plate of RC.

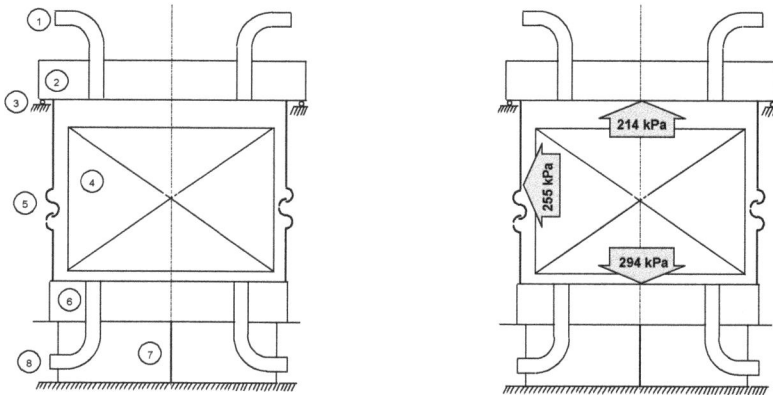

Fig. 3. Reactor cavity components and limit pressures [7]: 1 – upper Reactor Cavity Venting System (RCVS) pipes, 2 – upper plate, 3 – roller support, 4 – reactor core, 5 – casing, 6 – lower plate, 7 – support, 8 – lower RCVS pipes

At NPP with a full-scope containment, which covers all the piping of reactor cooling system, the coolant would be discharged to the containment, i.e. reinforced and leaktight building capable to withstand excess pressure of 500 – 700 kPa. At Ignalina NPP with the RBMK-1500 reactor, a part of RCS above the reactor core is located outside the reinforced compartments. The drum separators and part of downcomers are contained in the DS compartments, whish are connected to the reactor hall (see XI and XII in Figure 4). These compartments are called "reactor buildings" and they can withstand 24.5 kPa excessive pressure, that is a few times lower inside pressure than in the reinforced leaktight compartments I, II, III, IV, V and VI (see Figure 4). The compartments of the main RCS components (I) and corridor (II) can withstand 300 kPa, under-reactor compartments (III) and compartments of GDH and LWP (IV) – 80 kPa, bottom steam reception chambers (V) and vertical steam distribution shafts – 100 kPa of excessive pressure.

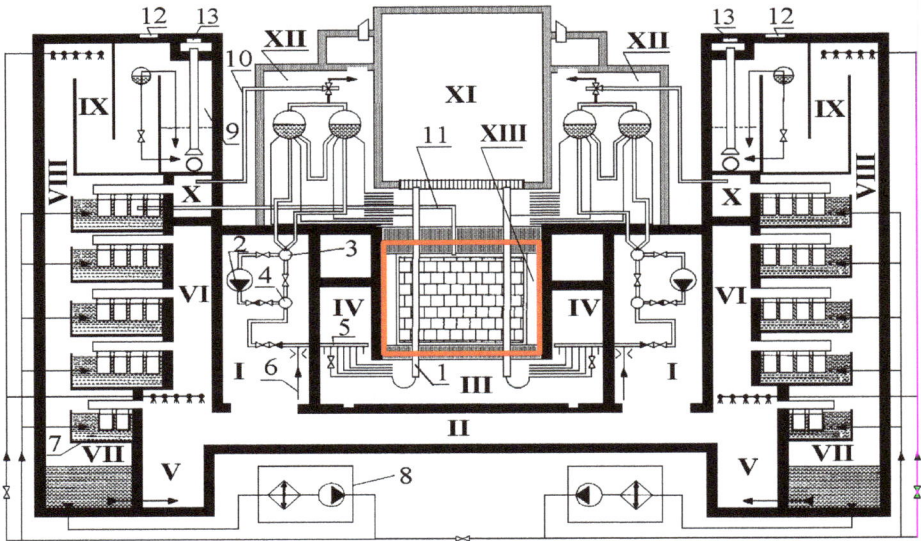

Fig. 4. RCS and Confinement of RBMK-1500: 1 – Fuel channel, 2 - Main Circulation Pump (MCP), 3 – Suction header, 4 – Pressure header, 5 – Group distribution header, 6 – Emergency Core Cooling System (ECCS) header, 7 – condensing pools, 8 – Condenser tray cooling system, 9 – Air release section, 10 – Steam release through main safety valve, 11 – Pipe of reactor cavity venting system, 12 – blow-down hatches, 13 – tip-up hatches. **Compartments**: I – Compartments of main RCS components (MCP, suction header, pressure header and downcomers), II – Corridor, III – Under-reactor compartment, IV – Compartments of Group Distribution Headers (GDHs) and Low Water Pipes (LWP), V – Bottom steam reception chambers, VI – Vertical steam distribution shafts, VII – Hot condensate chamber, VIII – Air venting channel, IX – Gas delay chamber, X – Top steam reception chambers, XI – Reactor hall, XII – Drum separator compartments, XIII – Reactor cavity

In case of an accident, these compartments have installed special valves or hatches that open to release the steam gas mixture to the environment. The part of steamlines and feedwater lines are contained in the turbine hall and deaerators compartments, respectively. If the rupture appears in these compartments then the release is not confined and the retention of fission products depends only on the natural sedimentation processes.

The ALS, RC and the other reactor buildings (DS compartments and reactor hall) of Ignalina NPP perform a function of containment, i.e. they are reinforced and leaktight, but due to its specifics it is usually called confinement. Therefore, in this chapter the term containment will be understood as a function rather than building.

3. Categorization of the beyond design basis accidents for RBMK reactors

Based on the presented specifics of RBMK, two categories of Beyond Design Basis Accidents (BDBA), i.e. core damage types in RBMK reactors were proposed by RBMK designers from Russia (Kurchatov Institute and Research and Development Institute of Power Engineering) [8]:

- damage of the core or its components with the reactor maintaining its overall structural integrity;
- total damage of the core, resulting in the loss of general structural integrity of the reactor system.

Such grouping of accident is very important regarding accident management. If the core or its components remain structurally intact (first category of core damage), then the controlling actions (accident management) for limiting and delaying damage of the core, as well as prevention of confinement damage and mitigation of fission products release are possible. If the general structural integrity of the reactor system is lost, then depending on the degree to which the general structural integrity of the reactor is maintained, (second category of core damage), the emergency plan has to be activated in order to protect the public (sheltering, evacuation, etc.).

The first category of core damage can be further subdivided into the following accident groups (Fig. 5):

- no severe damage of the core (1.1);
- severe core damage accompanied by containment of the core fragments in the reactor core, accident localization system or other reactor buildings (1.2).

Accidents, leading to a complete reactor core damage, with loss of structural integrity of the reactor can also be divided into two groups [9]:

- accidents when heat-up of the reactor core occurs during reactor operation or within the first seconds after the reactor shutdown, when decay heat is high (2.1);
- accidents when heat-up of the reactor core occurs after the reactor shutdown (2.2).

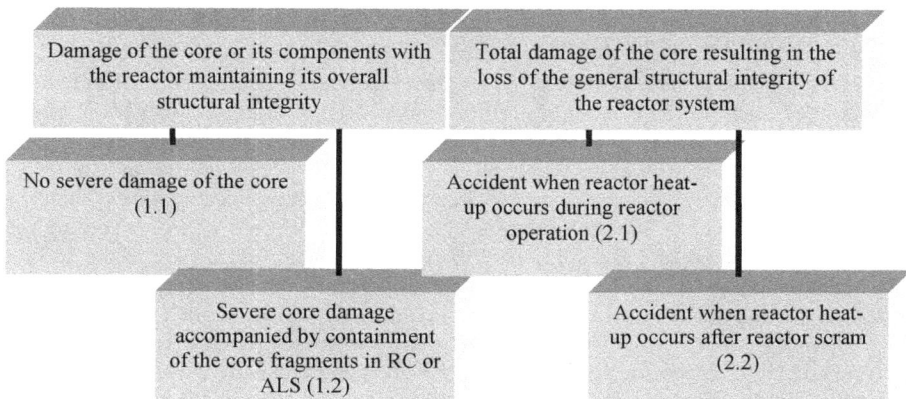

Fig. 5. Categories of core damage types in case of BDBA in RBMK [9]

As it was mentioned, such grouping of accidents is the starting point for the development of the measures for accident management. However, the development of accident management guidelines requires performing deterministic analysis of all possible accidents in each group. Based on this analysis the accident consequences, available time for possible operator actions and possible modifications of emergency systems may be determined.

In the next chapter the deterministic analysis of reactor core and reactor cooling system is presented, whereas the modeling of the process in the RBMK confinement is presented in the monograph [10].

4. Models for the deterministic analysis of BDBA

Models developed for the thermal-hydraulic analysis of processes in the reactor core and reactor cooling system are presented below. The models of RBMK-1500 are developed using system state-of-the-art code RELAP5 and RELAP/SCDAPSIM.

4.1 Reactor cooling system model of RBMK-1500

The RELAP5 computer code has been developed by Idaho National Engineering Laboratory [11]. It is a one-dimensional non-equilibrium two-phase thermal-hydraulic system code. The RELAP5 code has been successfully applied to PWR and BWR reactors. Since 1993 The RELAP5 model of the Ignalina NPP was used in the Lithuanian Energy Institute for the analyses of thermal-hydraulic response of the plant to various transients. The RELAP5/MOD3.2 model of the Ignalina plant (nodalization scheme) is presented in Figure 6.

The model consists of two loops. The left loop of RCS model consists of one equivalent core pass. Two drum separators are modeled as one "branch" type element (1). All downcomers are represented by a single equivalent pipe (2), further subdivided into a number of control volumes. The pump suction header (3) and the pump pressure header (8) are represented as branch objects. Three operating MCPs are represented by one equivalent element (5) with check and throttling-regulating valves. The stand-by MCP is not modeled. The bypass pipes (7) between the pump suction header and the pump pressure header is modeled with the manual valves closed. This is in agreement with a modification recently performed at the Ignalina NPP. All FCs of this left core pass are represented by an equivalent channel (12) operating at average power and coolant flow.

Compared to the model for the left loop, in the right one, the loop section between the pressure header and the DS is represented in a more detailed manner. The MCP system is modelled in more detail also (it is modelled with three equivalent pumps). The right loop model consists of three equivalent core passes. First core pass represents one single GDH with an equivalent FC of average power. Second core pass represents single GDH with failed to close check valve. A few equivalent channels of different power levels represent fuel channels, connected to this GDH. The other core pass represents the other 18 GDHs. The channels of this pass are simulated by an equivalent FC of average power. The steam separated in the separators is directed to the turbines via steam pipes (15). Two Turbine Control Valves (TCVs) organize steam supply to the turbines. The guillotine break of MCP pressure header (17) in the right loop model of RCS is modelled by a valve (18). The flow area of this valve is double of pressure header flow area. The valve (18) is connected to the volume (19), which represents the compartments covered by RCS pipelines.

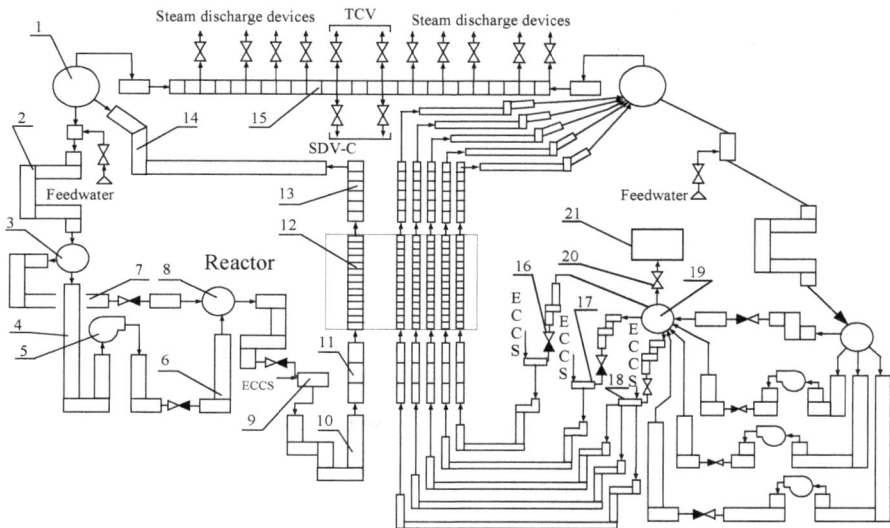

Fig. 6. RBMK-1500 model nodalization scheme: 1 - DS, 2 - downcomers, 3 - MCP suction header, 4 - MCP suction piping, 5 - MCPs, 6 - MCP discharge piping, 7 - bypass pipes, 8 - MCP pressure header, 9 - GDHs, 10 - lower water pipes, 11 - reactor core inlet piping, 12 - reactor core piping, 13 - reactor core outlet piping, 14 - steam-water pipes, 15 - steam pipes, 16 - check valve, 17 – single GDH, 18 – single GDH with failed to close check valve, 19 - ruptured pressure header, 20 - valve for break modeling, 21 – model of compartments, which surround the RCS pipelines

The fuel assemblies in reactor core are described as heat structure elements. The fuel channels with fuel assemblies were divided into a few (depending on the needs of modeling) equivalent groups according to the power and coolant flow rate values. For the core power of 4200 MW, the channel average power is assumed to be 2.53 MW, the maximum channel power is 3.75 MW and minimum channel power is 0.88 MW. It was assumed that approximately 95% of generated fission and decay power is generated in the fuel, and 5% in the graphite stack. More detail about the Ignalina NPP model, developed using RELAP5 code, is presented in [12, 13]. Model validation is performed by comparing calculation results and measurements using separate effect tests [14] and measurements at Ignalina NPP (integrate effects measurements) [13]. The experience of use of computer code for modeling of reactor cooling circuit in RBMK showed, that RELAP5 is very suitable for this task. One dimensional code is perfect for the modeling of thermal hydraulic and heat transfer processes in the RCS, which consists of many long pipelines without any cross flow.

For the analysis of processes, which occur in the reactor core (fuel channels of RBMK type reactors) at significant overheating of fuel assemblies up to fuel melting, specific computer tool for the analysis of processes during severe accident analysis should be used. For our purposes we used RELAP/SCDAPSIM code [15] that is an integrated, mechanistic computer code, which models the progression of severe accidents in light-water-reactor nuclear power plants. The entire spectrum of in-vessel severe accident phenomena, including reactor-coolant-system thermal-hydraulic response, core heat up, degradation and relocation, and lower-head thermal loads, is treated in this code in a unified framework for both boiling water reactors and pressurized water reactors. Unfortunately, the RELAP/SCDAPSIM code has some limitations, related to the application for RBMK-type reactors:

- SCDAPSIM gives a possibility to define PWR or BWR fuel bundles with a user-defined fuel enrichment, but does not give a possibility to include a plant specific core content (fuel with burnable erbium absorber, which is used in RBMK-1500);
- RELAP/SCDAPSIM code does not include transport of fission products from fuel through the RCS piping and their release to confinement;
- Heat generation is defined only once for all heat structures, i.e. fuel rod, fuel channel, graphite and Control & Protection System (CPS) channels. Therefore, the consideration of heat removal by CPS channels is very complicated.

These limitations of RELAP/SCDAPSIM code were taken into account in modelling of BDBA in RBMK reactors: (1) the BWR fuel rod type was used for the modeling of RBMK fuel rods, (2) the transportation of fission products from fuel through cooling circuit was not evaluated because code limitation, (3) in order to avoid the troubles with the modeling of heat transfer from one fuel channel to another through graphite columns in radial direction in the core – the single channel model was used. Such model is acceptable for a rough analysis. The nodalization scheme of such model is presented in Figure 7. In order to perform the analysis, the following boundary conditions should be assumed for the model elements, modeled as time dependent volumes and junctions:

- pressure and water temperature in the group distribution header (3);
- pressure in the drum separators (1);
- coolant flow rate through the fuel channel with average power (2).

This model is described in more detail in the papers [16, 17] and the monograph [18].

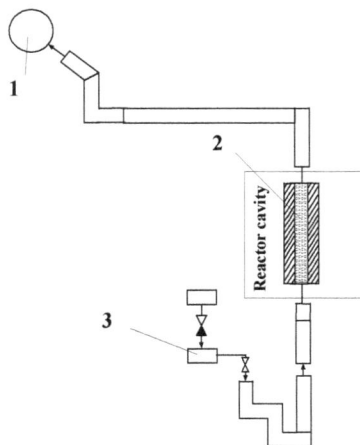

Fig. 7. Nodalization scheme of a simplified RBMK-1500 model (single fuel channel): 1 – drum separator and steam lines, 2 – fuel channel, 3 – GDH

5. Deterministic analysis of BDBA in RBMK-1500

5.1 Analysis of accidents with no severe damage of the core

As it was mentioned, the first category of BDBA (*damage of the core or its components with the reactor maintaining its overall structural integrity* see Figure 5) is divided into two groups: (1.1) *no severe damage of the core*; (1.2) *severe core damage accompanied by the containment of core fragments in the reactor core, accident localization system or other reactor buildings.*

The examples of accidents for RBMK-1500 in the first group (1.1, see Figure 5) are the failure of the final heat sink systems with the loss of their functions or staff error, failure of group of reactor Control & Protection System (CPS) rods, failure of one or more channels in the reactor Emergency Core Cooling System (ECCS) short-term or long-term subsystems, failure of the system which supplies feedwater during a prolonged period of time, and other failures. This group of accidents involves additional failures of the equipment in the safety systems (additional to failures considered by the single failure principle). Usually the accidents of first group (*no severe damage of the core*) are analyzed in order to justify the effectiveness of the functional backup, as well as to assess the conditions and the time available for the backup systems to be actuated. As a rule, this belongs to the field of the accident management on the basis of symptom-oriented emergency operating procedures.

Thus, in case of a postulated failure of CPS rods movement during reactor operation at power, the actions taken by the operators to shutdown the reactor and to hold it in a subcritical state were determined. The reactor can be shutdown and maintained subcritical by inserting one CPS rod into the core, decreasing the water temperature in the CPS cooling circuit or decreasing the temperature of the graphite stack [8]. At Ignalina NPP the Additional Hold-down System is implemented in case of CPS malfunction to prevent reactor re-criticality by injecting the liquid poison (Gadolinium Nitrate) into the CPS cooling circuit [19].

During reactivity initiated accidents the situation can occur when the group of control rods is withdrawn erroneously. Under adverse conditions it is possible that the signal for local Automatic control will not be generated and the local power increase can occur in a group of fuel channels. Validated calculations using the experimental data showed that this increase could reach 2-2.5 times of the nominal channel power, but it does not cause significant coolant flow decrease in the fuel channels (due to the increased channel resistance in the steam zone) and overheating of the fuel channels [9]. Since this chapter mainly deals with the thermal-hydraulics, more detailed examples with reactivity initiated accidents are not presented there.

Another example could be the loss of long-term cooling. The performed deterministic calculations showed that the reactor core cannot be damaged without the make-up by feedwater during any transient after full reactor shutdown in approximately 1.5 hours [8, 20]. One high pressure pump is sufficient to cool down the reactor for one hour after the reactor shutdown. If the water supply from one pump is re-established, all the parameters of RCS and reactor remain within safe operation limits. However, if the high pressure pumps are not available, the manual operators' actions are required. In [18] the optimal reactor core cooldown scenario for RBMK-1500 in case of station blackout was developed (see Figure 8 - Figure 10). The modeling of such scenario was performed using RELAP5 model presented in Figure 6.

In the analysis presented below, it is considered that the operator takes early actions: 15 minutes after the beginning of the accident the operator begins to supply cold water from ECCS hydro-accumulators into GDH of both RCS loops. After approximately 1.5 hours from the beginning of the accident, the peak fuel cladding temperature exceeds 400 °C (Figure 8). According this signal, the operator opens one steam relief valve to decrease pressure in RCS (Figure 9). At the same time the operator takes actions to maintain the water supply by gravity from deaerators and prepares the connection for water supply from the artesian water source. The activation of ECCS hydro-accumulators after 15 minutes from the beginning of the accident provides only a small amount of water due to equalization of pressures in hydro-accumulators and GDH. Additional amount of water from hydro-accumulators is injected when the pressure in RCS decreases (~2 h after the beginning of the accident). The water supply from ECCS hydro-accumulators and opening of one steam relief valve would result in the increase of water level in RCS (Figure 10). At the time moment t = 2.5 h, the water supply from ECCS hydro-accumulators stops. Approximately 170 m³ of water is injected from ECCS hydro-accumulators.

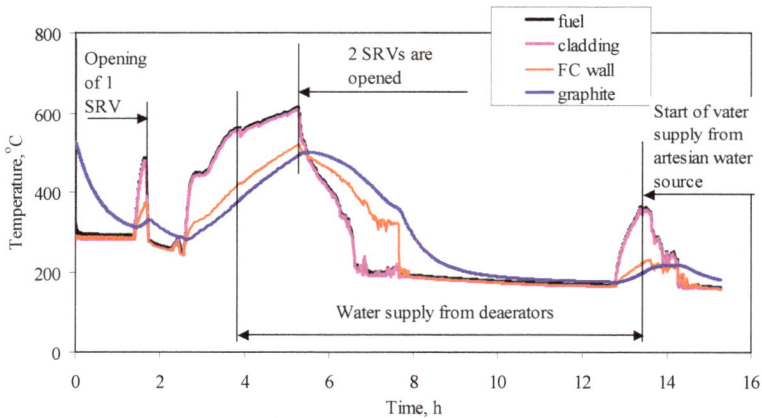

Fig. 8. RCS de-pressurization and water supply into reactor from ECCS hydro-accumulators, deaerators and artesian water source in case of station blackout. Temperatures of fuel, fuel rod cladding, fuel channel and graphite

When the pressure in RCS decreases down to 1.2 MPa (pressure in deaerators), the water flow from the deaerators begins (t = 3.8 h). There are four deaerators, which contain 480 m³ of water with temperature of ~ 190 °C. After the connection of deaerators to RCS, the pressure decrease leads to boiling of water in deaerators. The initial flow rate of water from deaerators does not warrant adequate cooling of the core – the temperature of core components is increasing at the time interval t = 3 – 5 h (Figure 8). To increase the flow rate of water from deaerators, at the time moment t = 5.3 h the operator opens one additional Steam Relief Valve (SRV). As it is seen from Figure 8, this action improves the core cooling conditions and the temperature of core components starts to decrease. The pressure in RCS decreases down to the pressure in artesian water system (~ 0.6 MPa) only more than 13 hours after the beginning of the accident. A complete connection of artesian water to RCS is permitted only after the decrease of pressure in RCS down to the level of pressure in the artesian water source. These measures should prohibit the injection of coolant from RCS into the pipeline of artesian water. After the connection of artesian water source to supply water into reactor, the water level in reactor core starts to increase, which means the success of core cooling. The mentioned operators' actions lead to a slow decrease of pressure in RCS, the fuel rods claddings and channels walls temperatures become not higher than 600 ᵒC.

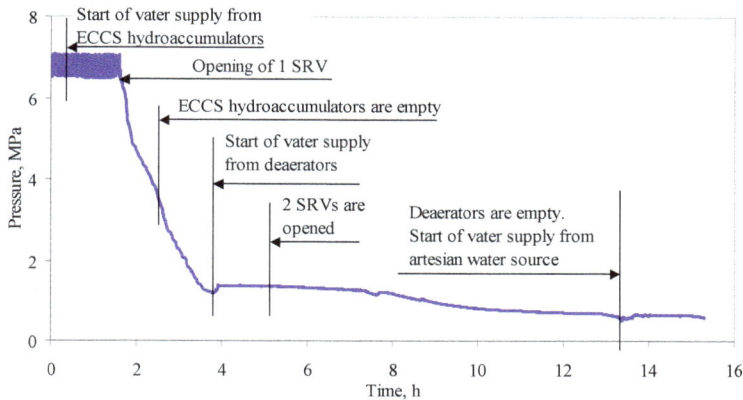

Fig. 9. RCS de-pressurization and water supply into reactor from ECCS hydro-accumulators, deaerators and artesian water source in case of station blackout. Pressure behavior in RCS

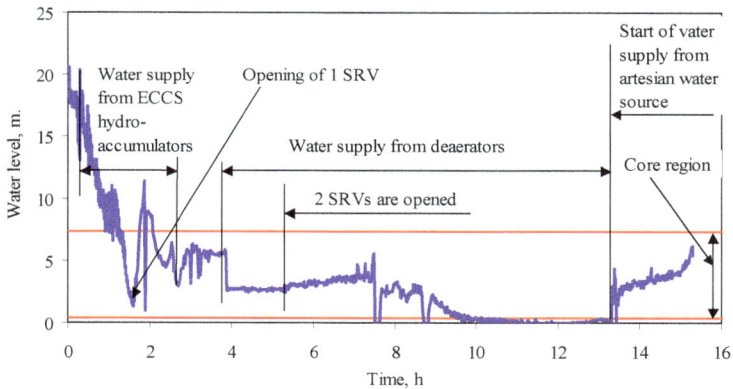

Fig. 10. RCS de-pressurization and water supply into reactor from ECCS hydro-accumulators, deaerators and artesian water source in case of station blackout. Calculated water level behavior in RCS

The results of the mentioned neutron-physical and thermal-hydraulic investigations have served as the basis for expanding the region where the accidents of the first group with multiple failures can be controlled using the symptom-oriented emergency operating procedures, and they have made it possible to determine the actions to be taken by personnel in order to prevent severe core damage.

5.2 Analysis of accidents leading to severe core damage accompanied by containment of core fragments in the RC, ALS or other reactor buildings

Accidents in the second group (1.2, see Figure 5) are analysed in order to develop the measures to preserve the structural integrity of the reactor or to determine the required means and time available for subsequent cooling of the reactor core. The accidents of this group are conventional severe accidents with core meltdown as a result of misbalance between energy source and heat sink. The development of such accidents in RBMK has much in common with overheating processes of vessel-type reactors, but it differs by the RBMK features mentioned above.

The heating and melting of a RBMK core can potentially occur as a result of misbalance between heat generated in the core and removed by reactor cooling system and emergency core cooling systems. A typical example of such accident is the damage of the boundaries of the circulation loop (LOCA type accident), accompanied by the failure of the ECCS or additional loss of feedwater.

The LOCAs, according the PSA terminology for RBMK-1500, were categorized according to the rupture size and location in the RCS. A large LOCA means a rupture of the biggest diameter pipes in RCS, i.e. pipes with diameter of 300 – 1000 mm. Medium LOCA is a rupture of pipes with diameter of 100 – 300 mm, whereas small and very small LOCA signifies a rupture of pipes with diameters of 50 – 100 mm and 30 – 50 mm respectively. The location of LOCAs in the RCS is groped into 4 zones (see Figure 11).

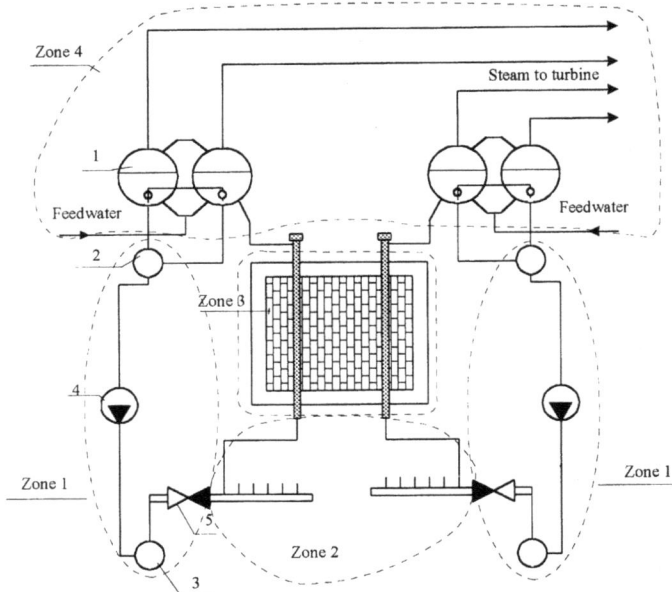

Fig. 11. Distribution of the zones in the RCS: 1 – DS, 2 – MCP suction header, 3 - MCP pressure header, 4 – MCP, 5 – check valve

Pipelines located in Zone 1 have large diameters: ~1000 mm – the main circulation pumps pressure and suction headers, ~600 mm – MCP connecting pipelines and ~300 mm – group distribution header piping. Therefore, large and medium LOCAs are possible in this zone. Pipelines located in Zone 2 have smaller diameter than the ones in Zone 1: ~300 mm – GDH piping and ~50 mm – lower water piping. Piping of Zone 4 (outside ALS compartments) consists of a part of downcomers (~600 mm diameter), feedwater lines of 500 mm diameter and steamlines of 600 mm diameter. Fuel channel break in the reactor cavity (LOCA in Zone 3) is a separate type of accident. The analysis of LOCAs in Zones 1, 2 and 4 was performed using RELAP5 model, presented in Figure 6.

5.2.1 LOCA in Zone 1

In case of break in Zone 1, a huge amount of coolant used for the reactor core cooling is discharged through the break. Consequently, the reactor core cooling is extremely worsened or terminated at all in the group of the fuel channels. GDH check valve prevent coolant water leaking from the core in the opposite direction. Depending on the location of the break, the cooling of FCs connected to one group distribution header (in the case of GDH break) or in all channels of one RCS loop (in the case of MCP pressure or suction header break) can be lost. The emergency protection (reactor shutdown) is activated within the first seconds due to the pressure increase in the reinforced leaktight compartments. Cooling of the reactor core is restored after the activation of ECCS. After 2 – 3 seconds the short-term subsystem of ECCS (two trains of hydro accumulators and one train from the main feedwater pumps) is activated. This subsystem starts to supply water into GDH downstream check valve and is designed to cool down the reactor within the first 10 minutes. Later the long-term subsystem of ECCS, which consists of 6 ECCS pumps and 6 auxiliary feedwater pumps, is activated. The long-term subsystem of ECCS supplies the water to both loops of RCS. The fuel cladding and fuel channel wall temperatures start to decrease after the ECCS activation.

In the case of large LOCA in Zone 1, the pressure upstream fuel channels decreases very fast. To prevent the reverse of coolant flow in the channels, the check valves are installed in each GDH. The failure of some of these valves effects the cooling conditions of channels connected to the affected GDH and may change the consequences of the accident. For example, in the case of the MCP pressure header break, the channels connected to the GDH with failed to close check valve are cooled by the reverse coolant flow from the drum separators (Figure 12). At the beginning of the accident, these FCs are cooled by the saturated water flow, but later (after DSs get empty) only by the saturated steam. Due to the worsened cooling conditions, fuel cladding temperatures in the channels connected to GDH with failed to close check valve increases higher than in the other channels of the affected RCS loop.

The feature of LOCA type accidents in Zone 1 is the heat-up of the fuel in the affected RCS loop during the first seconds of the accidents. It should be noted that the first fuel cladding temperature increase asserts only at the very beginning of the accident and takes a very short time: no more than 30 seconds. In the case of the MCP pressure header break with additional failures of short-term subsystem of ECCS, a short-term violation of an acceptance criterion for fuel rods cladding of 700 ℃ is observed in a considerable group of channels. Such increase of temperatures is related with stagnation of coolant flow rate after the GDHs check valves closing. This stagnation is terminated with the start of ECCS water supply. If the loss of the preferred AC power of the Unit does not occur simultaneously, so the stagnation is terminated after 10 s from the start of water supply from ECCS pumps. The excess of acceptance criteria for fuel rod cladding of 700 ºC is probable in FCs, initial power of which is higher than 2.5 MW (see Figure 13). There are 370 of such FCs in the affected RCS loop (see Figure 14). If the loss of preferred AC power takes place simultaneously, the stagnation is prolonged (ECCS pumps are started after the start of diesel generators). In this case the acceptance criterion for fuel rod cladding is violated in FCs which have the initial power higher than 2.0 MW. There are 670 of such FCs in the affected RCS loop (see Figure 14). However, the peak temperature of FCs walls is much below than the acceptance criterion for the FC wall (650 ºC).

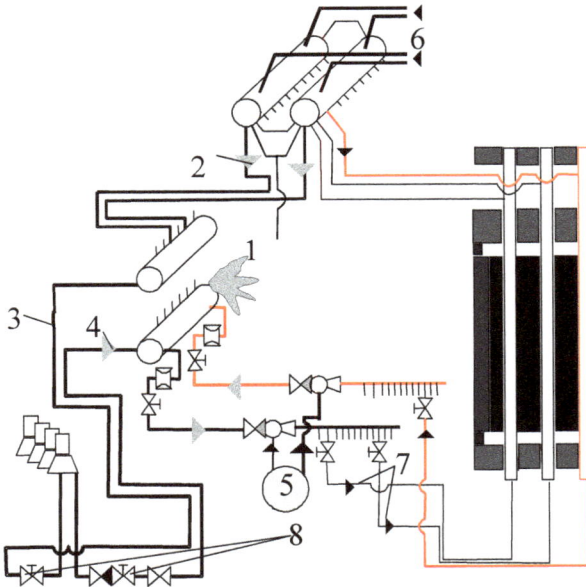

Fig. 12. MCP pressure header rupture with GDH check valve failure: 1 – break of MCP pressure header; 2 – overflow of coolant from DS; 3 – MCP suction piping; 4 – coolant discharge from MCP pressure piping; 5 – ECCS water supply; 6 – steam supply from intact loop of the RCS; 7 – coolant supply into reactor core; 8 – maintenance valves

The detailed analysis was performed to evaluate the possibility of failure of those fuel rods where cladding temperature exceeds 700 ºC. This analysis was performed using RELAP/SCDAPSIM model presented in Figure 7. The calculated pressure inside the fuel rods remains below pressure outside fuel (Figure 15). Thus, the ballooning of fuel rod claddings do not occur in the fuel channels with initial power less than 3.4 MW. The detailed analysis allows removing the surplus conservatism in the analysis (Figure 14).

Another fuel cladding temperature increase starts approximately 200 seconds after the beginning of the accident and is caused by the decrease of the reversed coolant flow, which in turn is due to the pressure decrease in DSs of the affected loop of RCS (Figure 12). A considerable temperature increase is possible only in case of operator non-intervention. Operator has a possibility to reduce coolant discharge through the break by closing the maintenance valves. These actions lead to the water level increase in the affected DS and improve cooling conditions of the fuel channels. The fuel channels of the intact RCS loop is reliably cooled with water supplied by the MCPs and ECCS long-term cooling subsystem.

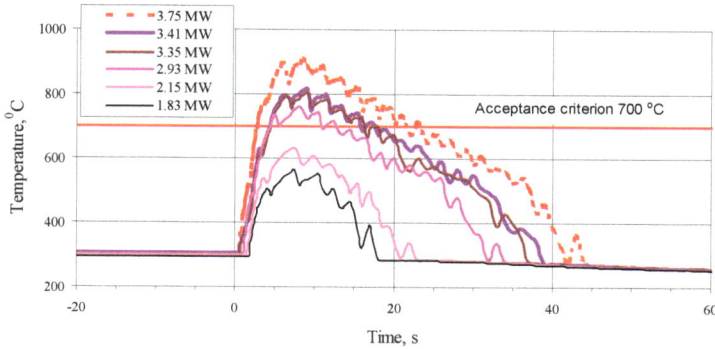

Fig. 13. Break of MCP pressure header with short-term ECCS failure. Temperature of fuel cladding in FC of the affected RCS

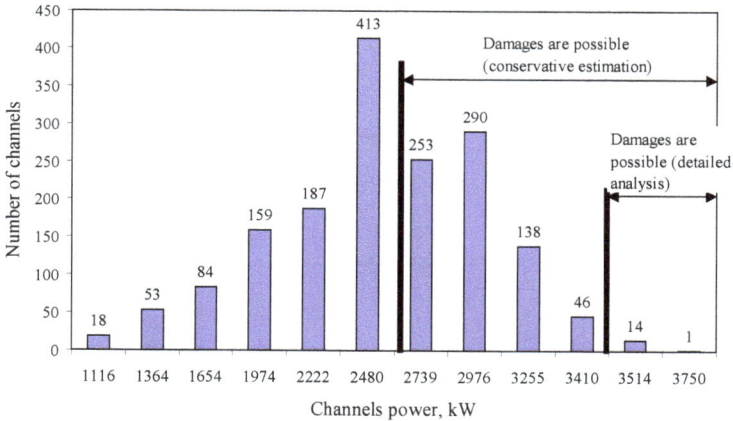

Fig. 14. Break of MCP pressure header with short-term ECCS failure. Estimation of the number of failed fuel channels (conservative estimation versus detailed analysis using RELAP/SCDAPSIM code)

For the case of medium LOCA in Zone 1 a detailed analysis is not carried out as the consequences this event is covered by medium LOCA in Zone 2. Short-term increase of temperatures of fuel rod cladding and FC walls is not observed at the initial stage of the accident at the GDH break upstream the check valve (medium LOCA in RCS Zone 1), but is traced at the break downstream the check valve (LOCA in Zone 2).

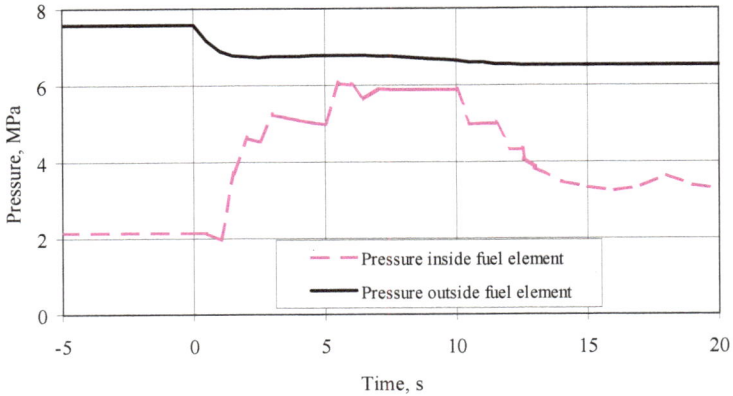

Fig. 15. Break of MCP pressure header with short-term ECCS failure. Pressure inside and outside the fuel rod (element) in the fuel channel with initial 3.4 MW power

5.2.2 LOCA in Zone 2

The consequences of LOCA in Zone 2 are similar to the consequences of LOCA in Zone 1. In both cases the break location is upstream the reactor core. Moreover, the phenomena in both cases are similar. In case of GDH break the coolant supply is terminated through 39 – 43 fuel channels connected to the distribution header of this group. In case of GDH guillotine break downstream check valve, the coolant in FCs connected to the affected GDH starts to flow in the opposite direction from DSs (Figure 16). Loss of the coolant from DS is compensated by

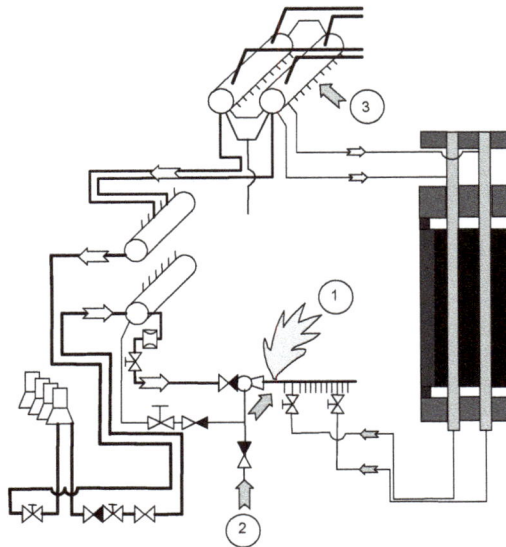

Fig. 16. GDH break downstream check valve. Structure of coolant flows: 1 – GDH break; 2 – ECCS water supply into the core; 3 – coolant supply through pressure header - ECCS bypass

ECCS water. Short-term fuel cladding and fuel channel wall temperatures increase is observed at the beginning of the accident due to the coolant flow direction change as in the case of LOCA in Zone 1. During partial breaks of GDH pipe, or in case of guillotine break of one GDH with the failure of check valve in the adjacent GDH (Figure 17), the coolant flow rate stagnation is possible in FCs connected to the affected GDH in this zone. In case of lower water piping break, coolant supply is terminated only into one FC.

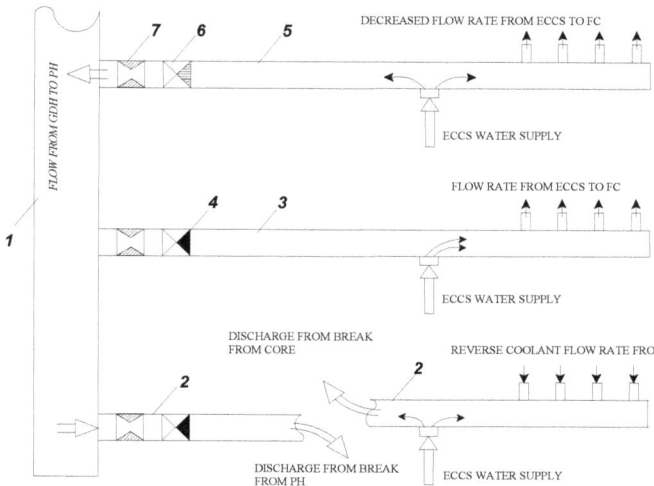

Fig. 17. Guillotine break of GDH downstream check valve at failure of check valve in adjacent GDH: 1 – MCP pressure header, 2- broken GDH, 3- normal GDH, 4 – check valve, 5 – GDH with fail to close check valve, 6 – fail to close check valve, 7 – flow limiting device

There are no pipelines with a diameter bigger than 300 mm, thus the large LOCAs in Zone 2 were not analyzed. For the medium LOCA in Zone 2, the following was considered: GDH guillotine break, GDH partial break resulting into stagnation of coolant flow rate and GDH guillotine break at failure to close the check valve in the adjacent GDH:

- The analysis of GDH guillotine break at complete ECCS failure and at operation of one, two or three ECCS pumps demonstrated that for reliable cooling of the core at a long stage two ECCS pumps are enough. With such number of ECCS equipment, after one hour from the beginning of the accident, the ECCS water flow rate starts to exceed water discharge through the break. However, short-term increase in temperatures of fuel rod cladding and FC walls at the initial stage of accident in FCs, connected to the broken GDH, is inevitable in any case. The excess of acceptance criterion for fuel rod cladding (700 ℃) is probable in 12 FCs. A more detailed analysis (using RELAP/SCDAPSIM model presented in Fig. 7) shows that this short temperature peak does not lead to the failure of any fuel rods.
- In the case of coolant flow rate stagnation in channels, multiple FCs breaks are probable after approximately 20 s from the beginning of the accident if the reactor is not shutdown on time. On the contrary, if a reactor is shutdown quickly (until the FCs heat up), the acceptance criterion for channels walls will not be violated and the channels

will remain intact. The conditions for the reactor long-term cooling remain similar to GDH break: the operation of two ECCS pumps is necessary.

- In the case of GDH break with a failure to close the check valve in the adjacent GDH, ECCS water supply worsens the cooling conditions of the channels connected to the GDH with failed to close check valve. It occurs that ECCS water interferes with the reverse coolant flow rate through these FCs. Stagnation of coolant flow rate is formed in these FCs. However, ECCS water supply helps to fill DSs and to ensure cooling of channels in the intact RCS loop and the channels connected to the 18 GDHs of the affected RCS loop. The analysis of GDH break with failure to close the check valve in the adjacent GDH is carried out at the operation of 1 - 4 ECCS pumps. The results of the analysis showed that for the reliable cooling of FC, connected to other 18 GDHs of the affected RCS loop, it is necessary to have not less than two operating ECCS pumps in long-term cooling subsystem. The channels connected to the GDH with failed to close check valve will be cooled because of radial heat transfer between the adjacent graphite blocks.

For the small LOCA in Zone 2, guillotine and partial breaks of lower water pipe were considered:

- The results of the analysis of guillotine break of the lower water pipe showed that the reactor core is reliably cooled during the first minutes after the accident in this case. One ECCS pump is enough for the reactor long-term cooling. The pump should be started during the first hour after the beginning of the accident.
- The performed analysis demonstrated that in the worst case, at partial break of the lower water pipe, the signal on the reactor shutdown on pressure increase in ALS compartments can not be generated. If the partial break causes stagnation of the coolant flow rate through the affected FC, it will lead to the heat up and break of this channel. The peak temperature of fuel in the affected channel will not reach the temperature of melting, i.e. 2800 ºC. After the break of the channel pipe, pressure in the reactor cavity increases, which results in the formation of a signal on the reactor shutdown. After the fuel channel wall break the conditions of flow stagnation will be destroyed and the broken parts of fuel channel and fragments of fuel assemblies below and above the break will be cooled by coolant flow from the top and bottom. The remaining intact fuel channels will also be reliably cooled. It is necessary to note that for RBMK type reactors the rupture of a single FC is a design basis accident. Such accident would correspond to a small breach in the reactor vessel of BWR. Steam-gas mixture from RC will be discharged through the reactor cavity venting system to the left tower of ALS. The steam will be condensed, fission products will be scrubbed in the condensing pool, but will not be discharged to the environment. Thus, the damaged fuel assembly will be contained in the reactor cavity.

5.2.3 LOCA in Zone 3

A single fuel channel break is the accident when the fuel channel is overheated during the reactor operation on power at nominal pressure. In case of other accidents that are included in this group (1.2, see Figure 5), the overheating of fuel can occur after the reactor shutdown at low pressure in RCS, because the strength of fuel channels at nominal pressure at the temperature margin of 650–800 ºC is limited [4]. Otherwise (if FC walls temperature exceeds this limit in few fuel channels during normal pressure in the circuit) multiple ruptures of fuel channels can occur. As design basis accident for RC is a single FC rupture, the accident with multiple ruptures of fuel channels is included into the group of accidents when the reactor core is completely damaged and the integrity of the reactor constructions is not preserved. The reactor cavity venting system at Ignalina NPP with RBMK-1500 reactor was improved providing the additional flow path to ALS, i.e. the RCVS capacity was increased to withstand a multiple rupture of fuel channels. Figure 18 presents the summary of the analysis performed to estimate the number of fuel channels that can be ruptured simultaneously in the beginning of the accident, i.e. at nominal RCS pressure and temperature, and reactor power of 4200 MW$_{th}$, the integrity of the reactor cavity would be maintained. A detailed analysis is presented in [21]. The coolant release rate was calculated using code RELAP5 (model presented in Figure 6) and the analysis of the Reactor Cavity and ALS response was performed using code CONTAIN. The performed analysis showed that making the most conservative assumptions, the reactor cavity could withstand simultaneous rupture of at least 11-16 FCs.

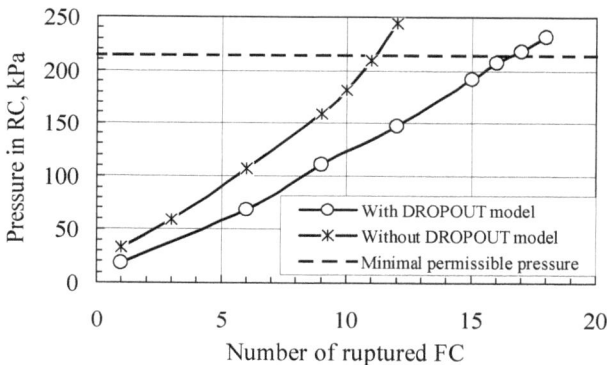

Fig. 18. Pressure in the reactor cavity as a function of a number of ruptured FCs [21]

5.2.4 LOCA in Zone 4

Pressure sharply decreases in DSs and RCS in the case of LOCA in Zone 4, but coolant circulation through the reactor core can be destroyed only in the latest stages of the accident, when pressure in RCS drops below 0.4 MPa. Thus, the characteristic fuel cladding and channel walls temperature peak within the first seconds of the accident (in case of LOCA in Zone 1 and Zone 2) is not met there. Guillotine breaks of one and two steamlines were considered for the large LOCA in RCS Zone 4:

- The results of the analysis showed that at LOCA in Zone 4 the operation of ECCS short-term subsystem is not necessary, i.e. the temperature of fuel rod cladding and FC walls is much lower than acceptance criteria without operation of this subsystem.
- For reliable cooling of the reactor core in long-term post-accidental period, it is necessary to have not less than two ECCS pumps in operation in the case of two steamlines break, and not less than one pump in the case of one steamline break.
- In case of breaks in Zone 4 without the reactor shutdown (break of steamlines), the temperature rises much more slowly (especially the temperature of FCs walls). This specifies that such breaks in Zone 4 are less dangerous, than breaks in Zone 1 (Fig. 19). In the case without the reactor shutdown, the melting of the core at low pressure in RCS is probable, but does not result in the immediate damage of the reactor cavity.

Pressure in DS	Peak temperatures of fuel cladding

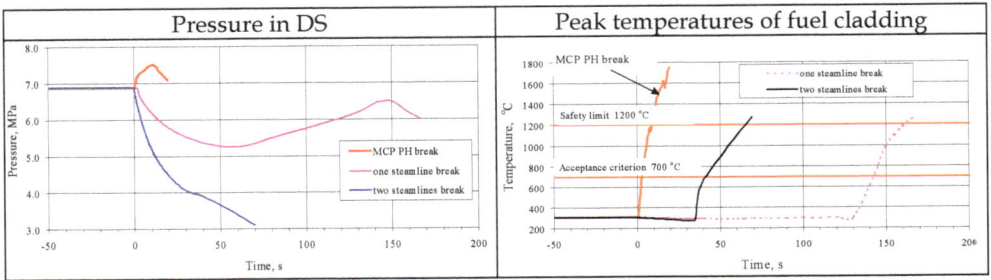

Fig. 19. Comparison of accident consequences in case of breaks in Zone 1 (MCP PH break) and Zone 4 (steamlines break)

5.3 Analysis of accidents when heat-up of the reactor core occurs during the reactor operation or within the first seconds after the reactor shutdown

The second category of core damage (see Figure 5), *accidents with loss of the general structure integrity of the reactor and ALS,* can potentially be due to the possibility of multiple ruptures of fuel channels at high pressure in RCS. The structural integrity of RBMK-1500 reactor depends on the integrity of the reactor cavity, which with a conservative strength margin was designed for conditions of an anticipated accident caused by the rupture of a single channel (in the nominal operating regime of a reactor). During operation of all nuclear power plants with RBMK reactors, there were three cases of fuel channel rupture, but the neighboring channels were not damaged [4]. This shows that the neighboring fuel channel–graphite cells have sufficient strength and elasticity and that the load caused by the rupture of a single channel is small. The *accidents, leading to the complete reactor core damage, with loss of the structural integrity of the reactor* can be divided into two groups [9]:

- *accidents when heat-up of the reactor core occurs during the reactor operation* or within the first seconds after the reactor shutdown, when decay heat is high (2.1, see Fig. 5);
- *accidents when heat-up of the reactor core occurs after the reactor shutdown* (2.2, see Fig. 5).

The first group of accidents (2.1) includes accidents when the heat-up of the reactor core occurs in the beginning of the accident (a few seconds after reactor scram activation) when decay heat in the core is high and the temperatures of fuel cladding and FC walls in the group of channels can reach the dangerous limits. An example of such accidents is the group of accidents, when the local flow stagnation occurs in the group of fuel channel during the LOCA. This situation is possible in the partial break case [12, 22]. It was mentioned earlier that partial rupture of lower water pipe could lead to the flow stagnation in the affected fuel channel. In the case of GDH partial rupture, the flow stagnation can occur in all fuel channels connected to this affected GDH. Under adverse conditions, the partial break of MCP pressure header can cause flow stagnation in the fuel channels of one affected loop. Since there is no sufficient time for actions of the operator in this case, the short-term accident management measures (automatic actuation of safety systems) are necessary. New reactor protection against coolant flow rate decrease through GDH generated signal for early ECCS activation is implemented in RBMK-1500 in this case. This short-term measure leads to the disturbance of the coolant flow rate stagnation in the group of fuel channels [12] and all parameters of RCS, thus, the reactor remains within safe limits.

Other examples of the first group of accidents (2.1) can also be the initiating events, namely:

- Anticipated Transients Without reactor Scram (ATWS);
- GDH blockage;
- Loss of natural circulation due to a sharp decrease of pressure in the RCS.

The analysis of ATWS (performed for the RBMK-1500 in 1996) demonstrated [3] the lack of inherent safety features in the RBMK design. The power is not reduced by means of inherent physical processes such as steam generation. The reactivity loss due to the fuel temperature rise (Doppler Effect) is not sufficient. The consequences of the accident for RBMK-1500 reactor, during which the loss of preferred electrical power supply and failure of automatic reactor shutdown occurred, are presented in Figure 20. The analysis was performed using RELAP5 model, presented in Figure 6.

Due to the loss of preferred electrical power supply all pumps are switched off (see Figure 20 (a)); therefore, the coolant circulation through the fuel channels is terminated. Because of the lost circulation, fuel channels are not cooled sufficiently and for this reason, the temperature of the fuel channel walls starts to increase sharply. As it is seen from Figure 20 (b), already after 40 seconds from the beginning of the accident, the peak fuel channel wall temperature in the high power channels reaches the acceptance criterion of 650 °C. It means that because of the further increase of the temperature in the fuel channels, plastic deformations begin, i.e. because of the influence of internal pressure, the channels can be ballooned and ruptured. During the first seconds of the accident, the main electrical generators and turbines are switched off as well. Steam generated in the core is discharged through the steam discharge valves, but their capacity is not sufficient Therefore, the pressure in the reactor cooling system increases and reaches acceptance criterion 10.4 MPa approximately after 80 seconds from the beginning of the accident (see Figure 20 (c)). Further increase of the pressure can lead to a rupture of pipelines.

Thus, the analysis of the anticipated transients without the shutdown demonstrated that in some cases the consequences can be quite dramatic for the RBMK-1500 reactors. Hence, in 1996 the priority recommendation was formulated as follows: to implement a second diverse shutdown system based on other principles of operation,. The implementation of such system requires much time and financial sources, thus at first it was decided to implement a compensating measure: a temporary shutdown system. This temporary system was called by the Russian abbreviation „DAZ" („Dopolnitelnaja avarijnaja začita" – „Additional emergency protection"). This system used the same control rods as well as design reactor shutdown system, however, signals for this system control were generated independently in respect of the design reactor shutdown system. The analysis performed to justify the selected set points for reactor scram activation showed that after the implementation of DAZ system, the reactor is shutdown on time and cooled reliably; moreover, the acceptance criteria are not violated even in case of transients when the design reactor shutdown system does not function. Figure 20 presents the behavior of the main parameters of the reactor cooling system in case of the loss of the preferred electrical power supply and simultaneous failure of the design reactor shutdown system [23]. In this case two signals for activation of DAZ system (reactor shutdown) are generated: on the increase of pressure in the drum-separators and on the decrease in the coolant flow rate through the main circulation pumps. In Unit 1 DAZ system was installed in 1999, in Unit 2 in 2000. Later (in 2004) the second diverse shutdown system was installed in the Ignalina NPP Unit 2. After these modifications the frequency of ATWS at Ignalina NPP became negligible (<10[-7]/year).

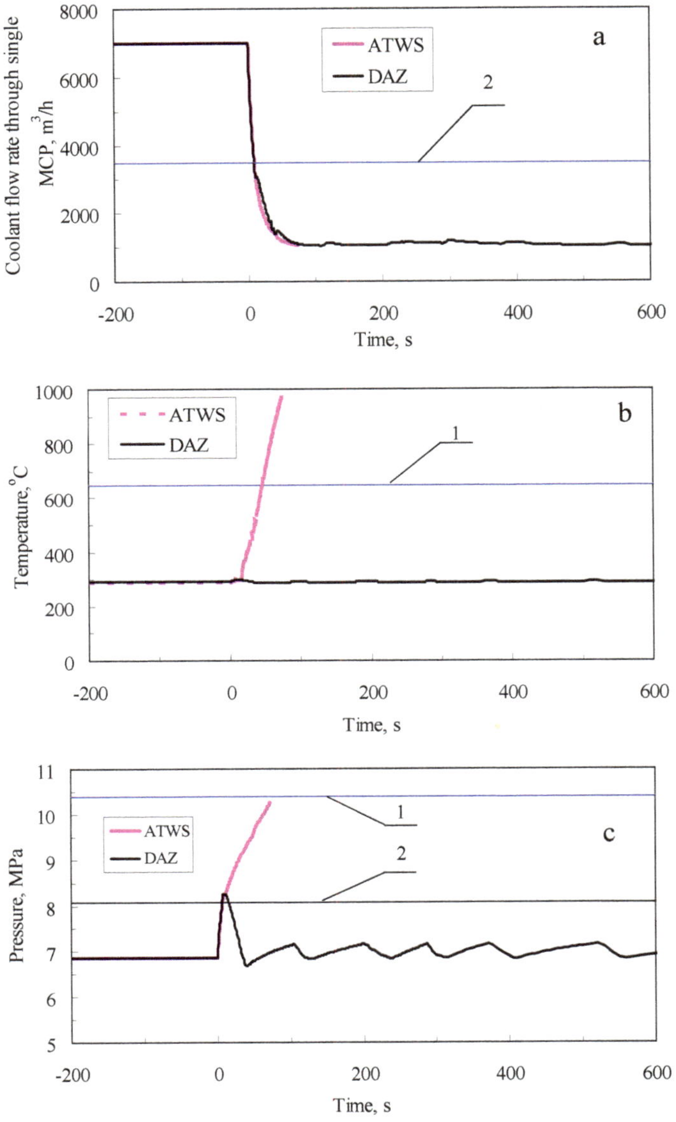

Fig. 20. Analysis of loss of preferred electrical power supply and simultaneous failure of design reactor shutdown system, when DAZ system was installed: a) coolant flow rate through one main circulation pump, b) the peak fuel channel wall temperature in the high power channel, c) pressure behaviour in drum - separators, 1 – acceptance criterion, 2 – set points of DAZ system activation (reactor shutdown)

The GDH blockage for RBMK-1500 also depends on such group of accidents when during normal operation a group of fuel channels is overheated and multiple rupture of FC can occur. It is shown [12] that the coolant flow through the ECCS bypass line is not enough to cool down the fuel channels connected to the blocked GDH. The critical heat flux would appear in some fuel channels and cause failure of fuel claddings and FC walls. In the year 2000 a new reactor scram (emergency shutdown) signal based on coolant flow rate decrease through GDH was implemented at Ignalina NPP. The new signal ensures the timely reactor shutdown so that the dangerous fuel cladding and FC walls temperatures are not reached [12]. Therefore, this accident is moved from the group of severe accidents into the group of accidents without core damage.

The accidents when the loss of natural circulation occurs due to a sharp decrease of pressure in the RCS (due to break of steamlines) are presented in section 5.2.4. Some parts of steamlines are located in the compartments without pressure gauges. Thus, there is no direct signal indicating that the steamline break occured in these compartments (as in the other cases the pressure increase in compartments indicates coolant discharge through the break). It means that signals for the reactor shutdown and ECCS activation will be generated with delay on the basis of secondary parameters (e.g., water level decrease in DSs). On the other hand, a sharp pressure drop in the RCS is a characteristic feature in the case of RBMK steamline break; it destroys the natural circulation of coolant through the core. The flow stagnation in the core together with the late reactor shutdown can cause overheating of group of fuel channels. This was mentioned in the safety analysis report of Ignalina NPP [3] and the review of safety analysis report [24]. In the year 1998 - 1999 a new reactor scram signal based on fast pressure decrease in DS was implemented at Ignalina NPP. This modification allowed avoiding the overheating of a group of fuel channels.

5.4 Analysis of accidents when heat-up of the reactor core occurs after the reactor shutdown

The second group of accidents from the second category of BDBA (*total damage of the core or its components with the reactor maintaining its overall structural integrity*) are the *accidents when heat-up of the reactor core occurs after the reactor shutdown* (2.2 – see Figure 5). The heat-up, damage and melting of the reactor core could occur in the late phase after the reactor shutdown due to loss of the long-term cooling. The results of the Level 1 probabilistic safety assessment of the Ignalina NPP showed that in the topography of the risk, transients dominate above the accidents with LOCAs and the failure of the core long-term cooling are the main factors of the frequency of core damage. The initiating events leading to the loss of long term cooling accident are such:

- loss of intermediate cooling circuit;
- loss of service water;
- station blackout.

The station blackout (the most likely initiating event) is the loss of normal electrical power supply for local needs with an additional failure on start-up of all diesel generators. In the case of loss of electrical power supply MCPs, the circulating pumps of the service water system and feedwater supply pumps are switched-off. The failure of diesel generators leads to the non-operability of the emergency long-term core cooling subsystem. It means the impossibility to feed RCS by water. The results of the analysis [20, 25] suggest that approximately 1.5 hours after the beginning of the accident, a dangerous heat-up of fuel rods and FC walls starts. Three ways of potential accident management for the loss of the long-term core cooling were discussed in [25]:

- decay heat removal by ventilation of DS compartments,
- decay heat removal by direct water supply into the reactor cavity,
- de-pressurisation of the reactor coolant system and water supply to the GDH from ECCS hydro-accumulators, deaerators or using non-regular means.

The results showed that the first two ways are inexpedient. The ventilation of DS compartments and the direct water supply into the RC are not sufficient to remove the decay heat from the core. However, the de-pressurisation of RCS and the following water supply from regular and non-regular means to the GDH in the case of loss of long-term cooling gives considerably better results compared with the other two measures. The performed analysis [25] demonstrated that the reactor core cooling by RCS depressurisation and water supply from deaerators give additional four hours for the operator to install water supply from the external (artesian) water source. Thus, this way of accident management is recommended to be included in the RBMK-1500 accident management programme.

In the case of loss of intermediate cooling circuit or loss of service water, the consequences of the accidents will be very similar. In these last cases there are no direct signals for the reactor shutdown; however, these initiating events lead to the loss of feedwater supply and MCP will be tripped due to the insufficient cooling. Reactor scram would appear on secondary parameters and further sequence of the accident will be the same as in case of the station blackout: due to decay heat the water in the core is evaporating and the core heat-up process starts. Core overheating can be avoided by using water stored in the ECCS hydro-accumulators, deaerators and other water sources located at NPP.

Figure 21 shows the behavior of fuel, claddings, channel walls, and graphite stack temperatures, calculated using RELAP5 and RELAP5/SCDAPSIM codes (models presented in Figure 6 and Figure 7), in case of RBMK-1500 reactor blackout, without any additional water supply. Such event, which is developing into a severe accident, serves as an example for the discussion of severe accident phenomena in RBMK type reactors. As it is presented in Figure 21, the failure of fuel channels could occur after ~3 hours. The failure of FCs is expected because the pressure in RCS is nominal, and the acceptance criterion for fuel channel (650 °C) will be reached (Figure 21). It is assumed that after ~3.8 hours the operator opens one steam relief valve to discharge steam from RCS. This action allows depressurizing RCS and prevents ruptures of FCs. If no operator actions were taken (no manual depressurization), then approximately 50 FCs with a higher power level could be ruptured. If all FCs ruptured within short time interval, the reactor cavity would be destroyed and the consequences of the accident would be similar to the Chernobyl accident.

Figure 22 shows the long-term behavior of fuel, claddings, fuel channels, and graphite stack temperatures, calculated using RELAP5/SCDAPSIM code (model presented in Figure 7), in case of RBMK-1500 reactor blackout, with operator intervention (opening one of steam relief valves for RCS depressurization). As the operator starts depressurization, the accident scenario continues at low pressure. Due to the pressure decrease the rest of coolant in pipelines below the reactor core starts boiling and steam cools down the core for a short time. As it is shown in Figure 22, the temperatures of the core components decrease for a short period of time. However, after 4.3 hours the second (repeated) heat-up of the reactor core elements begins. When the temperature of fuel cladding increases above 800 °C, the failure of fuel claddings occurs due to ballooning. The ballooning happens because at that time the pressure in RCS (outside fuel rods) is close to the atmospheric and the pressure inside fuel rods is high. The conditions for fast oxidation of claddings and fuel channels,

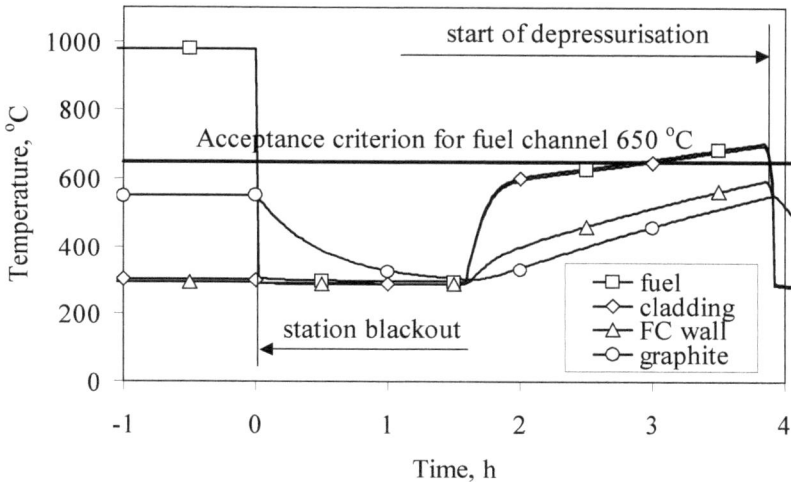

Fig. 21. Station blackout, when operators depressurizes RCS. Temperature of fuel, fuel cladding, fuel channel and graphite

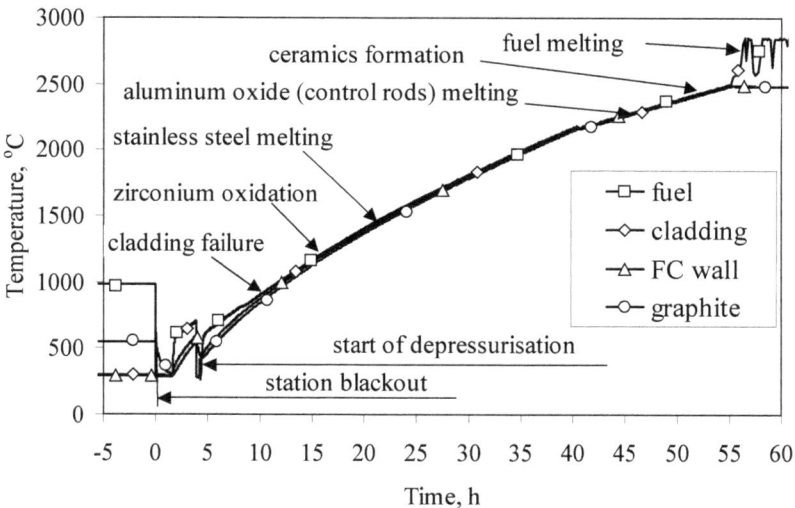

Fig. 22. Station blackout, when operators depressurizes RCS. Main consequences in case of station blackout

made from zirconium-niobium alloy, are reached after the fuel cladding and fuel channel temperatures exceed 1000 – 1200 °C (~15 hours after the beginning of the accident). Due to steam-zirconium reaction the generation of hydrogen starts. The oxidation and hydrogen generation processes terminated after the pressure in RCS decreases down to atmospheric. This indicates that there is no steam in RCS, thus the steam-zirconium reaction is impossible. Later the processes would continue at low pressure in RCS and RC would remain intact.

When the temperatures of fuel claddings and FC walls reach 1450 °C, the melting of stainless steel grids and zirconium starts at 1760 °C (Figure 22). Probably at the same time the fuel channels will fail. At temperature of 1930–2050 °C and 2330 °C the melting of aluminum oxide (control rods claddings) and boron-carbide (control rods elements) in the separate control rods channels starts. The formation of ceramic (U, Zr, ZrO$_2$) starts at temperature of 2600 °C. The analysis performed using RELAP5/SCDAPSIM code shows that fuel melting (melting of ZrO$_2$ and UO$_2$) starts at low pressure, approximately 50 hours after the beginning of the accident at temperatures of 2690 °C and 2850 °C respectively (Figure 22). Such comparably slow core heat up process is due to the high inertia of graphite stack, which provides a heat sink. Hence, the high pressure melt ejection and direct containment heating – the phenomena more related to PWR design – could not occur at RBMK-1500 reactor due to the limited space inside the reactor cavity. However, to cool down the reactor, it is necessary to start water supply into the fuel channels within the first 15 hours after the beginning of the accident. The water supply in later phases could lead to a fast steam-zirconium reaction and it could accelerate core damage processes.

6. Conclusion

In this paper the specifics of RBMK reactors design was presented. Based on the specific feature of RBMK, possible Beyond Design Basis Accidents were divided into four groups:

- accidents with no severe damage of the core;
- *accidents leading to a severe core damage accompanied by containment of the core fragments in the reactor cavity and accident localization system or other reactor buildings;*
- *accidents when heat-up of the reactor core occurs during the reactor operation or within the first seconds after the reactor shutdown;*
- *accidents when heat-up of the reactor core occurs after the reactor shutdown.*

The deterministic analysis of all these groups of BDBA was performed using a system of thermal hydraulic computer codes RELAP5 and RELAP/SDAPSIM. The consequences of these BDBAs and possible accident mitigation measures were discussed.

- For the first group of accidents (*accidents with no severe damage of the core*) it was showed: (1) In the case of erroneously withdrawn of a group of control rods, the local power increase appears in the adjacent fuel channels, but this do not lead to overheating of the fuel in these channels. The operators have possibility to compensate this local power increase by inserting remaining control rods. In the case the local power exceeds limits – the reactor will be shutdown automatically by activation of emergency shutdown system. (2) In the case of loss of long-term cooling, when there are no possibilities to inject water in the reactor using regular means, the operators can supply the water into reactor from ECCS hydro-accumulators, later to perform the RCS de-pressurization by opening manually steam relief valve. Finally, after the pressure in RCS is reduced, the low-pressure non-regular water sources can be used (deaerators and artesian water).

- The accidents in the second group (*accidents leading to a severe core damage accompanied by containment of the core fragments in the reactor cavity and accident localization system or other reactor buildings*) are initiated due to misbalance between energy source and heat sink. If the emergency core cooling system is not activated, or the amount of supplied water is less as required, the core meltdown can occur. Based on the performed deterministic analysis the setpoints for ECCS activation were selected. The capacity of reactor cavity venting system was increased to prevent failure of reactor cavity in case of multiple fuel channel rupture (up to simultaneous rupture of 16 fuel channels).

- The third group contains the *accidents when heat-up of the reactor core occurs during the reactor operation or within the first seconds after the reactor shutdown*, when decay heat is high. Because fast process of heat-up of fuel rods in this case – there is no time for operator actions in this case. The new algorithms for reactor shutdown and fast emergency core cooling system activation were proposed for RBMK-1500 to prevent overheating of fuel in local flow stagnation or flow blockage in the group of fuel channels cases. Also the new ECCS activation algorithm was developed to cooldown the reactor in the case of loss of natural circulation due to a sharp decrease of pressure in the RCS. To prevent the catastrophic core damage in the anticipated transients without reactor scram case (when main reactor shutdown system fails to shutdown reactor) the additional emergency protection was implemented in the RBMK-1500.

- The forth group of accidents – *the accidents when heat-up of the reactor core occurs after the reactor shutdown*. The performed analysis shown, that even in the case of failure of all design (regular) and non-regular means to cooldown the rector in the case of loss of long term core cooling, the core heat-up process is slow in RBMK-type reactors. In the station blackout case, to prevent failure of reactor cavity at high pressure, the operators are required to open the steam relief valve manually, to start RCS depressurization. Due to the high inertia of graphite stack, which provides a heat sink, the melting of fuel stats at low pressure not earlier as 50 hours after loss core cooling.

The analysis was performed for the RBMK-1500 reactor (Ignalina NPP, Lithuania), but the main ideas of the accident mitigation are also valid for the RBMK-1000, which are still operating in Russia.

7. Abbreviations

ALS   Accident Localization System
ATWS   Anticipated Transients Without reactor Scram
BDBA   Beyond Design Basis Accidents
BWR   Boiling Water Reactor
CPS   Control & Protection System
DAZ   Acronym for Russian – Additional emergency protection

DS   Drum Separator
ECCS   Emergency Core Cooling System
FC   Fuel Channel
GDH   Group Distribution Header
LWR   Light Water Reactor
LOCA   Loss Of Coolant Accident
LWP   Low Water Pipes
MCP   Main Circulation Pump
NPP   Nuclear Power Plant
RBMK   Acronym for Russian –graphite-moderated boiling water reactor type
PWR   Pressurized Water Reactor
RC   Reactor Cavity
RCS   Reactor Cooling System
RCVS   Reactor Cavity Venting System
SRV   Steam Relief Valve
TCV   Turbine Control Valve

8. References

[1] K. Almenas, A. Kaliatka, E. Uspuras, *Ignalina RBMK-1500. A Source Book. Extended and Updated Version*, Lithuanian Energy Institute, Kaunas, Lithuania (1998).

[2] Accident analysis for nuclear power plants with graphite moderated boiling water RBMK reactors, Safety Reports Series No. 43, IAEA, Vienna, 2005.

[3] In-depth safety assessment of Ignalina Nuclear Power Plant. Final Report. Ignalina NPP, Lithuania. 1996.

[4] O. Yu. Novoselsky, V. N. Filinov, Computational Assessment of RBMK Pressure Tube Rupture at Accident Heating. International Exchange Forum "Analytical Methods and Computational Tools for NPP Safety Assessment" Obninsk 1996.

[5] The analysis of steam-gas mixture release from the reactor cavity of RBMK-1500 reactor for determination of the boundaries. Phase 4, Results of the Analysis, Report No. 74.069, NIKIET, Moscow, 2000. (in Russian).

[6] Calculation of discharge capacity of the RCVS of INPP 1st stage, NIKIET, Report 4.161 Dated 1992.

[7] Rimkevicius S., Urbonavicius E., Cesna B. Safety margins of RBMK-1500 accident localisation system at Ignalina NPP // Safety margins of operating reactors. Analysis of uncertainties and implications for decision making. International Atomic Energy Agency. IAEA-TECDOC-1332, Vienna, January 2003. / 2003, p. 95-106.

[8] Vasilevskij V.P., Nikitin J.M., Petrov A.A., Potapov A.A., Tcherkashev J.M. Features of RBMK severe accidents development and approaches to such accidents management// Atomic energy. Vol. 90, Issue 6. Moscow, Russia. 2001. (In Russian).

[9] Kramerov A.J., Michailov D.A. About the approach to severe accident studying in channel boiling reactors (basically at overheating by decay heat) // Proc. of the 5th International Information Exchange Forum "Safety Analysis for NPPs of

VVER and RBMK Types Reactors". Obninsk, Russia .16-20 October 2000. 8 p. (In Russian).

[10] Rimkevicius S., Uspuras E. Modelling of thermal hydraulic transient processes in Nuclear Power Plants: Ignalina compartments / Ed. J. Vilemas // New York: Begell House Inc., 2007. Kaunas: Lithuanian Energy Institute, 2007. 197 p. ISBN 978-1-56700-247-8.

[11] RELAP5 Code Development Team, RELAP5/MOD3 Code Manual, Volume 1, Code Structure, System Models, and Solution Methods, NUREG/CR-5535, INEL-95/0174, 1995

[12] Kaliatka A., Uspuras E., Thermal-hydraulic analysis of accidents leading to local coolant flow decrease in the main circulation circuit of RBMK-1500, Nuclear Engineering and Design. ISSN 0029-5493, Vol. 217, N 1–2, 2002, pp. 91–101

[13] Uspuras E., Kaliatka A., Accident and transient processes at NPPs with channel-type reactors: monography // Kaunas: Lithuanian Energy Institute, 2006. Thermophysics: 28. 298 p. ISBN 9986-492-87-4.

[14] Urbonas R., Uspuras E., Kaliatka A., State-of-the-art computer code RELAP5 validation with RBMK-related separate phenomena data, Nuclear Engineering and Design, ISSN 0029-5493, Vol. 225, 2003, pp. 65-81.

[15] Allison C.M. and Wagner R.J., RELAP5/SCDAPSIM/MOD3.2 (am+) Input Manual Supplemental, Innovative Systems Software, LLC, 2001.

[16] Kaliatka A., Ušpuras E. Development and testing of RBMK-1500 model for BDBA analysis employing RELAP/SCDAPSIM code // Annals of Nuclear Energy. ISSN 0306-4549. 2008, Vol. 35, p. 977-992.

[17] Kaliatka A., Uspuras E. Specifics of RBMK core cooling in beyond design basis accidents // Nuclear Engineering and Design. ISSN 0029-5493. 2008, Vol. 238, p. 2005-2016.

[18] Urbonavicius E., Kaliatka A., Ušpuras E. Accident Management for NPPs with RBMK reactors. Monograph // New York: Begell House Inc., Kaunas: Lithuanian Energy Institute, 2010. 205 p. ISBN 978-1-56700-267-6.

[19] Final Safety Justification for Ignalina Nuclear Power Plant Diverse Shutdown System. Safety justification for Additional Hold-down System, DS&S Report XE405-TEC188_Appendix-E, Ignalina NPP, 2004.

[20] Afremov D.A., Solovjev S.L. Development and application of design-theoretical methods of the analysis of certain severe accidents for RBMK reactor // Heat-and-power engineering. No. 4. Moscow, Russia. 2001. (In Russian).

[21] Cesna B., Rimkevicius S., Urbonavicius E., Babilas E., "Reactor cavity and ALS thermal-hydraulic evaluation in case of fuel channels ruptures at Ignalina NPP, 2004," Nuclear Engineering and Design, Vol. 232, 2004, pp. 57-67.

[22] Dostov A.I., Kramerov A. J. Investigation of RBMK safety at the accidents initiated by partial breaks in main circulation circuit // Atomic energy. Vol. 92, Issue 1. Moscow, Russia. 2002. (In Russian).

[23] Kaliatka A., Ušpuras E. Development and evaluation of additional shutdown system at the Ignalina NPP by employing RELAP5 code // Nuclear Engineering and Design. ISSN 0029-5493. 2002, Vol. 217, N 1–2, p. 129–139.

[24] Uspuras E. Status of Ignalina's safety analysis reports // Intern. Conf. on the Strengthening of Nuclear Safety in Eastern Europe, 14-18 June 1999. - Vienna, Austria, 1999. - P. 415-442.

[25] Uspuras E., Kaliatka A., Vileiniskis V. Development of accident management measures for RBMK-1500 in the case of loss of long-term core cooling // Nuclear Engineering and Design. ISSN 0029-5493. 2006, Vol. 236, p. 47-56.

5

Evolved Fuzzy Control System for a Steam Generator

Daniela Hossu, Ioana Făgărăşan, Andrei Hossu and Sergiu Stelian Iliescu
University Politehnica of Bucharest, Faculty of Control and Computers
Romania

1. Introduction

Poor control of steam generator water level is the main cause of unexpected shutdowns in nuclear power plants. Such shutdowns are caused by violation of safety limits on the water level and are common at low operating power where the plant exhibits strong non-minimum phase characteristics. In addition, the steam generator is a highly complex, nonlinear and time-varying system and its parameters vary with operating conditions. Therefore, there is a need to systematically investigate the problem of controlling the water level in the steam generator in order to prevent reactor shutdowns.

Difficulties on designing a steam generator (SG) level controller arise from the following factors:

- *nonlinear plant characteristics*. The plant dynamics are highly nonlinear. This is reflected by the fact that the linearized plant model shows significant variation with operating power.
- *nonminimum-phase plant characteristics*. The plant exhibits strong inverse response behavior, particularly at low operating power due to the so-called *"swell and shrink"* effects.
- dynamics uncertainties,
- corrupted feed-water flow measurement signal with biased noises.

At low loads (less than 15% of full power) the non-minimum phase behavior is much more pronounced.

Various approaches have been reported in the literature: an adaptive PID level controller using a linear parameter varying model to describe the process dynamics over the entire operating power range (Irving et al. 1980); a model of the steam generator water level process in the form of a transfer function, determined based on first-principles analysis and expert experience has been presented in (Zhao et al., 2000); LQG controllers with "gain-scheduling" to cover the entire operating range (Menon & Parlos, 1992); a hybrid fuzzy-PI adaptive control of drum level, a model predictive controller to identify the operating point at each sampling time and use the plant model corresponding to this operating point as the prediction model (Kothare et al., 2000). Paper (Park & Seong, 1997) presents a self organizing fuzzy logic controller for the water level control of a steam generator. A nonlinear physical model with a complexity that is suitable for model-based control has been presented by Astrom and Bell (Ästrom & Bell, 2000). The model describes the behavior of the system over a wide operating range.

With the advent of the current generation of high-speed computers, more advanced control strategies not limited to PI/PID, can be applied (Hirota, 1993), (Pedrycz & Gomide, 2007), (Yen et al., 1995), (Ross, 2004).

Model predictive control (MPC) design technique has gained wide acceptance in process control applications. Model predictive control has three basic steps: output prediction, control calculation and closing the feedback loop (Camacho & Bordons, 2004), (Demircioglu & Karasu, 2000), (Morari & Lee, 1999).

In this chapter, we apply MPC techniques to develop a framework for systematically addressing the various issues in the SG level control problem.

Fuzzy models have become one of the most well established approaches to non-linear system modeling since they are universal approximations which can deal with both quantitative and qualitative (linguistic) forms of information (Dubois & Prade, 1997), (Zadeh, 2005), (Zadeh, 1989) This chapter deals with Takagi-Sugeno (T-S) fuzzy models because this type of model provides efficient and computationally attractive solutions to a wide range of modeling problems capable to approximate nonlinear dynamics, multiple operating modes and significant parameter and structure variations (Kiriakidis, 1999), (Yager & Zadeh, 1992), (Ying, 2000). Takagi-Sugeno (T-S) fuzzy models have a good capability for prediction and can be easily used to design model-based predictive controllers for nonlinear systems (Espinosa et al., 1999).

The objective of this chapter is to design, evaluate and implement a water level controller for steam generators based on a fuzzy model predictive control approach. The chapter includes simulations of typical operating transients in the SG. A new concept of modular advanced control system designed for a seamless and gradual integration into the target systems is presented. The system is designed in such a way to improve the quality of monitoring and control of the whole system. The project targets the large scale distributed advanced control systems with optimum granularity architecture.

2. Fuzzy model

Fuzzy models can be divided into three classes: Linguistic Models (Mamdani Models), Fuzzy Relational Models, and Takagi-Sugeno (TS) Models. Both linguistic and fuzzy relational models are linguistically interpretable and can incorporate prior qualitative knowledge provided by experts (Zadeh, 2008). TS models are able to accurately represent a wide class of nonlinear systems using a relatively small number of parameters. TS models perform an interpolation of local models, usually linear, by means of a fuzzy inference mechanism. Their functional rule base structure is well-known to be intrinsically favorable for control applications.

This chapter deals with Takagi-Sugeno (T-S) fuzzy models because of their capability to approximate a large class of static and dynamic nonlinear systems. In T-S modeling methodology, a nonlinear system is divided into a number of linear or nearly linear subsystems. A quasi-linear empirical model is developed by means of fuzzy logic for each subsystem. The whole process behavior is characterized by a weighted sum of the outputs from all quasi-linear fuzzy implication. The methodology facilitates the development of a nonlinear model that is essentially a collection of a number of quasi-linear models regulated by fuzzy logic. It also provides an opportunity to simplify the design of model predictive control. In such a model, the cause-effect relationship between control u and output y at the sampling time n is established in a discrete time representation. Each fuzzy implication is generated based on a system *step response* (Andone&Hossu, 2004), (Hossu et al., 2010) , (Huang et al. 2000).

$$IF\ y(n)\ is\ A_0^i,\ y(n-1)\ is\ A_1^i,....,y(n-m+1)\ is\ A_{m-1}^i,$$

$$and\ \ u(n)\ is\ B_0^i,\ u(n-1)\ is\ B_1^i,\ ...,u(n-l+1)\ is\ B_{l-1}^i \tag{1}$$

$$THEN\ y^i(n+1) = y(n) + \sum_{j=1}^{T} h_j^i \Delta u(n+1-j)$$

where:

A_j^i fuzzy set corresponding to output $y(n$-$j)$ in the ith fuzzy implication

B_j^i fuzzy set corresponding to input $u(n$-$j)$ in the ith fuzzy implication

h_j^i impulse response coefficient in the ith fuzzy implication

T model horizon

$\Delta u(n)$ difference between $u(n)$ and $u(n$-1)

A complete fuzzy model for the system consists of p fuzzy implications. The system output $y(n+1)$ is inferred as a weighted average value of the outputs estimated by all fuzzy implications.

$$y(n+1) = \frac{\sum_{j=1}^{p} \mu^j y^j(n+1)}{\sum_{j=1}^{p} \mu^j} \tag{2}$$

where

$$\mu^j = \bigwedge_i A_i^j \wedge \bigwedge_k B_k^j \tag{3}$$

considering

$$\omega^j = \frac{\mu^j}{\sum_{j=1}^{p} \mu^j} \tag{4}$$

then

$$y(n+1) = \sum_{j=1}^{p} \omega^j y^j (n+1) \tag{5}$$

3. Fuzzy model predictive control

3.1 Problem formulation

The goal in this chapter is to study the use of the feed-water flow-rate as a manipulated variable to maintain the SG water level within allowable limits, in the face of the changing steam demand resulting from a change in the electrical power demand. The design goal of an FMPC is to minimize the predictive error between an output and a given reference trajectory in the next N_y steps through the selection of N_u step optimal control policies.

The optimization problem can be formulated as

$$\min_{\Delta u(n),\Delta u(n+1),\ldots,\Delta u(n+N_u))} J(n) \tag{6}$$

$$J(n) = \sum_{i=1}^{N_y} \mu_i (\hat{y}(n+i) - y^r(n+i))^2 + \sum_{i=1}^{N_u} \upsilon_i \Delta u(n+i)^2 \tag{7}$$

where:

μ_i and υ_i	are the weighting factors for the prediction error and control energy;
$\hat{y}(n+i)$	i^{th} step output prediction;
$y^r(n+i)$	i^{th} step reference trajectory;
$\Delta u(n+i)$	i^{th} step control action.

The weighted sum of the local control policies gives the overall control policy:

$$\Delta u(n+i) = \sum_{j=1}^{p} \omega^j \Delta u^j (n+i) \tag{8}$$

Substituting (2) and (8) into (7) yields (9)

$$J(n) = \sum_{i=1}^{N_y} \mu_i \left(\sum_{j=1}^{p} \left(\omega^j \left(\hat{y}^j (n+i) - y^r(n+i) \right) \right) \right)^2$$
$$+ \sum_{i=0}^{N_u-1} \upsilon_i \left(\sum_{j=1}^{p} \omega^j \Delta u^j (n+i) \right)^2 \tag{9}$$

To simplify the computation, an alternative objective function is proposed as a satisfactory approximation of (9) (Huang et al., 2000).

$$\tilde{J}(n) = \sum_{j=1}^{p}\left[\left(\omega^j\right)^2\left(\sum_{i=1}^{N_y}\mu_i\left(\hat{y}(n+i)-y^r(n+i)\right)^2 + \sum_{i=0}^{N_u-1}v_i\Delta u^j(n+i)^2\right)\right] \quad (10)$$

The optimization problem can be defined as:

$$\min_{\Delta u(n),\Delta u(n+1),...,\Delta u(n+N_u-1)} \tilde{J}(n) =$$

$$\min_{\Delta u(n),\Delta u(n+1),...,\Delta u(n+N_u-1)} \sum_{j=1}^{p}\left(\omega^j\right)^2\tilde{J}^j(n) \quad (11)$$

$$\tilde{J}^j(n) = \sum_{i=1}^{N_y}\mu_i\left(\hat{y}^j(n+i)-y^r(n+i)\right)^2 +$$

$$+ \sum_{i=0}^{N_u-1}v_i\left(\Delta u^j(n+i)\right)^2 \quad (12)$$

Using the alternative objective function (12), we can derive a controller by a hierarchical control design approach.

3.2 Controller design

1. *Lower Layer Design:* For the jth subsystem, the optimization problem is defined as follows:

$$\min_{\Delta u(n),\Delta u(n+1),...,\Delta u(n+N_u-1)} \tilde{J}^j(n) \quad (13)$$

$$R_j:\quad \begin{aligned}&IF\ y(n+k-1)\ is\ A_0^j,\cdots,y(n+k-m)\ is\ A_{m-1}^j\\&THEN\quad y^j(n+k) = y^j(n+k-1)+\sum_{i=1}^{T}h_i^j\Delta u(n+k-i)+\varepsilon^j(n+k-1)\end{aligned} \quad (14)$$

where $\varepsilon^j(n+k-1)$ serves for system coordination and it is determined at the upper layer.

2. *Upper Layer Design:* The upper layer coordination targets the identification of globally optimal control policies through coordinating $\varepsilon^j(n+k-1)$ for each local subsystem.
3. *System Coordination:* The controller is designed through a hierarchical control design (Figure 1). From the lower layer, the local information of output and control is transmitted to the upper layer. The whole design is decomposed into the derivation of p local controllers. The subsystems regulated by those local controllers will be coordinated to derive a globally optimal control policy.

The objective function defined in (11) can be rewritten in a matrix form:

$$R_j:\quad \begin{aligned}\tilde{J}^j(n) &= \left(\hat{Y}_+^j(n)-Y^r(n)\right)^T W_1^j\left(\hat{Y}_+^j(n)-Y^r(n)\right)+\\&+ \left(\Delta U_+^j(n)\right)^T W_2^j\left(\Delta U_+^j(n)\right)\end{aligned} \quad (15)$$

where:

$$\hat{Y}_+^j(n) = \left(\hat{y}^{\,j}(n+1)\, \hat{y}^{\,j}(n+2) \cdots \hat{y}^{\,j}\left(n+N_y\right) \right)^T \tag{16}$$

$$Y^r(n) = \left(y^r(n+1)\, y^r(n+2) \cdots y^r\left(n+N_y\right) \right)^T \tag{17}$$

$$\Delta U_+^j(n) = \left(\Delta u^j(n)\, \Delta u^j(n+1) \cdots \Delta u^j\left(n+N_u-1\right) \right)^T \tag{18}$$

$$W_1^j = diag\left\{ \mu_1^j, \mu_2^j, \cdots, \mu_{N_y}^j \right\} \tag{19}$$

$$W_2^j = diag\left\{ v_1^j, v_2^j, \cdots, v_{N_u}^j \right\} \tag{20}$$

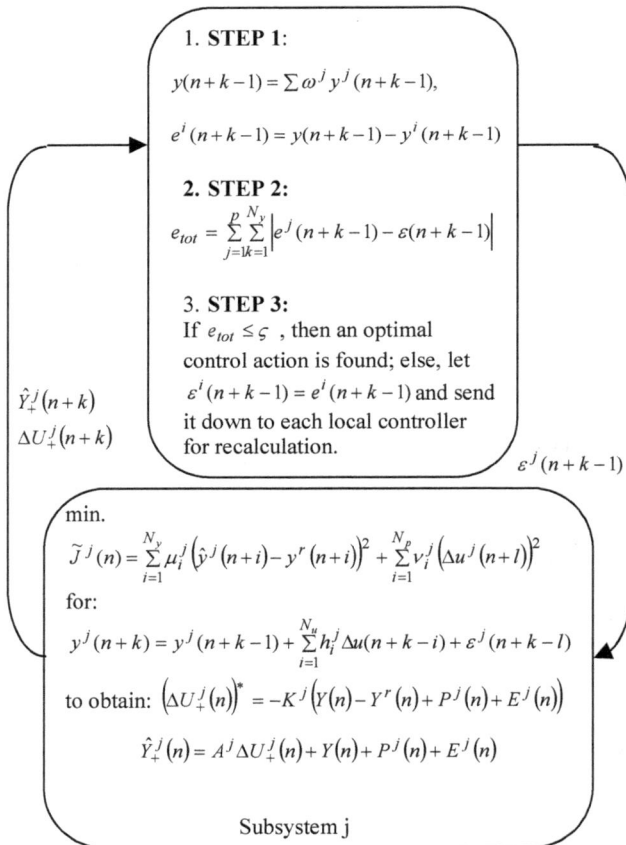

1. STEP 1:

$$y(n+k-1) = \sum \omega^j y^j(n+k-1),$$

$$e^i(n+k-1) = y(n+k-1) - y^i(n+k-1)$$

2. STEP 2:

$$e_{tot} = \sum_{j=1}^{p} \sum_{k=1}^{N_y} \left| e^j(n+k-1) - \varepsilon(n+k-1) \right|$$

3. STEP 3:

If $e_{tot} \le \varsigma$, then an optimal control action is found; else, let $\varepsilon^i(n+k-1) = e^i(n+k-1)$ and send it down to each local controller for recalculation.

$\hat{Y}_+^j(n+k)$
$\Delta U_+^j(n+k)$

$\varepsilon^j(n+k-1)$

min.

$$\tilde{J}^j(n) = \sum_{i=1}^{N_y} \mu_i^j \left(\hat{y}^j(n+i) - y^r(n+i) \right)^2 + \sum_{i=1}^{N_p} v_i^j \left(\Delta u^j(n+l) \right)^2$$

for:

$$y^j(n+k) = y^j(n+k-1) + \sum_{i=1}^{N_u} h_i^j \Delta u(n+k-i) + \varepsilon^j(n+k-l)$$

to obtain: $\left(\Delta U_+^j(n) \right)^* = -K^j \left(Y(n) - Y^r(n) + P^j(n) + E^j(n) \right)$

$$\hat{Y}_+^j(n) = A^j \Delta U_+^j(n) + Y(n) + P^j(n) + E^j(n)$$

Subsystem j

Fig. 1. Hierarchical controller design

The N_y - step prediction of the output by the jth FI can be rewritten as follows:

$$\hat{Y}_+^j(n) = A^j \Delta U_+^j(n) + Y(n) + P^j(n) + E_+^j(n) \tag{21}$$

where:

$$A^j = \begin{bmatrix} a_1^j & 0 & 0 & \cdots & 0 \\ a_2^j & a_1^j & 0 & & 0 \\ a_3^j & a_2^j & a_1^j & & 0 \\ \vdots & & & & \vdots \\ a_{N_y}^j & a_{N_y-1}^j & a_{N_y-2}^j & \cdots & a_{N_y-N_u+1}^j \end{bmatrix} \tag{22}$$

$$a_i^j = \sum_{k=1}^{i} h_k^j \tag{23}$$

$$Y(n) = \left(y_{D_{al}}(n) \, y_{D_{al}}(n) \cdots y(n)_{D_{al}} \right)^T \tag{24}$$

$$P^j(n) = \left(P_1^j(n) \, P_2^j(n) \cdots P_{N_y}^j(n) \right)^T \tag{25}$$

$$E_+^j(n) = \left(0 \sum_{k=1}^{2} \varepsilon^j(n+k-1) \cdots \sum_{k=1}^{N_y} \varepsilon^j(n+k-1) \right)^T \tag{26}$$

$$P_i^j(n) = \sum_{k=1}^{i} \sum_{l=k+1}^{T} h_l^j \Delta u(n+k-l) \tag{27}$$

The resulting control policy for the jth subsystem can be derived as

$$\begin{aligned} \tilde{J}^j(n) &= \left(\Delta U_+^j(n) \right)^T \left(A^{j^T} W_1^j A + W_2^j \right) \Delta U_+^j(n) + \\ &+ \left(\Delta U_+^j(n) \right)^T A^{j^T} W_1^j Z^j(n) + \\ &+ \left(Z^j(n) \right)^T W_1^j A^j \Delta U_+^j(n) + \left(Z^j(n) \right)^T W_1^j Z^j(n) \end{aligned} \tag{28}$$

where:

$$Z^j(n) = Y(n) - Y^r(n) + P^j(n) + E_+^j(n) \tag{29}$$

Minimizing (26) yields

$$\begin{aligned} \frac{\delta \tilde{J}^j(n)}{\delta \Delta U_+^j(n)} &= 2 \left(A^{j^T} W_1^j A^j + W_2^j \right) \Delta U_+^j(n) + \\ &+ 2 A^{j^T} W_1^j Z^j(n) = 0 \end{aligned} \tag{30}$$

The control law by the jth FI can be identified as

$$\left(\Delta U_+^j(n)\right)^* = -K^j Z^j(n) \tag{31}$$

$$K^j = \left(A^{j^T} W_1^j A^j + W_2^j\right)^{-1} A^{j^T} W_1^j \tag{32}$$

The optimal global control policies can be derived at the upper layer.

$$\Delta U_+(n) = \left(\Delta u(n)\ \Delta u(n+1)\cdots \Delta u(n+N_u-1)\right)^T \tag{33}$$

3.3 Parameter tuning

In controller design, the difficulty encountered is how to quickly minimize the upper bound of the objective function so that the control actions can force a process to track a specified trajectory as close as possible.

There has no rigorous solution to the selection of optimal control horizon (N_u) and prediction horizon (N_y).

The model horizon is selected so that $T\Delta t \geq$ open loop settling time.

The ranges of weighting factors W_1^j and W_2^j can be very wide, the importance is their relative magnitudes. The following procedure to tune the weighting factors is proposed:

- Select a value for W_1 and assign it to all local controllers. Determine W_2^j independently for each local controller in order to minimize the objective function for that subsystem
- Identify the largest W_2 and assign it to all subsystems.

Examine the system's closed-loop dynamic performance. Reduce the value of W_2 gradually until the desirable dynamic performance is identified.

3.4 Simulations

Process Modeling: The main problem in setting up a signal flow diagram for a level controlled system in a SG can be found in the inhomogeneous contents of the SG.

The filling consists of water at boiling temperature, pervaded by steam bubbles.

Since the volume fraction of the steam bubbles is quite considerable, the mean specific weight of the contents is very strongly dependent on the proportion of steam.

This, of course, means that the steam content also strongly influences the level in the SG. The steam content itself depends, in turn, on the load factor, on the changes in feed-water flow, and on feed-water temperature.

The presence of steam below the liquid level in the SG causes the *shrink-and-swell* phenomenon that in spite of an increased supply of water, the water level initially falls. Figure 2 shows responses of the water level to steps in feed-water and steam flow-rates at different operating power levels (Irving et al., 1980).

Particularly it is difficult to control automatically a steam generator water level during transient period or at low power less than 15% of full power because of its dynamic characteristics.

The inverse response behavior of the water level is most severe at low power (5%).

The changing process dynamics and the inverse response behavior significantly complicate the design of an effective water level control system.

A solution to this problem is to design local linear controllers at different points in the operating regime and then applies gain-scheduling techniques to schedule these controllers to obtain a globally applicable controller.

Consider a step in feed-water flow rate at 5% operating power. For this system, a fuzzy convolution model consisting of four fuzzy implications is developed as follows:

For j=1 to 4:

$$R^j : if \ y_{D_{al}}(n) \ is \ A^j$$

$$then \quad y_{D_{al}}^j(n+1) = y_{D_{al}}^j(n) + \sum_{i=1}^{200} h\_D_{al}{}^j u(n+1-i) \tag{34}$$

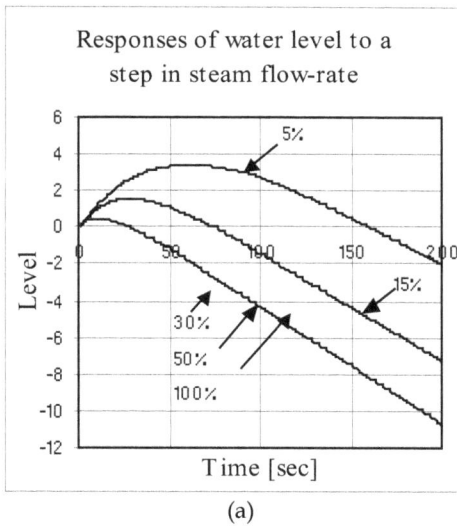

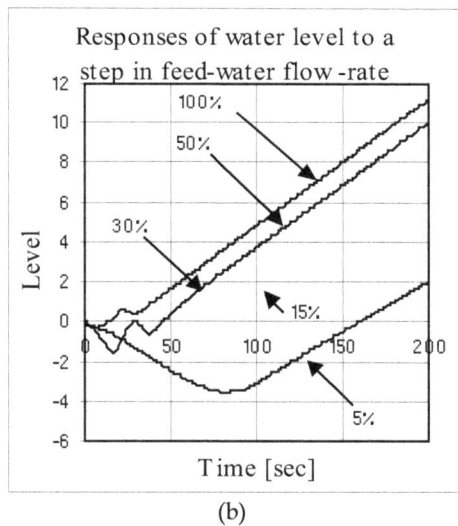

Fig. 2. Responses of water level at different operating power (indicated by %) to (a) a step in feed-water flow –rate. (b) a step in steam flow-rate.

Figure 3. shows the response of water level at 5% operating power to a step in feed-water flow –rate. In Figure 4 the system is decomposed into 4 subsystems: $y_{D_{al}}^1$, $y_{D_{al}}^2$, $y_{D_{al}}^3$, $y_{D_{al}}^4$.

Figure 5 shows the impulse response coefficients for $y_{D_{al}}^1 , y_{D_{al}}^2 , y_{D_{al}}^3 , y_{D_{al}}^4$ subsystems and Figure 6 shows the definition of fuzzy sets A^1, A^2, A^3 and A^4. Consider a step in steam flow rate at 5% operating power.

For this system, a fuzzy convolution model consisting of four fuzzy implications is developed as follows:

For j=1 to 4:

$$R^j : if \ y_{D_0}(n) \ is \ A^j$$

$$then \quad y_{D_0}^j(n+1) = y_{D_0}^j(n) + \sum_{i=1}^{200} h\_D_{0i}^j u(n+1-i) \tag{35}$$

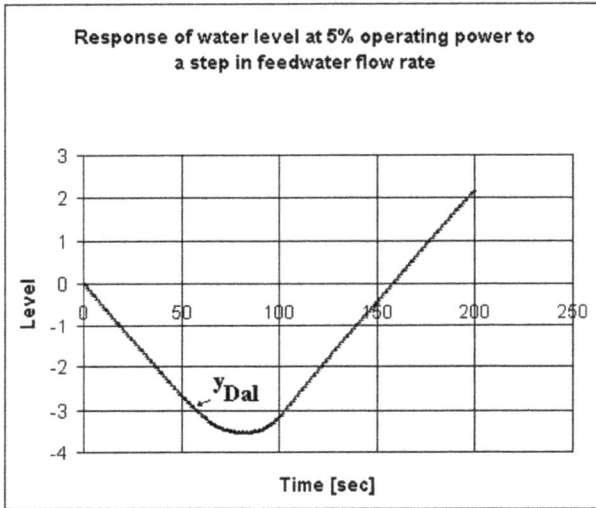

Fig. 3. Response of water level at 5% operating power to a step in feed-water flow –rate

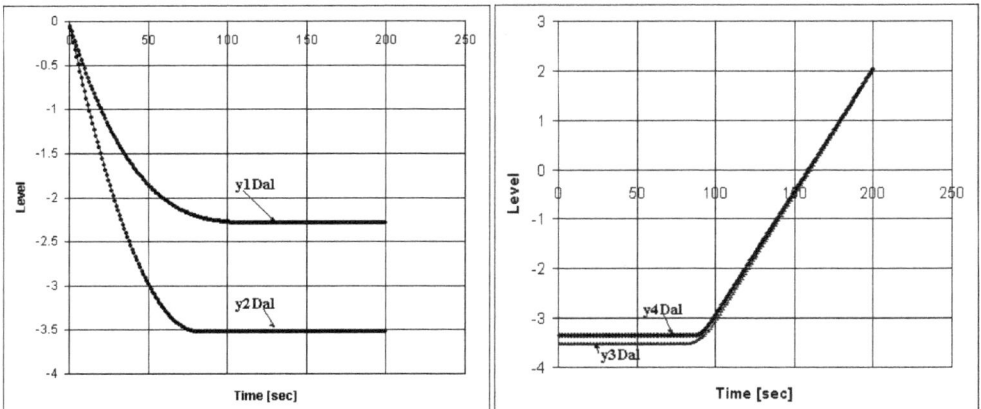

Fig. 4. The system is decomposed into 4 subsystems: $y_{D_{al}}^1$, $y_{D_{al}}^2$, $y_{D_{al}}^3$, $y_{D_{al}}^4$

	h1_Dal	h2_Dal	h3_Dal	h4_Dal
1	-0.058	-0.08	0	0
2	-0.057	-0.08	0.003	0.005
3	-0.056	-0.08	0.005	0.01
4	-0.055	-0.08	0.008	0.015
5	-0.054	-0.08	0.011	0.02
6	-0.053	-0.08	0.013	0.025
7	-0.052	-0.08	0.016	0.03
8	-0.051	-0.08	0.019	0.035
9	-0.05	-0.08	0.021	0.04
10	-0.049	-0.07	0.024	0.045
11	-0.048	-0.07	0.027	0.05
12	-0.047	-0.07	0.03	0.05
13	-0.046	-0.07	0.032	0.05
14	-0.045	-0.07	0.035	0.05
15	-0.044	-0.07	0.038	0.05
16	-0.043	-0.07	0.04	0.05
17	-0.042	-0.07	0.043	0.05
18	-0.042	-0.07	0.046	0.05
19	-0.041	-0.07	0.048	0.05
20	-0.04	-0.06	0.051	0.05
21	-0.039	-0.06	0.051	0.05
22	-0.038	-0.06	0.051	0.05
23	-0.037	-0.06	0.051	0.05
24	-0.036	-0.06	0.051	0.05
25	-0.036	-0.06	0.051	0.05
26	-0.035	-0.06	0.051	0.05

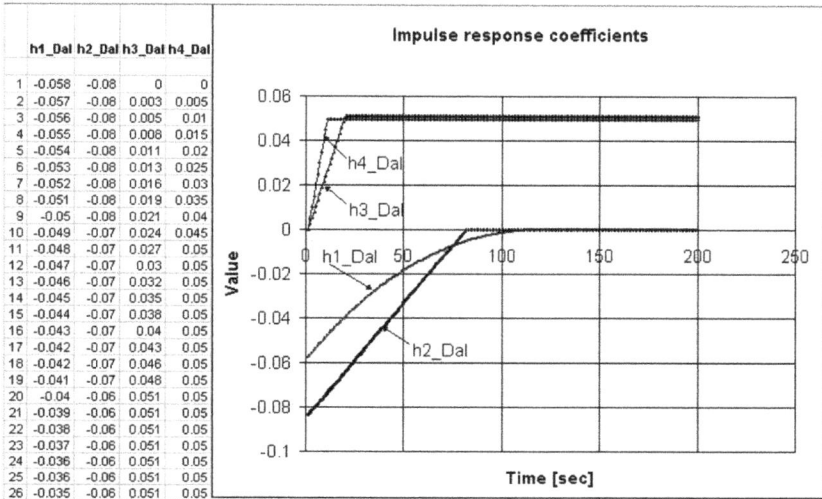

Fig. 5. The impulse response coefficients for $y_{D_{al}}^1$, $y_{D_{al}}^2$, $y_{D_{al}}^3$, $y_{D_{al}}^4$ subsystems

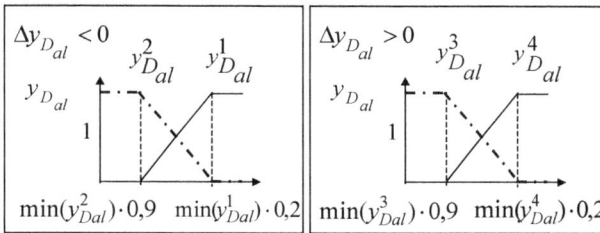

Fig. 6. Definition of fuzzy sets A^1, A^2, A^3 and A^4 for FI R^1, R^2, R^3 and R^4 respectively.

Figure 7 shows the response of water level at 5% operating power to a step in steam flow –rate.

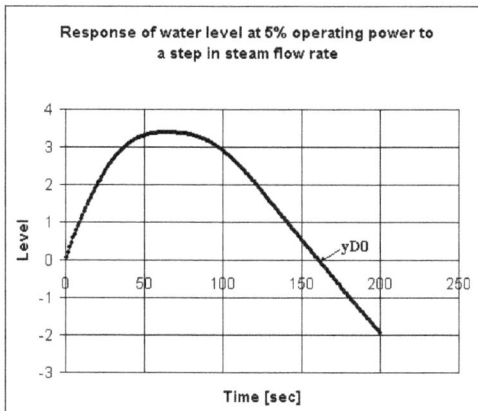

Fig. 7. Response of water level at 5% operating power to a step in steam flow –rate;

In Figure 8 the system is decomposed into 4 subsystems: $y_{D_0}^1$, $y_{D_0}^2$, $y_{D_0}^3$, $y_{D_0}^4$.

Figure 9 shows the impulse response coefficients for $y_{D_0}^1$, $y_{D_0}^2$, $y_{D_0}^3$, $y_{D_0}^4$ subsystems, Figure 10 shows the definition of fuzzy sets A^1, A^2, A^3 and A^4.

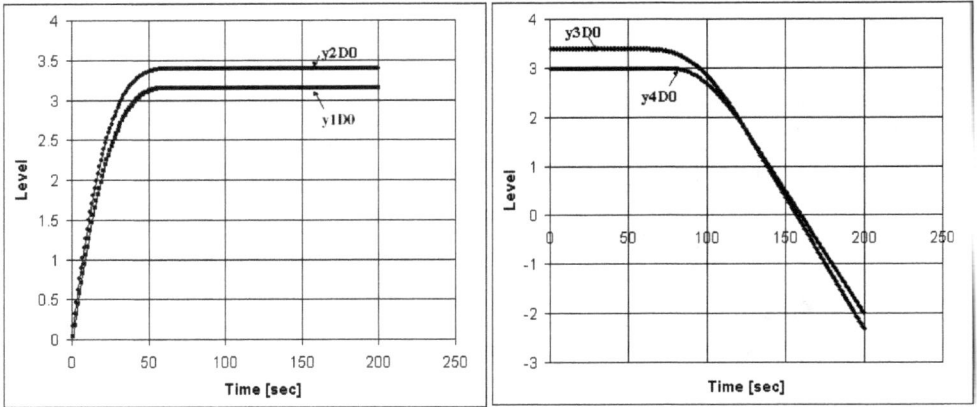

Fig. 8. The system is decomposed into 4 subsystems: $y_{D_0}^1$, $y_{D_0}^2$, $y_{D_0}^3$, $y_{D_0}^4$.

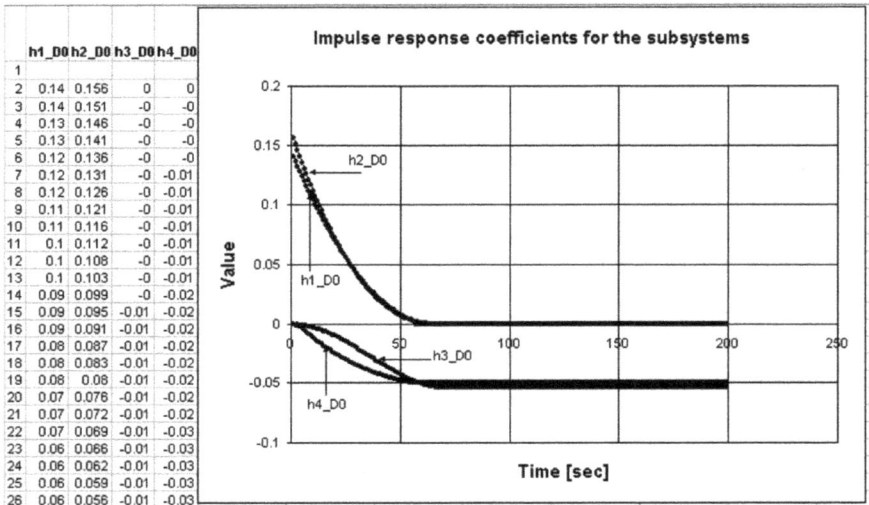

Fig. 9. The impulse response coefficients for $y_{D_0}^1$, $y_{D_0}^2$, $y_{D_0}^3$, $y_{D_0}^4$ subsystems

Controller Design: The goal is to study the use of the feed-water flow-rate as a manipulated variable to maintain the SG water level within allowable limits, in the face of the changing steam demand resulting from a change in the electrical power demand.

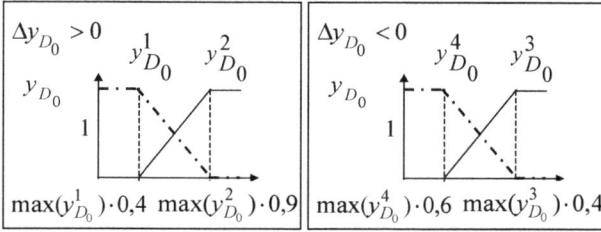

Fig. 10. Definition of fuzzy sets A^1, A^2, A^3 and A^4 for FI R^1, R^2, R^3 and R^4 respectively .

The simulations are organized around two different power transients:

- a step-up in power from 5% to 10% (Figure 11);
- a ramp-up in power from 5% to 10% (Figure 12)

The model horizon is $T=200$. Increasing N_y results in a more conservative control action that has a stabilizing effect but also increases the computational effort.

The computational effort increases as N_u is increased. A small value of N_u leads to a robust controller.

For both power transients the controller responses are very satisfactory and not very sensitive to changes in tuning parameters.

We can see that the performance is not strongly affected by the presence of the feed-water inverse response, only a slight oscillation is visible in the water level response.

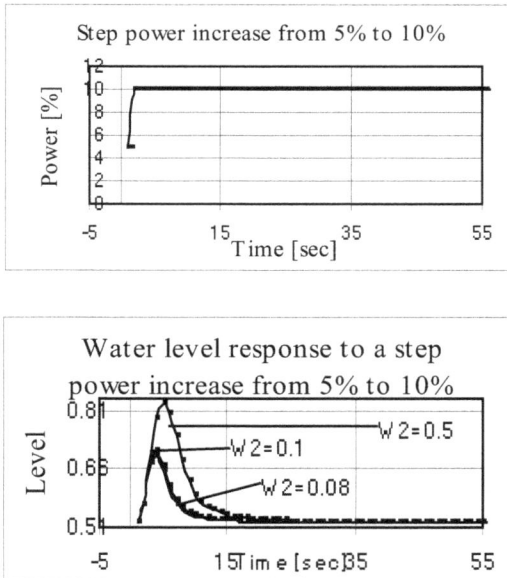

Fig. 11. Water level response to a step power increase from 5% to 10% (Nu=2, Ny=3, W₁=1)

(a)

(b)   (c)

(d)

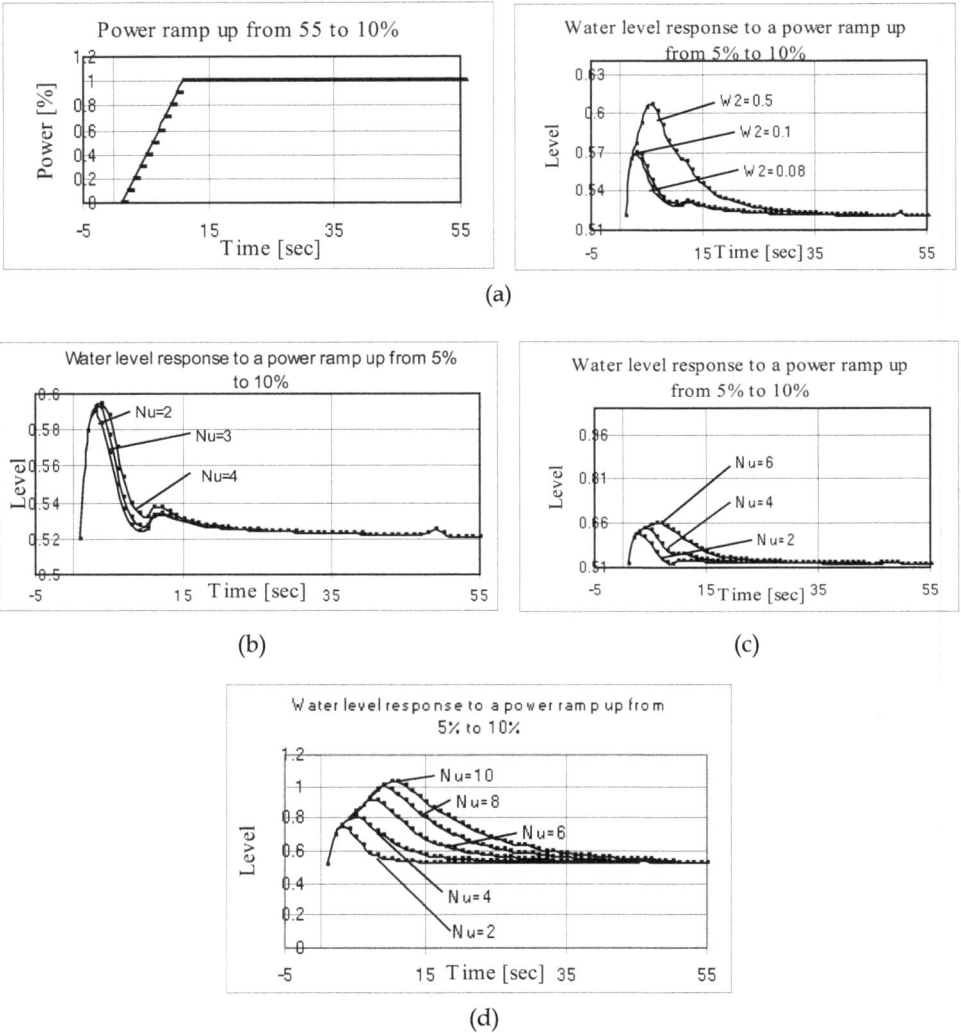

Fig. 12. Water level response. a) to power ramp up from 5% to 10% (Nu=2, Ny=3, W_1=1); b) to power ramp up from 5% to 10% (W2=0.1, Ny=3, W_1=1); c) to power ramp up from 5% to 10% (W2=0.1, Ny=7, W_1=1); d) to power ramp up from 5% to 10% (W2=0.1, Ny=11, W_1=1)

3.5 Evolved controller client/server architecture

An original concept of modular *evolved* control system, seamless and with gradual integration into the *primary* control system is proposed.

The aim of the application is to integrate the concepts of *evolved control algorithms, portability of software modules, real time characteristics of the application.*

The target systems are the *large scale distributed control systems* with *optimum granularity architecture*.

The first part of the life cycle phases of the new control system, from conception to validation stage, the new control system lives "hiding in the shadow" of the control system it will replace, and after validation the old system will be replaced by the new one.

The identification, modeling, control and validation stages of the life cycle of the system, will be done *on-line* (the new system uses a real image of the I/O process data), without affecting the existing control system.

Because of high level of *interconnectivity between system components*, it is necessary to provide the *highest independence between communication modules on one-hand and the control modules on the other hand*. In order to obtain high ability of integration, the communication modules have to cover the widest possible area of industrial communication interfaces and protocols.

One item of the application is to offer a *unified API of extended generality and extendibility in order to unify access and information retrieval from various wireless and wired technology and communication interfaces* (RS 232, RS 485, fieldbus: Profibus / Interbus, Ethernet IP, TCP/IP, etc).

Applications could properly adapt to changes in the network connections.

The design and implementation of a solution *to hide the embedded communication network problems from the application system programmers* is included.

One of the main objectives of the application is to supply *an integrated solution of systems, which should support all the phases of the life cycle: modeling, simulation, development and implementation*.

For *parameter tuning, for validation and also for embedding a large number of industrial communication protocols, multi-disciplinary simulation environments are developed which generate instruments for control, I/O data consistency check, and defect detection*.

In the end, real-time advanced control applications are developed, with seamless and gradual integration into the existing distributed control system.

A software package for *evolved control* includes a method based on *fuzzy model predictive control*.

By using the basic concept of decomposition-coordination in a large-scale system theory, the *fuzzy model predictive controller* design can be accomplished through a *two-layer iterative design process*.

The design is decomposed into the derivation of local controllers. The subsystems regulated by those local controllers will be coordinated to derive a *globally optimal control policy*.

In order to provide the real-time characteristic, we choose a multitasking environment for the application (WINDOWS Operating System).

From structural point of view we propose a Client / Server architecture for fuzzy Controller (FC) (Andone et al., 2006):

Client - is a Windows application representing the implementation of the graphical user interface (GUI). The Client enables the operator to control the system in two modes:

manual/automatic, to monitor the system response, etc. The Client has also the ability to connect and communicate with the Server application.

Server – is an ActiveX EXE application containing the implementation of the Fuzzy Controller (FC) kernel.

The Server includes a collection of objects, these objects cover the tasks of both data processing and the communication between dedicated applications for input and output data.

The Client application will have a thread pool architecture.

The Server application will have a real multithreading architecture (each active object having assigned its own execution thread).

The Server have also a multi-layer structure: at the higher level are implemented upper FC and the communication classes (using different transmission mechanisms – DDE, OPC, HLI, ActiveX, Winsocket, Pipes), at the lower level are implemented the controllers for the subsystems corresponding to the low level FC.

The Server's application as real multithreading architecture, provides the FC Kernel the real-time response characteristic, required for the industrial process control.

4. Conclusions

Control of SG water level strongly affects nuclear power plant availability.

The control task is difficult for a number of reasons, the most important among them being the nonlinear plant dynamics and the non-minimum phase plant characteristics.

There has been a special interest in this problem during low power transients because of the dominant reverse thermal dynamic effects known as shrink and swell.

The SG level control problem was viewed as a single input/single output control problem with the feed-water as the manipulated variable, the level as the controlled variable and the turbine steam demand as disturbance.

It has been shown that in the case of nonlinear processes, the approach using fuzzy predictive control gives very promising results.

The process non-linearity was addressed by scheduling the model (and the controller) with the power level.

The SG system is modeled by Takagi-Sugeno's fuzzy modeling methodology, where the system output is estimated based on gradient. The complex shrink and swell phenomena associated with the SG water level are well captured by the model.

The predictive controller based on fuzzy model is designed in a hierarchical control design.

An original concept of modular evolved control system, seamless and gradual integration into the existing distributed control system is proposed in the chapter.

A unified API of extended generality and extendibility in order to unify access and information retrieval from various wireless and wired technology and communication interfaces is developed in order to ensure independence between communication and control modules of the designed systems.

A Client / Server architecture for evolved controller that runs on the Windows environment, with real-time characteristics is proposed.

5. Acknowledgment

Parts of this chapter are reprinted from Hossu, D. Fagarasan, I., Hossu, A., Iliescu, S. St., Evolved fuzzy control system for a steam generator, *Int. Journal of Computers, Communications and Control*, IJCCC, ISSN 1841-9836, E-ISSN 1841-9844, Vol. V (2010), No.2, pp. 179 – 192.

6. References

Andone, D. ,Dobrescu, R., Hossu, A. & Dobrescu, M. (2006) Application of fuzzy model predictive control to a drum boiler , *ICAE - Integrated Computer-Aided Engineering*, IOSPress, vol.13, nr.4, ISSN 1069-2509, pp. 347-361.

Andone, D.& Hossu, A. (2004) Predictive Control Based on Fuzzy Model for Steam Generator, *2004 IEEE International Conference on Fuzzy Systems, Proc. FUZZ-IEEE 2004*, vol. 3, IEEE Catalog Number 04CH37542, ISBN 0-7803-8353-2, ISNN 1098-7584, Budapest, Hungary; pp. 1245-1250; July 25-29.

Hossu, D., Fagarasan, I., Hossu, A. & Iliescu, S. St. (2010) Evolved fuzzy control system for a steam generator, *Int. Journal of Computers, Communications and Control*, IJCCC, ISSN 1841-9836, E-ISSN 1841-9844, Vol. V , No.2, pp. 179 – 192.

Äström, K. & Bell, R. (2000) Drum-boiler dynamics, *Automatica* 36, pp. 363-378.

Camacho, E. & Bordons, C. (2004) *Model Predictive Control*, Springer-Verlag, London.

Demircioglu, H.& Karasu, E. (2000) Generalized Predictive Control – A Practical Application and Comparison of Discrete and Continuous-Time Versions, *IEEE Control Systems*, Oct 2000, vol 20, nr. 5, pp 36-44.

Dubois, D.& Prade, H. (1997) *Fuzzy Sets and Systems: Theory and Applications*, Academic Press, Inc., Orlando, FL.

Espinosa, J., Hadjili, M. L.& Wertz, V., J.(1999) *Predictive control using fuzzy models – Comparative study*, European Control Conference, Karlsruhe, Germany, Sept. 1999.

Hirota, K. (1993) *Industrial Applications of Fuzzy Technology*, Springer-Verlag, New York.

Huang, Y., Lou Helen, H., Gong, J.P.& Edgar, Th. F. (2000) Fuzzy Model Predictive Control, *IEEE Trans. On Fuzzy Systems*, vol. 8, no. 6, Dec. 2000, pp. 665-668.

Irving, E., Miossec, C.& Tassart, J. (1980) Toward efficient full automatic operation of the PWR steam generator with water level adaptive control, *Proc. Int. Conf. Boiler Dynamics Contr. Nuclear Power Stations*, London, U.K., pp. 309-329.

Kiriakidis, K. (1999) Non-linear control system design via fuzzy modeling and LMIs, *International Journal of Control*, Vol. 72, no. 7, pp. 676-685.

Kothare, M., Mettler, B., Morari, M., Bendotti, P.& Falinower, C. (2000) Level Control in the Steam Generator of a Nuclear Power Plant, *IEEE Trans. On Control Systems Technology*, vol. 8, no. 1, Jan. 2000, pp. 55-69.

Menon, S.K. & Parlos, A.G. (1992) Gain-scheduled nonlinear control of U-tube steam generator water level, *Nuclear Sci. Eng.*, vol. 111, pp. 294-308.

Morari, M. & Lee, J. H. (1999) Model predictive control: Past, present, and future, *Computers & Chemical Eng.*, pp. 667-682.

Park, G. Y. & Seong, P. H. (1997) Application of a self-organizing fuzzy logic controller to nuclear steam generator level control, *Nuclear Engineering and Design.*, vol. 167, pp. 345-356.

Pedrycz, W. & Gomide, F. (2007) *Fuzzy Systems Engineering: Toward Human-Centric Computing*, Wiley-IEEE Press.

Ross, T.J. (2004) *Fuzzy Logic with Engineering Applications*, second ed. Wiley & Sons.

Yager, R., R. & Zadeh, L. (1992) A., *An Introduction to Fuzzy Logic Applications in Intelligent Systems*, Kluwer Academic Publishers, Norwell, MA.

Yen, J., Langari, R. & Zadeh, L.(1995) A., *Industrial Applications of Fuzzy Logic and Intelligent Systems*, IEEE Press, Piscataway, NJ.

Ying, H. (2000) *Fuzzy Control and Modeling: Analytical Foundations and Applications*, Wiley-IEEE Press.

Zadeh, L. (2008) A New Frontier in Computation - Computation with Information Described in Natural Language, *International Journal of Computers Communications and Control*, Volume:3, Supplement: Suppl. S, pp. 26-27, 2008.

Zadeh, L. A. (2005) Toward a generalized theory of uncertainty (GTU): an outline, *Information Sciences-Informatics and Computer Science: An International Journal*, v.172 n.1-2, p.1-40, 9 June 2005.

Zadeh, L. A. (1989) Knowledge Representation in Fuzzy Logic, *IEEE Transactions on Knowledge and Data Engineering*, v.1 n.1, p.89-100, March 1989.

Zhao, F., Ou, J. & Du, W. (2000) Simulation modeling of nuclear steam generator water level process – a case study, *ISA Transactions 39*, pp. 143-151.

6

The Gap Measurement Technology and Advanced RVI Installation Method for Construction Period Reduction of a PWR

Do-Young Ko
Central Research Institute, Korea Hydro & Nuclear Power Co., Ltd.
Republic of Korea

1. Introduction

A nuclear power plant takes approximately 52–58 months from the first concrete construction until the completion of the performance test. Many research groups throughout the world have studied ways to shorten the construction period of nuclear power plants to 50 months. Therefore, the construction period of a nuclear power plant is one of the most important factors to make a company competitive in international nuclear energy markets. There are many advanced construction methods to decrease the construction period for new nuclear power plants. This chapter is related to the modularization of reactor vessel internals (RVI) that one of the most effective methods to reduce the construction period of nuclear power plants (Ko et al., 2009) (Ko & Lee, 2010) (Ko, 2011) (Korea Hydro & Nuclear Power Co., Ltd., 2009).

2. Development of reduced-scale model system for measurement system

Generally, the RVI comprise three components: the core support barrel (CSB), the lower support structure (LSS)/core shroud (CS), and the upper guide structure (UGS). The existing method of assembly is very complicated and requires approximately 8 - 10 months to complete (Korea Electric Power Research Institute, 1997) (Korea Hydro & Nuclear Power Co., Ltd., 2002). The installation of the reactor vessel (RV) is a critical process during the construction period.

This part describes the RVI installation method using the RVI modularization which can shorten the construction period by a minimum of two months compared to the existing method. In order to modularize the RVI, gaps between the CSB snubber lug and RV core-stabilizing lug must be measured using a remote method at outside the RV. Therefore, this part includes explanation on a measuring system to measure gaps between the RV and the CSB remotely with the aim of RVI modularization. The remote measurement system was developed for use at actual construction sites of nuclear power plants using a measurement sensor, a threaded connection jig, and a zero-point adjustment device. With these, a reduced-scale model system was validated. With the remote measurement system, experiments and analyses were performed using mockups for both the RV and the CSB to verify the applicability of the described system in a construction project. From the data acquired by the remote measurement system, shims were separately made and adjusted.

After installing the shims on RV core-stabilizing lugs, the gaps satisfied requirements within the permissible range of 0.381 – 0.508 mm. The reliability and applicability of the remote measurement method were evaluated and it was concluded that the remote measurement system enables RVI modularization with a significantly reduced construction period.

Fig. 1 shows an existing nuclear reactor installation method and the developed modularization method by remote gap measurement.

RVI is classified on a large scale into three categories: CSB, LSS/CS and UGS. When a nuclear power plant is built, the materials are delivered and assembled according to which category they fall under. It is, however, possible to modularize the installation of the CSB and LSS/CS (Ko et al., 2009) (ABB-CE, 1995).

Gaps between CSB snubber lug and RV core-stabilizing lug can be measured at outside the RV by a remote method in order to modularize the RVI. If the CSB module (CSB and LSS/CS) is installed into RV, access to measuring the gaps is cut off by the LSS/CS. Therefore, the gaps must be measured remotely at outside the RV.

Fig. 2 shows a picture of the second step in Fig. 1(a)

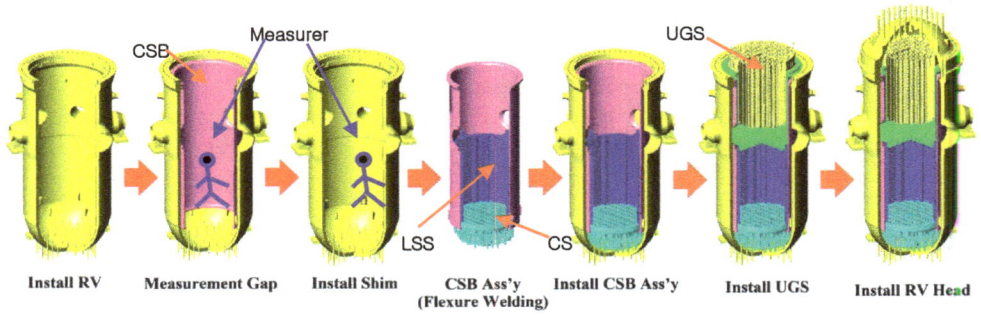

a) Existing nuclear reactor installation method

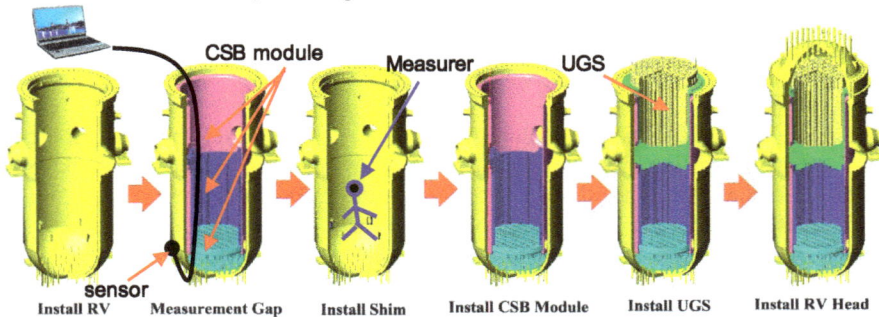

(b) Proposed modularization method

Fig. 1. Existing method and developed modularization method by gap measurement remotely

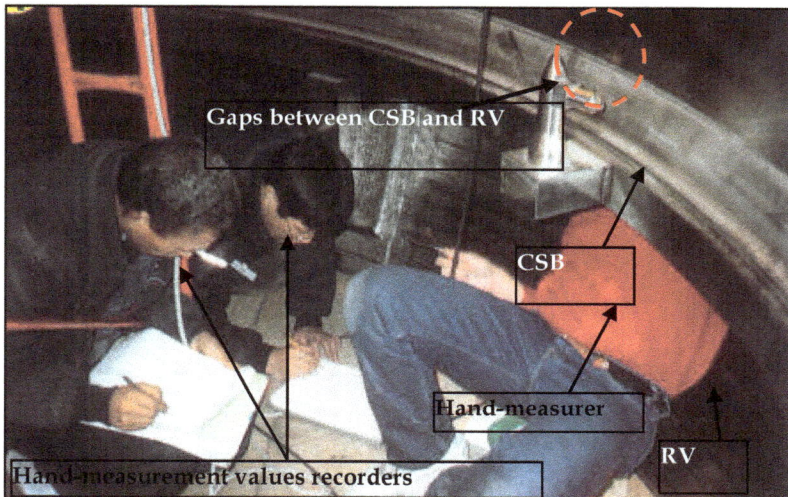

Fig. 2. Picture of the second step in Fig. 1 (a), showing manual gap measurement.

The hand-measurement of Fig. 2 takes a lot of measurement times and it occurs measurement errors by measurers. Also, a measurement space is small and narrow, and environment to measure gaps between RV and CSB is uncomfortable.

Fig. 3 shows timescales of existing installation method and developed method.

As shown in Fig. 3, developed modularization method by gap measurement remotely can reduce the critical path of the construction period by approximately 8-12 weeks. If the construction period is reduced, a construction budget expected to be saved the minimum $ 176 million in Korea.

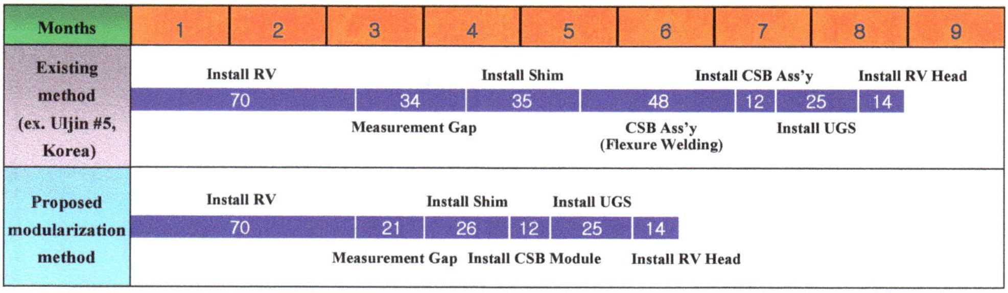

Fig. 3. Timescales of existing installation method and proposed method

Fig. 4 shows a section of the RV core-stabilizing lug and the CSB snubber lug that shall be measured remotely for the modularization of the RVI.

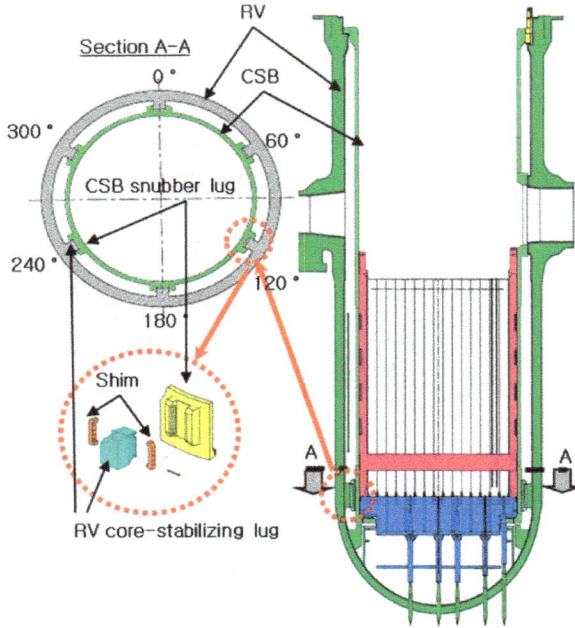

(a) Measurement parts between RV core-stabilizing lug and CSB snubber lug

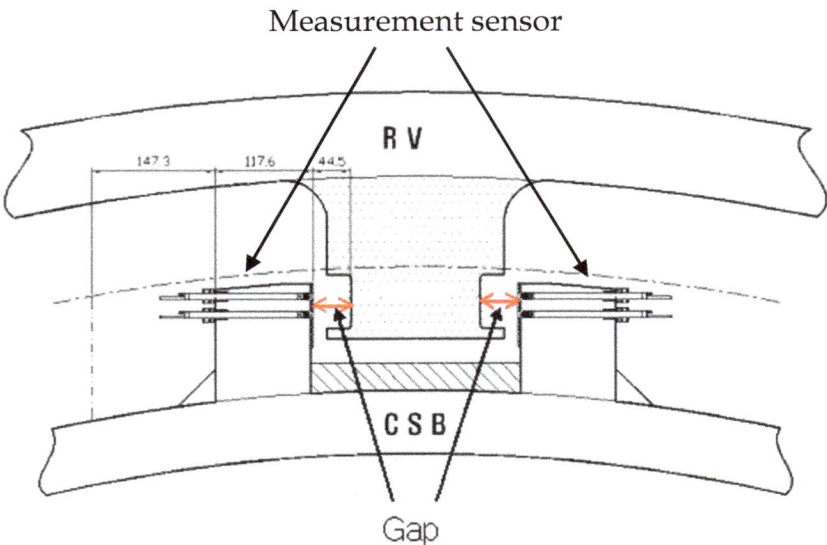

(b) Gaps between an RV core-stabilizing lug and a CSB snubber lug

Fig. 4. Measurement parts and gaps between the RV Core-stabilizing lug and the CSB snubber lug

2.1 Design of a reduced-scale model system for the physical simulation of gap measurement

2.1.1 Design

The purpose of designing a reduced-scale model system for the modularization of the reactor internals is to confirm the performance and application conditions of a remote distance measurement sensor selected. The sensor should be able to measure gaps between the CSB snubber lug and the RV core-stabilizing lug in the range 0.381–0.508 [mm].

This system was designed to be used at narrow space and small probe hole conditions and to be light easy to handle by using aluminum material. A threaded connection jig allows the remote measurement sensor to be assembled into the CSB snubber.

The block diagram in Fig. 5 shows the reduced-scale model system. The servomotors provide the movement in the up/down and front/rear directions for the reduced-scale models of the CSB snubber lug with remote measurement sensors connected. The air compressor drives the measurement sensors and the obtained data are displayed on the computer screen through the interface modules and network cables.

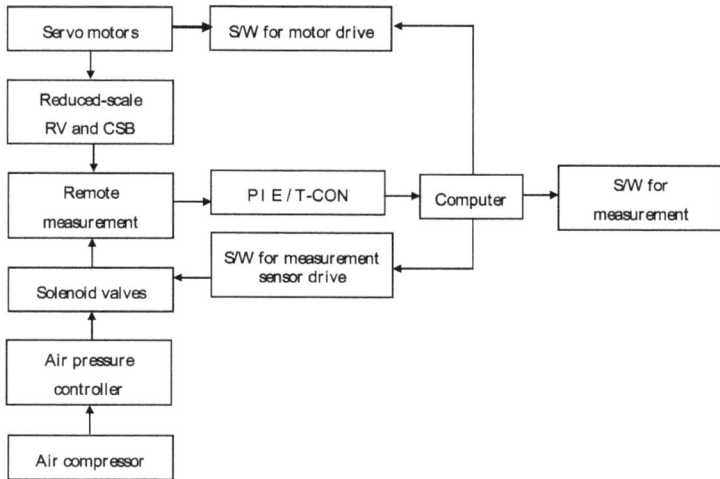

Fig. 5. Block diagram of reduced-scale model system

2.1.2 Fabrication

The picture in Fig. 6 shows the reduced-scale model system that consists of the reduced-scale CSB snubber lug , the reduced-scale RV core-stabilizing lug, 4 remote distance measurement sensors, the air supply device, 2 servo motor devices, the data interface modules and a laptop computer.

The components of the reduced-scale model system were fabricated as follows.

-    The model pieces of the CSB snubber lug and RV lug were machined to precisely represent the surface and inside of the remote distance measurement sensors' hole. The material is aluminum alloy.

- The threaded connection jig of the remote distance measurement sensor was fabricated.
- The zero point adjustment device was fabricated.
- The block gauge was used to test the reliability of the remote distance measurement sensors.

(a)

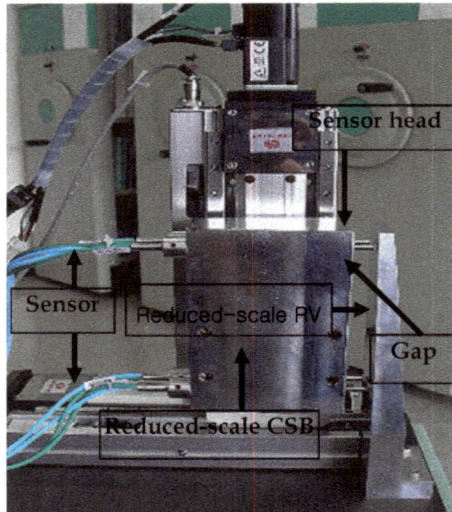

(b)

(a) Overall system, (b) Enlarged circle of Fig. 6 (a) to simulate measurement of gaps between CSB snubber lug and RV core-stabilizing lug

Fig. 6. Reduced-scale model system for the modularization of reactor internals

2.2 Measurement sensor

Many essential factors in selecting a measurement sensor for the RVI modularization were studied. Namely, the measurement environment, the measured object, size of the sensor, the weight of a sensor, measurement range, drive force of the sensor, accuracy and resolution were investigated

First, contact sensors and non-contact sensors were compared (Ko et al., 2009) (ABB-CE, 1995).

Contact sensors are used to directly measure distance by moving the sensors toward the measured object. Non-contact sensors are used to indirectly measure comparatively very small distances by laser, high frequency and eddy current(Figliola & Beasley, 2000)(Beckwith et al., 1993).

Contact sensors are more suitable than non-contact sensors. Non-contact sensors are not ideal because their outside diameter is bigger than the hole diameter 8 [mm] of the CSB snubber lug, the measurement range is smaller than it is with contact sensors, and, most importantly, design changes to the CSB snubber lug are bigger in non-contact sensors than in contact sensors.

The most important principle can be directly applied without design changes to the existing reactor internals. The contact sensor of a remote distance measurement sensor must be inserted into the measurement hole of CSB snubber lug. Therefore, an essential condition of a sensor must be a probe type because a measurement sensor must pass through the measurement hole of the CSB snubber lug. The outside diameter must be 8 [mm] or less since the measured diameter of the hole in the CSB snubber lug is 8 [mm]. The length of the probe head must be 117.6 [mm] or more because the end point of the measurement hole in the CSB snubber lug must reach the RV after a zero point adjustment. Also, a sensor must span the distance from the CSB snubber lug to the RV when the RV and CSB are assembled, thus requiring the backward probe head to be 147.3 [mm] or less. Measurement range can be measured from 50 [mm].

The previous measurement was done by hand-measurement and the measured maximum value was 48.92 [mm] (Uljin #5 nuclear power plant in Korea). The resolution must be 0.0254 [mm] or below as the sensor has to measure until 0.0254 [mm] (1/1000″) in the case of gap measurement between the CSB snubber lug and the shim on the RV core-stabilizing lug, after the shim was assembled onto the CSB snubber lug by cap screws.

A sensor was investigated that remote measurement was possible in at least 25 [m] or more. The following items were considered when selecting a sensor: material, space and outward shape of the reactor internals. Also, no additional devices should be installed around the RV core stabilizing lug and the CSB snubber lug in order to perform the remote measurement.

Table 1 shows the suitable specifications of a sensor that have been researched to measure gap. The shape of a sensor is probe type of a contact sensor and the measurement method is digital.

Finally, the SOLARTRON (UK) sensor (DT/20/P) was selected for the reduced-scale model system (Solartron-metrology, 2006).

Shape of sensor	Probe
Type of sensor	Contact, Digital
Outside diameter of probe head	8 [mm] or below
Length of probe head	117.6 [mm] or over
Backward space of CSB snubber lug hole	From hole entrance 147.3 [mm] or below
Measurement range	50 [mm] or over
Resolution	25.4 [um] or below
Accuracy	±12.7 [um] or below
Operating temperature	No relation
Distance of remote measurement	25 [m] or over
Driving force	Electric or Pneumatic
Numbers of synchronous measurement	72 points and over
Operating tool	Computer-based

Table 1. Suitable specifications of a remote sensor for gap measurement

2.3 Experiment and result

The designed system was tested to confirm the performance and application conditions of a remote distance measurement sensor selected.

The tests were carried out repeatedly to confirm reliability, consistency, accuracy and stability of the measurement.

The reliability test method of the sensors is as follows. First, a sensor was fixed to an anchor of granite comparator stand. Second, the probe of the sensor is positioned so as to be close to the granite comparator stand at suitable heights. Third, the probe was extended so that it touches on the face of stand. This state is the zero point of the sensor. Forth, put gauge blocks of 5 [mm], 10 [mm], 15 [mm], and 18 [mm] on the stand, and measure the reliability of the sensor. Reliability test results were obtained as shown in Fig. 7.

As the size of a gauge block is large, the error of measurement sensor was increased from 0.0037 [mm] to 0.0119 [mm]. However, the reliability test was satisfied because the sensors did not exceed the maximum allowable error (0.0254 [mm]). And the measurement errors under the pressure of 0.8 [bar] and 2 [bar] are similar.

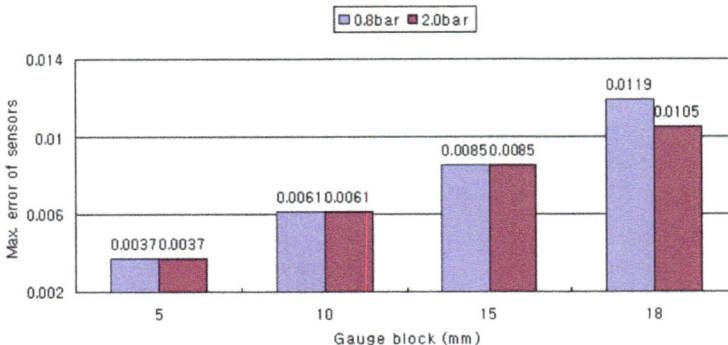

Fig. 7. Reliability test results of remote distance measurement sensors using gauge blocks

The Gap Measurement Technology and Advanced RVI Installation Method for Construction Period Reduction
of a PWR
121

The consistency test method of connection jigs for sensors is as follows. First, a sensor was inserted into a threaded connection jig and fixed firmly. Second, insert a sensor and a connection jig to the hole of the minimized model of the CSB snubber lug, and fix it firmly by threaded connection jig. Third, when sensors have received air pressure of 0.8[bar], 1.4[bar], and 2.0[bar], the sensors measure the distance five times. Fourth, the process measurements are repeated three times and attached again after removing the connection jigs of the remote distance measurement sensors. Consistency test results under 1.4[bar] at each gauge blocks were obtained as shown in Fig. 8.

The measurement errors of sensor 2 (0.0024 [mm]) and 4 (0.0025 [mm]) were large compared with sensor 1 and 3, but consistency test were satisfied because the error did not exceed the maximum allowable error.

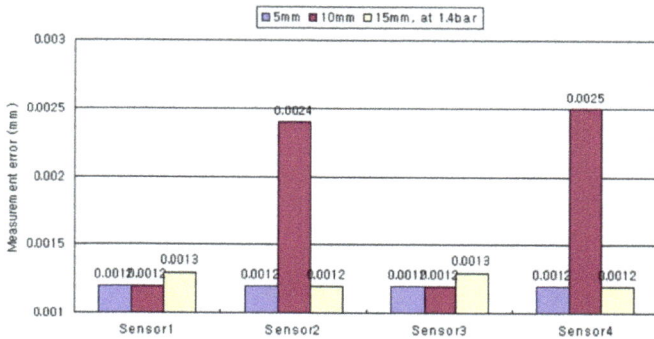

Fig. 8. Consistency test results of connection jigs for remote sensors

The accuracy test method for the zero point adjustment device for the sensors is as follows. First, a zero point adjustment device that was designed and made, binds the right and left of the minimized model of the CSB, and is fixed. Second, run the probes of the sensors installed into the minimized model of the CSB, and remove the zero point adjustment device on the minimized model of the CSB after repeating the distance measurement five times. Third, compare the distance measurements from three repetitions of the process. Accuracy test results were obtained as shown in Fig. 9.

The measurement errors of sensor 2 and 4 were large compared with sensor 1 and 3, but accuracy test were satisfied because it did not exceed the maximum error.

Switching noises and EMI (electromagnetic interference) might occur because of the use of electric lamps and transceivers to build nuclear power plants (Ko & Bae, 2006).

Therefore, the stability tests were done at no disturbance noise and in a switching noise environment, and at EMI environment.

The stability test method of sensors at no disturbance noise is as follows. The air pressure to 1.4[bar] on the remote sensors of the reduced-scale model system was set up and tested five times, repeatedly. Stability test results at remote distance were obtained as shown in Fig. 10.

There were no errors in all sensors except sensor 4. The error was only 0.0003 [mm], but it was in allowable range.

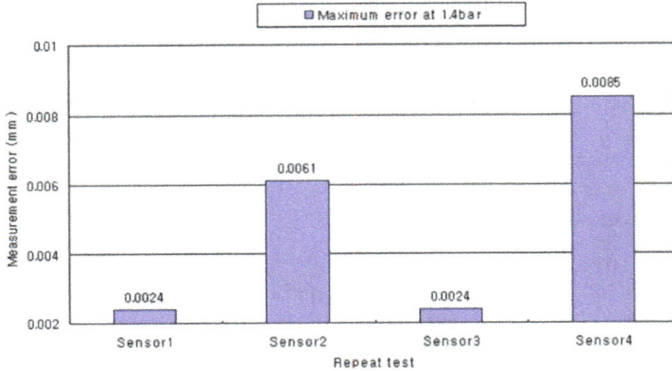

Fig. 9. Accuracy test results of a zero point adjustment device for sensors

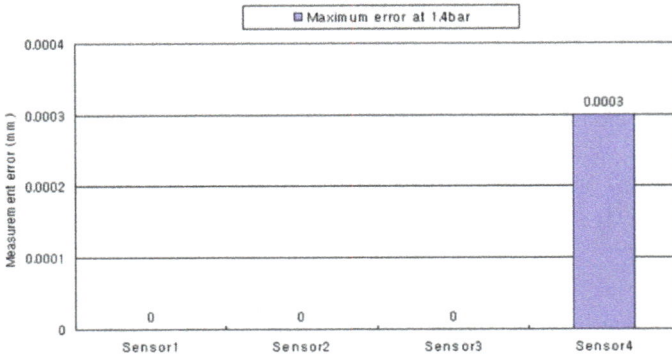

Fig. 10. Stability test results of sensors at no disturbance noise

The stability test method of the sensors in a switching noise environment is as follows. First, a 27[W] desk lamp (220V/60Hz) at 10 [cm] from the sensor probe was installed. Second, five times while turning the power on the lamp on and off were measured. During this test, the supplied air pressure was 0.8[bar] to the the sensor probe. The results at switching noise environment were obtained as shown in Fig. 11.

The measurement errors of sensor 2 and 4 were 0.0012 [mm] and 0.0008 [mm], but it were satisfied because the sensors did not exceed the maximum error (0.0254 [mm]).

Another stability test methods of the sensors at EMI environment is as follows. First, a VHF/UHF FM radio transceiver (TM-V7A/KENWOOD) and an antenna 30 [cm] away from the probes were installed. Second, five times each for the cases of 144[MHz/5W], 144[MHz/10W], 439[MHz/5W] and 439[MHz/10W] were repeatedly measured. During this test, the supplied air pressure was 0.8[bar] to the probes of the sensors. The test results at EMI environment were obtained as shown in Fig. 12.

In the environment of 144 [MHz/10W], the measurement error was 0.0003 [mm], but it was in allowable range also.

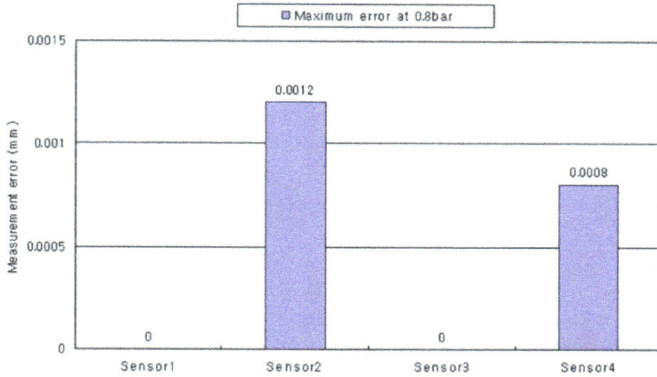

Fig. 11. Stability test results at switching noise environment

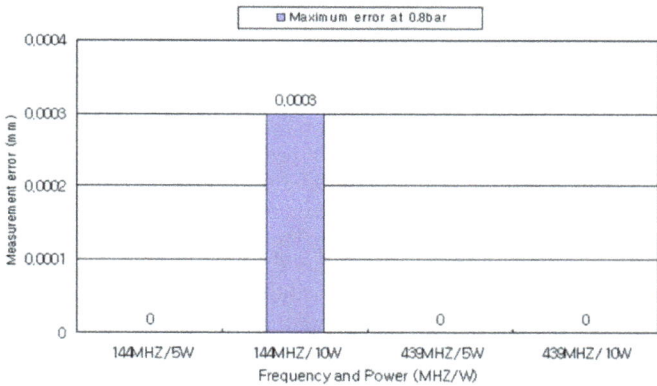

Fig. 12. Stability test results at EMI environment

As shown in the above test results, the selected sensor can be used in a remote measurement system for the modularization of reactor internals since the sensor errors did not exceed 0.0254 [mm] (1/1000″).

In the experiments and results, they were found that the measurement errors of sensor 2 and sensor 4 are bigger than sensor 1 and sensor 3. We judged that these results were occurred by something problem from self-characteristics of sensors.

2.4 Conclusion

From these results, the technology of remote measurement for the modularization of reactor internals may be advanced by design and development of the reduced-scale model system.

Also, the reduced-scale model system designed may be used as a gap measurement training system for a modularization method of reactor internals. And it needs the more suitable sensor under the consideration of the special conditions and environments in reactor internals.

3. Development of measurement system for RVI modularization

An RV mockup and a CSB mockup were also manufactured to evaluate and verify the reliability and applicability in construction sites of the developed remote measurement system. This part explains the development of the remote measurement system, including its design, fabrication and related experiments (Ko & Lee, 2010).

3.1 Development of a remote measurement system for gap measurements

3.1.1 Design

The purpose of this remote measurement design is to measure the gaps of between the RV core-stabilizing lug and the CSB snubber lug for RVI-modularization. The reason for the measurement of gaps is to set a permissible range of 0.381 – 0.508 mm between the RV core-stabilizing lug and the CSB snubber lug. The permissible range, when a nuclear reactor operates, ensures a margin by thermal expansion. This is identical to that used in current power water reactors.

For these gaps, adjustment of the shims was machined at the construction site and assembled in the RV after measurement of the gaps of the RV core-stabilizing lugs and the CSB snubber lugs.

Fig. 4 shows the placement of the six gap-measurement locations between the RV core-stabilizing lug and the CSB snubber lug. The RV core-stabilizing lugs located inside the RV and the CSB snubber lugs located outside the CSB are set at angles of 0°, 60°, 120°, 180°, 240°, and 300°. The angular positions correspond to locations 1 through 6, respectively (see Fig. 4). Also, each of six RV core-stabilizing lugs corresponds to a pair of (left and right) the CSB snubber lugs, as shown in Fig. 4. One side of the CSB snubber lug has six holes for hand-measurements in the current RVI-installation method. Therefore, one CSB snubber lug has twelve holes, and six CSB snubber lugs have a total of 72 holes, which are from the external RVI; the lengths of which should be measurable simultaneously, as shown in Fig. 13.

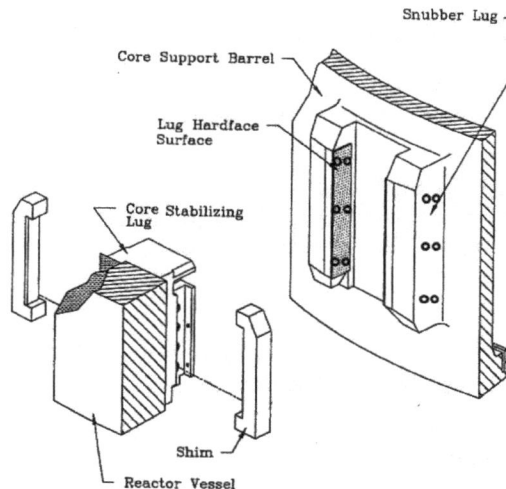

Fig. 13. RV core-stabilizing lug and CSB snubber lug

Many essential factors were studied before selecting the measurement sensors used for RVI-modularization.

Specifically, the measurement environment, the measured object, the size of the sensor, the weight of the sensor, the measurement range, the driving force of the sensor, the accuracy, and the resolution were investigated. Finally, the SOLARTRON (UK) sensor (DT/20/P) was selected for the remote measurement system (Ko et al., 2009).

The DT/20/P sensor was tested to confirm its performance and application conditions in a reliability test using gauge blocks. A consistency test was also done to check the connection jig, and an accuracy test was done for the zero-point adjustment device. Finally, a stability test was done to check the switching noise environment and the EMI (electromagnetic interference) using a reduced-scale model system. All test results were satisfactory for a sensor of a remote measurement system (Ko et al., 2009).

3.1.2 Fabrication

A remote measurement system was developed to measures the gaps between the RV core-stabilizing lug and the CSB snubber lug using a DT/20/P digital probe sensor. The major characteristics of the remote measurement system are as follows:

- The remote measurement system consists of a measurement sensor section, a pneumatic supply and control section, a power supply section, and a remote control computer and software program.
- The measurement sensor section is intended to measure gaps between the RV core-stabilizing lug and the CSB snubber lug. Those sensors, placed at 0° and 60°, are measured by 24 sensors with a signal cable connected to the channel box #1. Those sensors placed at 120° and 180° are measured by 24 sensors with a signal cable connected to the channel box #2, and those sensors placed at 240° and 300° are measured by 24 sensors with a signal cable to connected the channel box #3. A measurement sensor section is composed of 72 digital probes. This system is able to measure 72 points at once and operate by pneumatic actuation.
- The pneumatic supply and control section is intended to supply air to actuate the sensors by a remote control computer and a software program. The pneumatic supply section consists of an air compressor, an air filter, an air pressure regulator and an air tube. The pneumatic control section is composed of a flow control valve, a solenoid valve, a solenoid valve manifold, a USB orbit module and a T-CON. The solenoid valve and the solenoid valve manifold control operation of measurement sensors and the USB orbit module and the T-CON receive signal data of the digital probe, DT/20/P.
- The power supply section supplies electric power to the electric equipment, including the T-CON.
- The remote control computer and the software program consist of a laptop computer and the software to control digital probes and to process and store the measurement results.
- The channel box contains 24 digital probes, the T-CON, 24 solenoid valves, and four solenoid valve manifolds. The channel is designed for three channel boxes. These boxes were very easy to handle due to their suitable weight and size.
- All of the cables and air tubes are easily connected to the channel boxes.
- The network of the remote measurement system is very stable and causes no disturbance to the EMI environment.

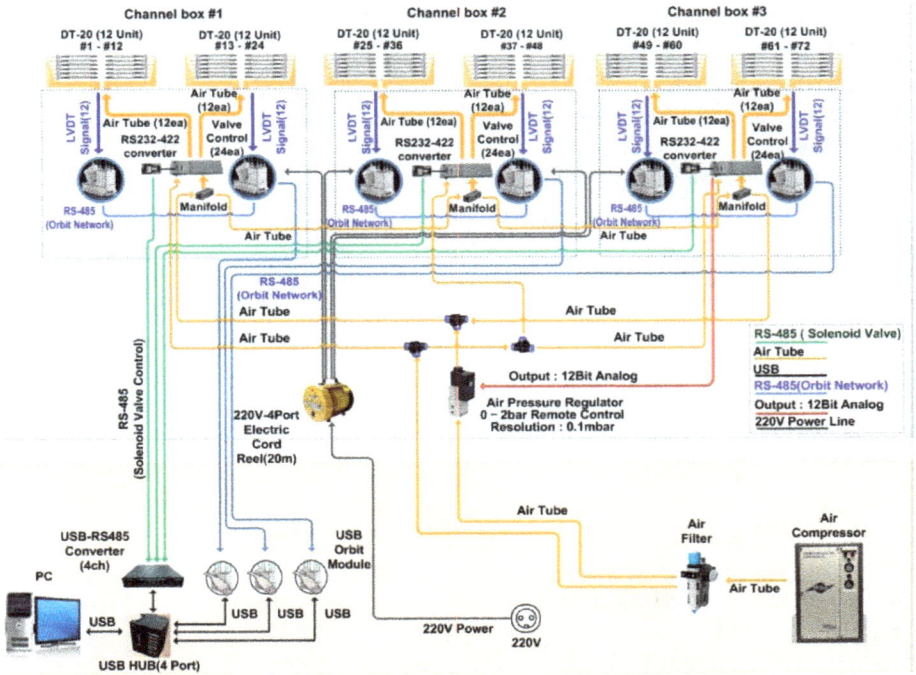

Fig. 14. Block diagram of developed remote measurement system

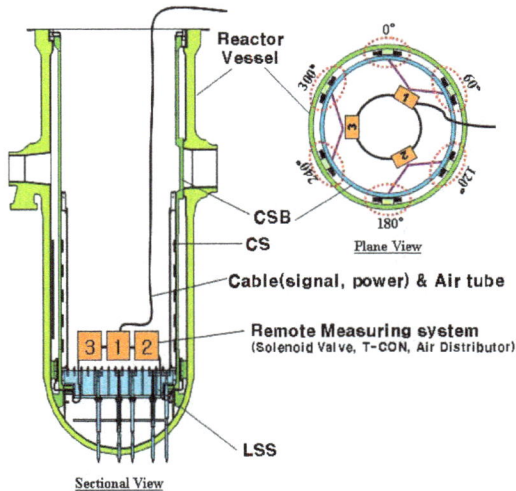

Fig. 15. Setup position of the remote measurement system

Table 2 presents a list of the parts of the remote measurement system, and Fig. 14 shows a block diagram of the developed remote measurement system.

List of Parts	Amount
DT/20/P digital probe	72
USB orbit module	3
AC-PSIM power supply	6
T-CON	75
Flow control valve	72
Threaded connection jig	72
0-Point adjustment jig	6
Solenoid valve	72
Solenoid valve manifold	12
2-line RS-485 signal cable & reel	3
220V-4port electric cable & cord reel	2
Air tube	500 m
4-port air manifold	1
I/O board SMPS	9
4-port USB hub	1
USB to RS-485 converter	1
Air clean unit	2
Air compressor	1
Marking tool	1
Electric lamp & cord reel	2
Air tube one touch reel(20 m)	1
System storage box	1

Table 2. List of parts for remote measurement system

As shown in Fig. 15, the channel boxes are located above the LSS in the CSB. The digital probes should be set on the CSB snubber lugs before assembly with the RV. The channel box should be connected to the air compressor, to the remote control computer, and to the electric power source after assembly of the RV and the CSB.

3.2 Design of RV and CSB mockup

An RV mockup and a CSB mockup were designed and manufactured because the remote measurement system should be subjected to a test to verify its applicability in construction projects. This was done using the RV mockup and the CSB mockup developed here.

3.2.1 Design concept of the RV and the CSB mockup

To design the RV mockup and the CSB mockup, it was necessary to follow a number of principles. First, the measurement parts of the RV core-stabilizing lugs and the CSB snubber lugs should be designed to simulate the actual size of the construction site. Second, the 72 points of the RV core-stabilizing lugs and the CSB snubber lugs should be measured simultaneously. The factors not related to any measurement part should be designed to be as simple as possible.

The RV mockup and the CSB mockup were designed to evaluate the setting suitability of remote measurement system. In particular, the RV core-stabilizing lug of the RV mockup and the CSB snubber lug of the CSB snubber lug were designed to match the RV and the CSB of an actual nuclear power plant.

Other parts of the actual RV and CSB were simply designed for usability of the experiment and out of concerns for the manufacturing budget. Thus, the thickness of cylinders was designed to be thinner than that of an actual RV and an actual CSB.

3.2.2 Design of the RV mockup

The RV inner diameter is identical to that of the original. However, the RV outer diameter, not related to the measurement, was designed so as to be thinner, making it a light RV mockup. Thus, it was designed to be different from the actual size. The RV upper flange was simply designed.

The RV upper flange-face was assembled with the CSB and was applied to actual conditions in the same manner. The RV cylinder was designed with a thin plate, including a RV core-stabilizing lug. The height of the RV mockup was minimized to 2,665 mm.

The assembled parts of the RV core-stabilizing lugs, the RV core stop lug, and the shim were designed to be identical to their real-life counterparts.

Six supports of 70 cm in height were attached to the bottom of the RV mockup to monitor the condition of the measurement sensors. The lifting lugs were welded onto the RV upper flange to facilitate assembly and separation from the CSB mockup.

Fig. 17 and Fig. 18 show the designed 3D model and an image of the manufactured RV mockup.

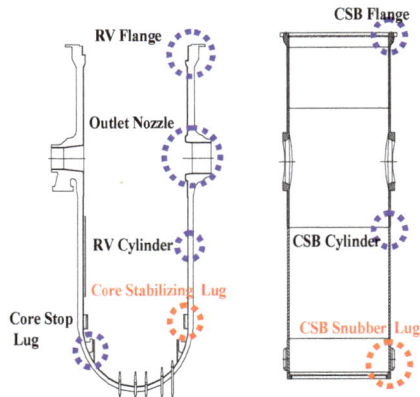

Fig. 16. Important parts of the design of the mockup

Other parts of the actual RV and CSB were simply designed for usability of the experiment and out of concerns for the manufacturing budget. Thus, the thickness of cylinders was designed to be thinner than that of an actual RV and an actual CSB.

Fig. 16 shows the important parts in the design of the RV mockup and the CSB mockup.

3.2.3 Design of the RV mockup

The RV inner diameter is identical to that of the original. However, the RV outer diameter, not related to the measurement, was designed so as to be thinner, making it a light RV mockup. Thus, it was designed to be different from the actual size. The RV upper flange was simply designed.

The RV upper flange-face was assembled with the CSB and was applied to actual conditions in the same manner. The RV cylinder was designed with a thin plate, including a RV core-stabilizing lug. The height of the RV mockup was minimized to 2,665 mm.

The assembled parts of the RV core-stabilizing lugs, the RV core stop lug, and the shim were designed to be identical to their real-life counterparts.

Six supports of 70 cm in height were attached to the bottom of the RV mockup to monitor the condition of the measurement sensors. The lifting lugs were welded onto the RV upper flange to facilitate assembly and separation from the CSB mockup.

Fig. 17 and Fig. 18 show the designed 3D model and an image of the manufactured RV mockup.

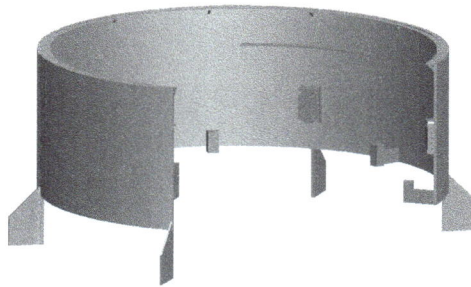

Fig. 17. 3D model of the RV mockup

Fig. 18. Image of the manufactured RV mockup

3.2.4 Design of the CSB mockup

The CSB outer diameter is identical to that of its real-life counterpart. However, the CSB inner diameter, not related to the measurement, was designed to be thinner, making it a light CSB mockup. Thus, it was designed to be different from the actual size.

The conditions of the CSB flange in contact with the RV upper flange were designed according to actual conditions. The CSB cylinder was designed with a thin plate including the CSB snubber lug. The height of the CSB mockup was minimized to 1,618 mm. The CSB snubber lugs were both designed in keeping with their actual shapes and were welded onto the CSB. The lifting lugs were welded onto the CSB flange to facilitate assembly and separation from the RV mockup. A stand for the installation of the remote measurement system was designed on the inside of the CSB mockup.

Fig. 19 and Fig. 20 show the designed 3D model and an image of the manufactured CSB mockup.

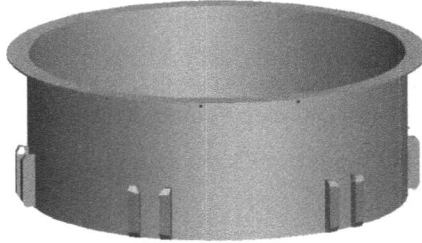

Fig. 19. 3D model of the CSB mockup

Fig. 20. Image of the manufactured CSB mockup

3.3 Experiment and result

3.3.1 Experiment

One difference between the remote measurement method and the hand-measurement method is that measurement had to be done using the remote measurement system after assembly of the CSB on the RV. An additional difference was that it was necessary to attach a gauge block to the RV core-stabilizing lug before assembly of the CSB to overcome the length-measurement limit of the remote measurement system.

The measurement value of the gap between the RV core-stabilizing lug and the CSB snubber lug is typically measured to a maximum 49 mm in the construction sites of nuclear power plants, but the maximum measurement range of the measurement sensor in the remote measurement system was 20 mm.

Therefore, the remote measurement method required twelve gauge blocks of 35 mm to measure 72 points simultaneously.

The experimental procedure of the remote measurement system using the RV mockup and the CSB mockup is as follows:

- Widths of six RV core-stabilizing lugs and twelve gauge blocks of 35 mm made using a micrometer were measured and attached to twelve gauge blocks to the right and the left of the six RV core-stabilizing lugs.
- The inside lengths of the CSB snubber lugs, the locations of the contact of the RV core-stabilizing lug with the CSB snubber lug, were measured using a cylinder gauge.
- The remote measurement system was installed on a stand inside the CSB mockup. The stand is used to simplify the LSS in RVI.
- 72 sensors assembled with 72 threaded connection jigs were inserted and fixed at 72 holes for measurement of the CSB snubber lugs.
- Three channel boxes are connected to air-hoses, signal cables, and electric power cables.
- The software program for remote measurement should be tested and measurement sensors should be confirmed as being capable of normal operation using the remote control computer.
- To set a start-point of the measurement sensors, six zero-point adjustment devices were installed between the left and the right areas of the CSB snubber lugs and were adjusted to the zero point. After the zero-point adjustment, the six zero-point adjustment devices were disconnected from the CSB snubber lugs.
- After the CSB mockup of the RV mockup was assembled, the gaps between the RV core stabilizing lugs and the CSB snubber lugs were measured using the remote measurement system.

The measurement test, using the remote measurement system, was carried out with the assembly and disassembly of the RV mockup and the CSB mockup three times in succession.

Fig. 21 shows a computer screen detailing the gaps as acquired by the remote measurement system.

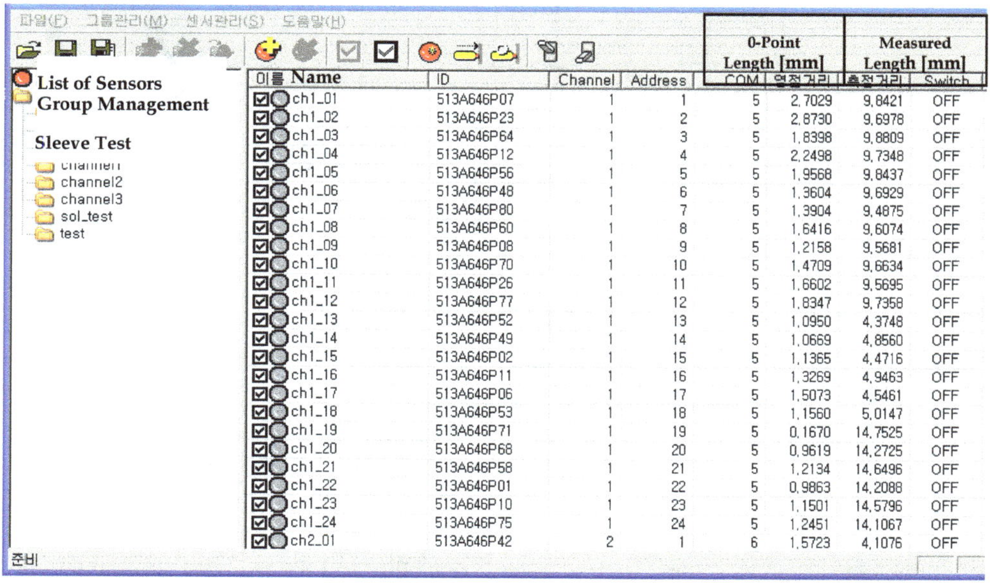

이름 Name	ID	Channel	Address	0-Point Length [mm] COM	0-Point Length [mm] 영점거리	Measured Length [mm] 측정거리	Switch
ch1_01	513A646P07	1	1	5	2,7029	9,8421	OFF
ch1_02	513A646P23	1	2	5	2,8730	9,6978	OFF
ch1_03	513A646P64	1	3	5	1,8398	9,8809	OFF
ch1_04	513A646P12	1	4	5	2,2498	9,7348	OFF
ch1_05	513A646P56	1	5	5	1,9568	9,8437	OFF
ch1_06	513A646P48	1	6	5	1,3604	9,6929	OFF
ch1_07	513A646P80	1	7	5	1,3904	9,4875	OFF
ch1_08	513A646P60	1	8	5	1,6416	9,6074	OFF
ch1_09	513A646P08	1	9	5	1,2158	9,5681	OFF
ch1_10	513A646P70	1	10	5	1,4709	9,6634	OFF
ch1_11	513A646P26	1	11	5	1,6602	9,5695	OFF
ch1_12	513A646P77	1	12	5	1,8347	9,7358	OFF
ch1_13	513A646P52	1	13	5	1,0950	4,3748	OFF
ch1_14	513A646P49	1	14	5	1,0669	4,8560	OFF
ch1_15	513A646P02	1	15	5	1,1365	4,4716	OFF
ch1_16	513A646P11	1	16	5	1,3269	4,9463	OFF
ch1_17	513A646P06	1	17	5	1,5073	4,5461	OFF
ch1_18	513A646P53	1	18	5	1,1560	5,0147	OFF
ch1_19	513A646P71	1	19	5	0,1670	14,7525	OFF
ch1_20	513A646P68	1	20	5	0,9619	14,2725	OFF
ch1_21	513A646P58	1	21	5	1,2134	14,6496	OFF
ch1_22	513A646P01	1	22	5	0,9863	14,2088	OFF
ch1_23	513A646P10	1	23	5	1,1501	14,5796	OFF
ch1_24	513A646P75	1	24	5	1,2451	14,1067	OFF
ch2_01	513A646P42	2	1	6	1,5723	4,1076	OFF

Fig. 21. Screen showing measurement results of the remote measurement system

3.3.2 Result

The ground plan for the gap between the RV and the CSB as measured by the remote measurement system is shown in Fig. 22.

Table 3 shows a comparison of the data by the remote measurement system and the hand-measurements.

The "Total" value, data by the remote measurement system, denotes the sum of the left gap (LG) and the right gap (RG), the left block-gauge length (LB) and the right block-gauge length (RB), and the width of the RV core-stabilizing lug (D). The "J" value represents the hand-measurement data using a micrometer. Therefore, the "Total – J" value denotes the error between the remote measurement system data and the hand-measurement data.

Table 3 now shows that the remote measurement system has high accuracy as the maximum error is only 0.0669 mm. The maximum error is smaller than 0.127 mm of acceptance criteria.

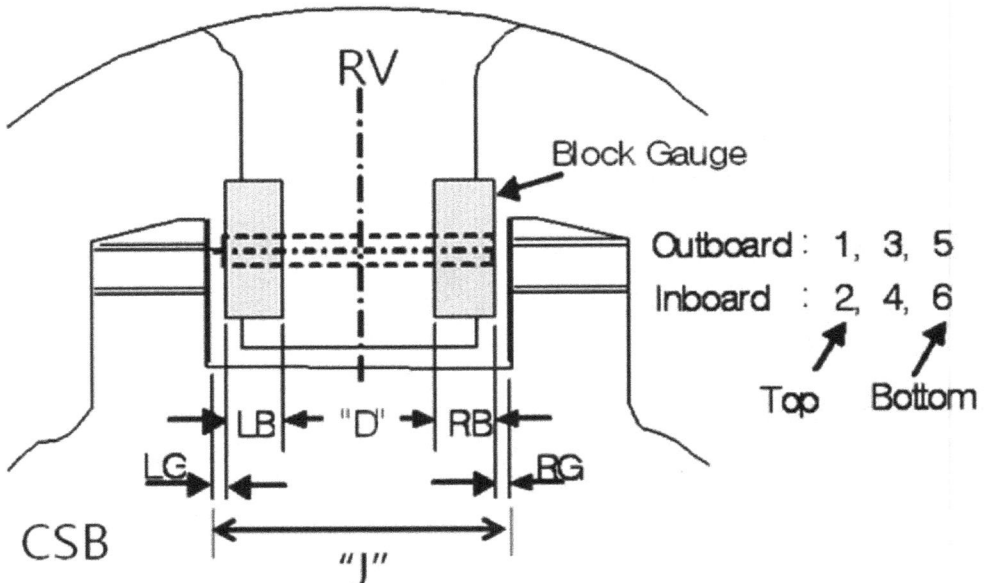

Fig. 22. Ground plan for the gap of between the RV and the CSB

The differences between the Total values and the J values are shown in Fig. 23.

On the other hand, on the basis of the permissible range, which was set to 0.381 – 0.508 mm, if there is a difference in the width that is within 0.127 mm, the requirements are considered to be satisfied. Errors in the measurement data that was used in the remote measurement system in the experiment and analyses did not exceed 0.127 mm.

Therefore, through this experiment and as demonstrated by the results, the remote measurement system was demonstrated to have reliability and the required applicability for deployment in the construction sites of nuclear power plants.

Location		LG +	LB +	D	+ RB	+ RG	=	Total	J	Total - J (Error)
0°	1	9.9007	34.885	152.150	34.895	9.5612	=	241.3919	241.3250	0.0669
	2	9.7528	34.900	152.130	34.895	9.6764	=	241.3542	241.3050	0.0492
	3	9.8675	34.890	152.120	34.905	9.5639	=	241.3464	241.3250	0.0214
	4	9.6571	34.965	152.110	34.905	9.6678	=	241.3049	241.3100	-0.0051
	5	9.8546	34.855	152.125	34.875	9.6056	=	241.3152	241.3250	-0.0098
	6	9.7016	34.865	152.100	34.875	9.7631	=	241.3047	241.3200	-0.0153
60°	1	4.3944	34.885	152.390	34.900	14.7650	=	241.3344	241.3650	-0.0306
	2	4.8742	34.895	152.385	34.900	14.2862	=	241.3404	241.3500	-0.0096
	3	4.4628	34.900	152.420	34.910	14.6466	=	241.3394	241.3700	-0.0306
	4	4.9950	34.905	152.390	34.905	14.1700	=	241.3650	241.3450	0.0200
	5	4.5724	34.875	152.435	34.865	14.6027	=	241.3501	241.3650	-0.0149
	6	5.0382	34.885	152.410	34.875	14.1214	=	241.3296	241.3450	-0.0154
120°	1	4.0993	34.900	152.250	34.890	15.1549	=	241.2942	241.3200	-0.0258
	2	3.8661	34.895	152.240	34.895	15.3632	=	241.2593	241.2800	-0.0207
	3	4.1366	34.905	152.240	34.895	15.1916	=	241.3682	241.3400	0.0282
	4	3.8756	34.905	152.210	34.900	15.3849	=	241.2755	241.2900	-0.0145
	5	4.1408	34.875	152.220	34.870	15.1761	=	241.2819	241.3300	-0.0481
	6	3.9372	34.875	152.205	34.880	15.4613	=	241.3585	241.3000	0.0585
180°	1	7.0775	34.905	152.145	34.890	12.3671	=	241.3846	241.3250	0.0596
	2	7.1105	34.900	152.125	34.875	12.2970	=	241.3075	241.3250	-0.0175
	3	7.2694	34.900	152.175	34.900	12.1571	=	241.4015	241.3500	0.0515
	4	7.2975	34.895	152.150	34.885	12.0841	=	241.3116	241.3275	-0.0159
	5	7.5123	34.850	152.160	34.855	11.9887	=	241.3660	241.3650	0.0010
	6	7.5585	34.850	152.140	34.845	11.9393	=	241.3328	241.3350	-0.0022
240°	1	3.0665	34.895	152.220	34.890	16.2133	=	241.2848	241.3200	-0.0352
	2	3.4621	34.900	152.090	34.895	15.9112	=	241.2583	241.2800	-0.0217
	3	3.1894	34.905	152.205	34.900	16.1000	=	241.2994	241.3150	-0.0156
	4	3.5021	34.905	152.080	34.905	15.8700	=	241.2621	241.2950	-0.0329
	5	3.3227	34.865	152.195	34.865	16.0655	=	241.3132	241.3350	-0.0218
	6	3.6515	34.865	152.085	34.875	15.8272	=	241.3037	241.3000	0.0037
300°	1	6.6439	34.875	152.265	34.895	12.6747	=	241.3536	241.3450	0.0086
	2	6.5915	34.895	152.130	34.900	12.8177	=	241.3342	241.3250	0.0092
	3	6.6015	34.890	152.260	34.905	12.7303	=	241.3868	241.3500	0.0368
	4	6.5128	34.905	152.145	34.905	12.8487	=	241.3165	241.3350	-0.0185
	5	6.5463	34.860	152.240	34.860	12.8112	=	241.3175	241.3600	-0.0425
	6	6.5746	34.870	152.125	34.865	12.8949	=	241.3295	241.3350	-0.0055

Table 3.Comparison of data by remote measurement system and hand-measurement

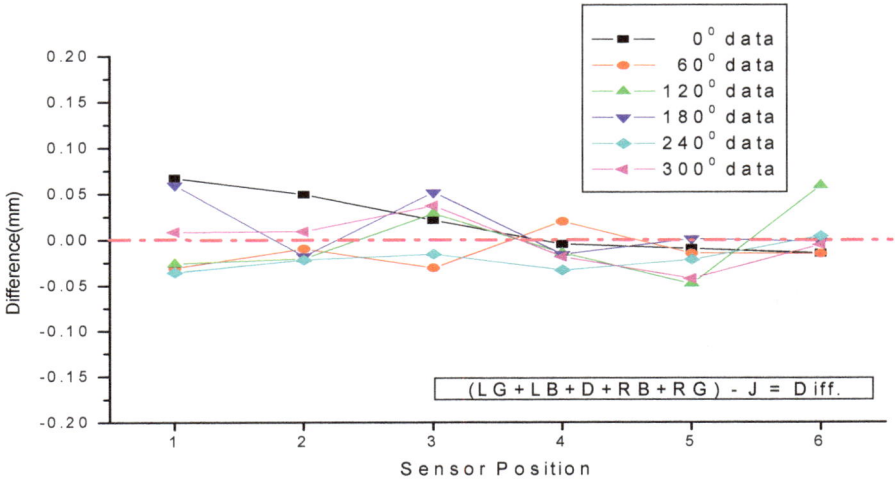

Fig. 23. Differences between the Total values and the J values

3.4 Conclusion

A system was successfully developed that can measure the gaps between the RV core-stabilizing lugs and the CSB snubber lugs remotely for RVI-modularization. To confirm the reliability of this method and its applicability to construction sites, the measurement system was designed, manufactured, and tested it using an RV mockup and CSB mockup.

It is therefore concluded that the remote measurement system reduces the construction period by nearly two months or more compared to the existing method of hand measurements.

As the developed remote measurement system can be used at actual construction sites.

4. Development of an improved installation procedure and schedule of RVI modularization

4.1 Development of an improved installation procedure for RVI modularization

Under the existing installation procedure, six snubber shims were assembled between the RV core-stabilizing lugs and the CSB snubber lugs after alignment of the CSB and the RV. Subsequently, assembly and flexure welding of the LSS and the CS were conducted. The new and improved installation procedure for RVI modularization was developed allowing assembly and flexure welding of the LSS and the CS to be performed before the main installation process. Alignment of the CSB assembly and the RV and assembly of the snubber shims were undertaken during the main installation process.

The improved installation procedure for RVI modularization (Ko, 2011) appears in Table 4 as order 1 and order 3. A detailed explanation is given below of the flexure welding of the LSS and the CS in the CSB as well as of the alignment and installation of the snubber shims of the CSB assembly and the RV.

Order	RVI installation procedure	Remark
1	Assembly of the LSS and the CS in the CSB, Flexure welding	Improvement
2	Welding of the flow baffle in the RV	
3	Alignment of the CSB assembly and the RV, Installation of snubber shims	Improvement
4	Installation of the CSB assembly in the RV	
5	Installation and alignment of the UGS	
6	Installation and alignment of the RV head	
7	Installation of alignment keys, dowel pins and guide lugs insert	
8	Final alignment, Installation of head down ring and HJTC tube	

Table 4. Improved installation procedure for RVI modularization

4.1.1 Alignment of the CSB and RV and calculation of the dimensions of the snubber shim

1. The dimensions between the RV outlet nozzle and the reactor coolant loop (RCL) were measured after the welding of the RCL. They were recorded along with the measured dimensions before the welding. The widths of the RV core-stabilizing lugs were also measured.
2. The measured positions were marked on RV keyways and the relative positions of the keyways on the RV centerline were measured using the widths and vertical degrees.
3. The target-hole positions of the RV flange on the RV centerline and the dimensions of the gauge blocks of the RV core stabilizing lug were also measured. Here, the dimensions of the required measurements were the heights and widths of the gauge blocks.
4. Before installation of the gauge blocks, the integrity of the RV core-stabilizing lugs and the cap screws was checked. Neolube was applied twice on both the threaded surfaces on the cap screws and the surfaces of the bearings.

4.1.2 Installation of the gauge block

1. Gauge blocks were installed and the cap screws were then tightened. They needed to be tightened according to a three-step tightness method, as follows: 160, 213 and 266 ft-lbs. The final torque was 256 - 276 ft-lbs and the tightness sequence was as follows:
2. The number of cap screws was 1, 2, 3, and 4 from top to bottom. The tightness sequence of step 1 was 2, 3, 1, and 4, and the torque was 160 ft-lbs.
3. The tightness of step 2 was identical to that of 2) and the torque was 213 ft-lbs.
4. The tightness of step 3 was identical to that of 2) and the torque was 266 ft-lbs.
5. The cap screws were unscrewed in the reverse sequence of the tightness sequence: 4, 1, 3, and 2.
6. One more time, the tightness was carried out by the sequence of 2), 3), and 4).
7. The remaining gauge blocks were installed on the CSB snubber lugs according to the sequence of 2), 3), 4), 5), and 6). After installation of the gauge blocks, the gaps between the upper part and the lower part were uniformly maintained at 0.1016 mm.

4.1.3 Dimensions of the CSB Snubber Lug

1. After installation of the gauge blocks, all widths of the gauge blocks at the same six intervals were measured. Subsequently, the dimensions of the CSB snubber lugs were measured. The measured parts of the CSB snubber lugs were the inside widths of both of their surfaces. In total, six holes of the CSB snubber lugs were measured.
2. The measured positions of the CSB keyway were marked and the widths measured.
3. Four dummy alignment keys (DAKs) were installed on the RV keyways and adjusted so that they could be positioned within 0.254 mm of the RV centerline. The vertical degree of an RV head seating surface of a DAK was adjusted so that it could be positioned within 0.0254 mm/ft. The position of the DAK and the vertical degree were measured again.

4.1.4 Installation of the remote measurement system

1. In order to set up the remote measurement system inside the CSB assembly, three channel boxes were mounted in the CSB assembly. As shown in Fig. 15, the channel boxes were opened and 72 digital probes were taken out of the CSB assembly through spaces between the CSB and the LSS.

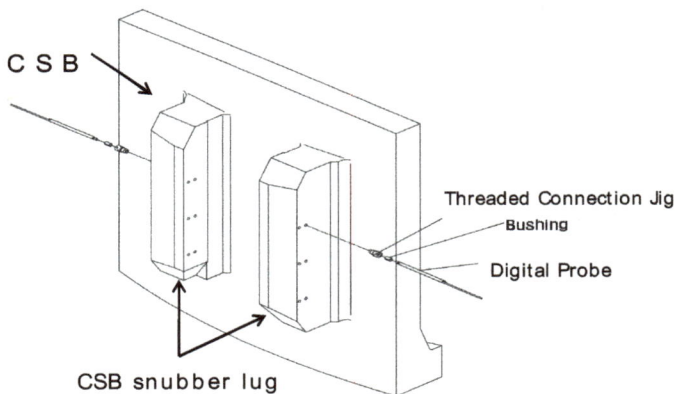

Fig. 24. Assembly of digital probe and threaded connection jig

2. As shown in Fig. 24, all 72 digital probes were installed the holes of the CSB snubber lugs after the digital probes were assembled with the threaded connection jigs. For this step, the digital probes were necessarily placed at 1 mm or more from the measured holes of the CSB snubber lug. Bolt tightening was done using a hexagonal wrench. The bolts numbered 1, 2, 3, and 4 were clockwise and the tightening order was 1, 3, 2, and 4. All remaining sensors were installed via the methods described above.
3. The electric power cords, air hoses, and signal cables between the channel boxes were connected and taken out of the RV. The air hoses were connected with an air compressor.
4. The signal cables were linked with a USB-orbit module and an RS-485 converter. The USB ports of the USB-orbit module were linked with a USB hub. The USB hub and USB ports of the RS-485 converter were connected to the USB ports of a remote measurement computer. Table 5 shows the remote measurement system guideline.

Object	Status	Operation	Object	Status	Operation
Air-compressor		Electric power turned on and pressure adjusts to 0.8MPa; air valve open	RS-232 to 485 Converter		Solenoid valve links with RS-232 to 485 converter
RS-485 Converter		RS-485 converter links with solenoid valve cables	Air regulator		Air regulator links with solenoid valve
USB orbit module		Orbit cables link with USB orbit module	Solenoid valve & AC adapter		Channel boxes supply electric power
USB hub		USB ports of orbit module link to USB hub	RS-485 Converter		Computer power on
Remote measurement computer		USB ports of computer link to USB ports of RS-485 converter and USB hub	USB Hub		Computer power on
Solenoid valve & AC adapter		Solenoid valve link AC adapter	Software program of remote measurement system		Software program executes

Table 5. Remote measurement system guideline

5. Electric power supplies were checked in order with the remote measurement system first, followed by the air compressor, and then the remote measurement computer to verify whether or not the connections worked exactly according to the operating guide.

6. The air pressure was set to 0.8 – 1 bar using a software program on the remote measurement computer. At this stage, the digital probes had to be checked using a software program to verify whether or not they operated normally. Fig. 14 shows the configuration of the remote measurement system.

4.1.5 Zero-point adjustment of the remote measurement system and marking

1. In order to attach a zero-point adjustment device firmly onto the CSB snubber lug, a zero-point adjustment plate and a connector on the gap control section were tightened. The remaining zero-point adjustment devices adhered to the CSB snubber lug as described above. Fig. 25 shows the zero-point adjustment device mounted on the CSB snubber lug.

2. Length measurement was done using the software program of the remote measurement computer. Measurements were taken five times. The average values were then used to set the zero-point data. When the setting of the zero-point data was complete, the zero-point data were saved and recorded.

3. The zero-point adjustment device was detached from the CSB snubber lug and a marking tool was attached to the CSB snubber lug. Digital probes were stained with a red stamping ink and their correct operation was confirmed using the software program of the remote measurement computer. The remaining digital probes were executed in the same way.

4. The marking tool was removed from the CSB snubber lug and the channel boxes in the CSB assembly and air compressor were then separated from the air hoses, electric power cords, and signal cables connected to the remote measurement computer. The disconnected cords and cables had to be arranged so that a disturbance did not result from the combination of the RV and the CSB assembly. Fig. 26 shows the marking tool assembly attached onto the CSB snubber lug.

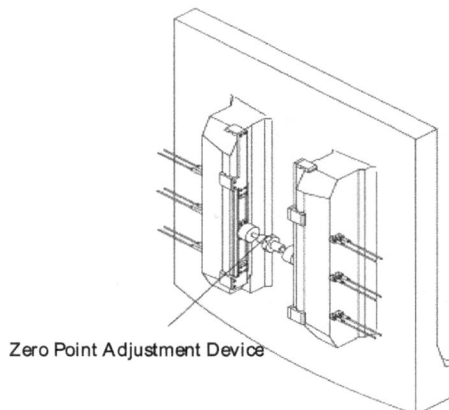

Zero Point Adjustment Device

Fig. 25. Installation of zero point adjustment device

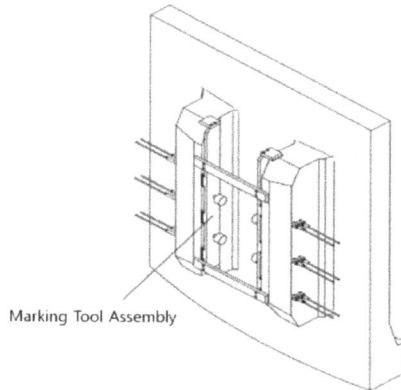

Marking Tool Assembly

Fig. 26. Installation of a marking tool

4.1.6 Combination of the CSB assembly and the RV

1. The CSB assembly was aligned to the RV centerline and the CSB assembly was inserted in the RV. The CSB assembly was turned at 45° and was lowered to prevent damage to the DAK. It was then combined after the CSB assembly was turned to the original position before ensuring a 50 cm interval between the CSB assembly and the RV.

2. When the CSB assembly was installed, the load measured by a hydra-set was continuously checked. In addition, when the CSB assembly was at a height of approximately 30 cm from the RV head seating surface, the bottom surface (datum "B") of the CSB upper flange was used to stop the descent of the CSB assembly. A basis surface (datum "B") of the CSB upper flange was used for a parallel adjustment to within 0.381 mm of the RV head seating surface.

3. The load of the hydra-set was decreased to 10,000 lb and was checked given that the CSB came in contact with the RVI installation surface. The RV centerline and the CSB centerline were aligned within 0.0254 mm by CSB position devices (8 EA).

4. The vertical degree for the CSB keyway and the datum hole were measured and their relative positions on the CSB centerline were confirmed.

5. The gaps between the RV head seating surface and the upper surface of the CSB flange were measured in 45° intervals. The gaps (2.1336 – 2.9464 mm) of the RV outlet nozzle and the temperatures of the nozzle area were also measured.

6. The alignment of the RV/CSB centerline and the requirements of the nozzle gap were checked. If the requirements were not satisfied, it was necessary to repeat this procedure. If the position of the DAK changed before and after the installation of the CSB due to the checking of the position of the DAK, the measurements had to be done again and the existing checklist was invalidated. All installation requirements were met; the final adjustment conditions and the variation of the CSB centerline on the RV centerline were measured and recorded.

4.1.7 Length measurement using the remote measurement system

1. The channel box of the internal CSB and the air hose, compressor, electric power cord, power supply and signal cables of the external CSB were connected. Electric power was then supplied.

2. Using a software program running on a remote measurement computer, the length was measured a total of five times. The pressure for the measurement was adjusted to 0.8 ~ 1 bar; this was set to have a zero length to ensure that the data were entered correctly. The average value of the measured lengths was used as the data. Once the measurement was completed, the data were stored and the measured length was recorded.

3. After the gap measurements were completed, all electric power was turned off. After the RV and the CSB assembly were detached, the air hose, electric power cord and signal cable were respectively separated from the compressor, electric power, USB hub and RS-485 converter. Once separated, the air hose, electric power cord and signal cable were temporarily fixed in the CSB assembly to ensure that they would not interfere with the disassembly of the CSB assembly.

4.1.8 Separation of the CSB Assembly and RV

1. The CSB assembly was separated from the RV and set on a storage stand. The CSB assembly was lifted at 45° turns after vertical lifting of approximately 50 cm in order to prevent the DAKs from being damaged. The CSB assembly was checked continuously via a load measured by a hydra-set.

2. After the checked positions of the gauge blocks were marked using digital probes, the widths of the marked positions were measured. After removing the gauge blocks, the gaps between the CSB snubber lugs and the RV core-stabilizing lugs were used to calculated the processing dimensions of the snubber shims; the gaps were calculated at 0.381-0.508 mm as a permissible range.

4.1.9 Installation and measurement of the shim on the RV core-stabilizing lug

1. The dimensions of the refined shims were measured and a penetration test was conducted. Before installation of the shims, the RV core stabilizing lugs and the cap screws were checked to confirm the integrity of each screw. Neolube, a dry film lubricant, was applied twice to the threaded surfaces of the cap screws and bearing surface. Fig. 13 shows the assembly of the snubber shims on the RV core-stabilizing lug.

2. When the shims were installed, the cap screws were assembled by hand and tightened according to a three-step tightness method: 160, 213 and 266 ft-lbs. It was important that after the shims were installed, the upper gaps and lower gaps were maintained as constant (0.1016 mm).

3. After the shims were installed, their full widths were measured in six positions to measure the equal intervals. The DAKs were adjusted to vertical degrees at 0.0254 mm/ft for the RV head seating surface. At this point, the vertical degrees of the DAKs and the status of the installed positions were recorded.

4. According to section 2.6 "Combination of the CSB assembly and the RV," the CSB assembly was installed in the RV. When the RV core-stablizing lugs and the RV snubber shims were connected, they were lowered using the hydra-set. The RV centerline and the CSB centerline were adjusted within 0.0254 mm using the DAKs, and the CSB assembly was completely lowered. If the positions of the DAKs changed, the measurements had to be done again.

5. The gaps between the RV and the CSB outlet nozzles were recorded and the offsets of the CSB keyways regarding the DAKs were recorded as well. The offsets of the CSB centerline in relation to the RV centerline were also calculated and recorded.

6. The air hoses, electric power cords and signal cables of the remote measurement system
    were reconnected to the compressor, electric power, USB hub and RS-485 converter,
    and electric power was supplied. The length of the snubber shims was measured five
    times. Once the measurement was complete, the measured data were stored and the
    measured lengths were recorded. The measured lengths of the shims were confirmed to
    be within a permissible range (0.381 - 0.508 mm). If the measured lengths of the shims
    exceeded the permissible range, they would be used after reprocessing.
7. All electric power was turned off, and the air hoses, electric power cords, and signal
    cables were respectively separated from the compressor, electric power, USB hub and
    RS-485 converter. Separated air hoses, electric power cables and signal cables were
    temporarily fixed in the CSB assembly when the RV and CSB assembly were detached
    in order to avoid interference with cables and pieces of equipment. Finally, the gaps
    between the CSB and the RV core-stop lugs were measured.

4.1.10 Separation and confirmation of the remote measurement system

1. The CSB assembly was separated from the RV and set down on a storage stand. The
    channel boxes and digital probes, threaded connection jigs, air hoses, electric power
    cables, and signal cables were completely removed from the CSB assembly.
2. After removing the CSB assembly, snubber shims were confirmed in the combined
    state. All heads of the cap screws were dug with holes of Φ 3.0226 mm at a depth of
    19.05 ± 0.762 mm.
3. All holes of the heads of the cap screws had pins inserted and their installation status
    was checked. Plugs (Φ 12.7 mm) installed to fix the pins were inserted; after welding the
    plugs, penetration tests were carried out.

4.2 Development of improved installation schedule for RVI modularization

Table 6 presents a comparison of the existing RVI installation process at the Shin-kori #1
nuclear power plant and the modularization installation process developed.

Compared with the existing method, it was found that the developed installation process
using RVI modularization can shorten the installation period to about 67 days in the critical
path. This can reduce the construction period, as follows: RV & CSB dimension check (15
days), CSB alignment & gap measurement (13 days), RV & CSB & LSS/CS alignment (13
days), flexure welding (20 days) and CSB assembly installation & alignment check (6 days).

The RV & CSB dimension check (15 days), flexure welding (20 days) and CSB assembly
installation & alignment check (6 days) are conducted through a concurrent process before
the determination of the critical path in the existing installation process. In addition, the CSB
module alignment & gap measurement (13 days) and RV & CSB module alignment (13
days) are correspondingly reduced using the remote measurement system and the
improved installation procedure.

Fig. 27 shows the existing RVI installation schedule in the critical path at the Shin-kori #1
nuclear power plant in Korea. It consists of the following steps: (1) the RV & CSB dimension
check (2) the CSB alignment & gap measurement (3) snubber shim machining & installation
(4) the RV & CSB & LSS/CS alignment (5) flexure welding (6) the CSB assembly installation
& alignment check and (7) the upper guide structure (UGS) & RV head installation &
alignment check. Therefore, the RVI installation period at the Shin-kori #1 nuclear power
plant should be required approximately 129 days in the critical path.

Unit. Day

	Existing Method (Ex. Shin-kori #1, Korea)	Proposed Modularization Method
RV & CSB Dimension Check	15	-
CSB Alignment & Gap Measurement	18	CSB Module Alignment & Gap Measurement (5)
Snubber Shim Machining & Installation	21	21
RV & CSB & LSS/CS Alignment	28	RV & CSB Module Alignement (15)
Flexure Welding	20	-
CSB Assembly Installation & Alignment Check	6	-
UGS & RV Head Installation & Alignment Check	21	21

Table 6. Comparison of existing RVI installation period and modularization installation period

Fig. 28 shows the RVI modularization installation schedule. To use RVI modularization in an actual construction project, modularization installation schedule was developed. The RVI modularization schedule in the critical path consists of the following steps: (1) CSB module alignment & gap measurement (2) snubber shim machining & installation (3) RV & CSB module alignment (4) and UGS & RV head installation & alignment check. Therefore, it was determined that the RVI modularization installation period should require about 62 days in the critical path.

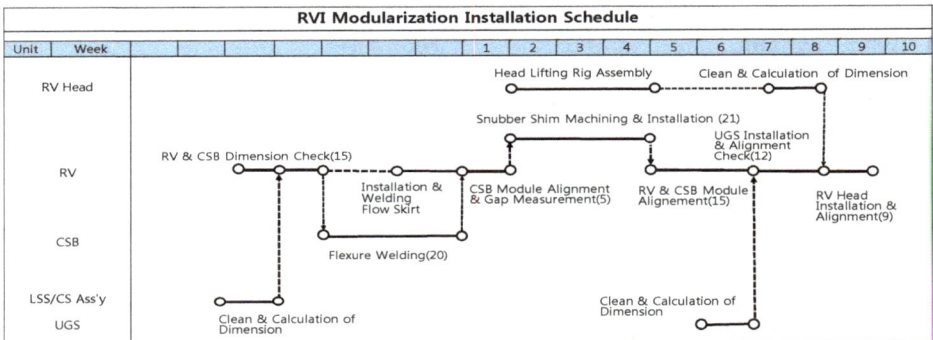

Fig. 27. Existing RVI installation schedule

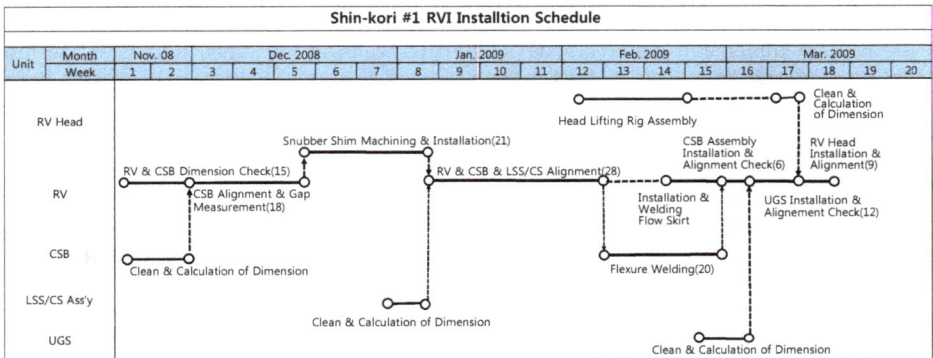

Fig. 28. Developed RVI modularization installation schedule

4.3 Results

An improved installation procedure and schedule for RVI modularization were developed. These developments facilitated a RV & CSB dimension check, flexure welding of the LSS and the CS in the CSB and a CSB assembly installation & alignment check before the main installation process. The new procedure and schedule also facilitated a CSB alignment & gap measurement and a RV & CSB module alignment, as undertaken during the main installation process.

According to the improved installation procedure for RVI modularization, the gaps between the RV core-stabilizing lug and the CSB snubber lug of the RVI mockup are measured by a remote measurement method. The results measured by these methods are analyzed with design shims attached to the assembly between the RV core-stabilizing lug and the CSB snubber lug. After the shims on the RV core-stabilizing lug were installed, The measured gap values satisfied the requirements within the permissible range, 0.381 – 0.508 mm, were found.

4.4. Conclusion

The development of an improved installation procedure and schedule is one of the most important technologies for RVI modularization. The improved installation procedure and schedule developed here can be used at nuclear power plant construction sites. The improved installation procedure through an experiment using a verified remote measurement system was validated and then manufactured mockup.

On the basis of these studies, technologies for RVI modularization using the improved installation procedure and schedule will be applied to the construction project.

5. Acknowledgements

Korea Hydro and Nuclear Power Co., Ltd. applied for international patents on all R&D results in this chapter.

6. References

ABB-CE, (1995). Support Work Agreement Work Order Delivery H-4, Korea Electric Power Corporation.

Ko, D. Y. & Bae, B. H. (2006). A Study on the EMC for Application of Wireless Communication System in Nuclear Power Plants. Conference on Information and Control System, Korea.

Ko, D. Y., Lee, J. G., Kang, Y. C., & Kim, S. H. (2009). Development of a measurement system of gap between CSB and RV to shorten a nuclear reactor installation period. Nuclear Engineering and Design 239, pp.(495-500).

Ko, D. Y. & Lee, J. G. (2010). Development of a Remote Measurement System for the gap between RV and CSB for RVI-modularization. Nuclear Engineering and Design 240, pp. (2912-2918).

Ko, D. Y. (2011). Development of an Improved Installation Procedure and Schedule of RVI Modularization for APR1400. Nuclear Engineering and Technology, Vol. 43 No. 1, pp.(89-98).

Korea Electric Power Research Institute. (1997). Construct Ability Improvement for Nuclear Power Plants, TR-95ZJ02-97-41.

Korea Hydro & Nuclear Power Company. (2002). Construction Study Report for Development of Next Generation Reactor, TR.A99NJ13.P2002.Shin2, pp. 633-672.

Korea Hydro & Nuclear Power Company. (2009). The Development of Modularization Technology for Reactor Internals, TR-S05NJ02-J2009-58, pp. 26-135.

Rechard S. Figliola & Donald E. Beasley. (2000). Theory and Design for Mechanical Measurements, Wiley, pp.(575-580).

Solartron-metrology. (2006). Digital technology digital probes, Solartron-metrology.

Tomas G. Beckwith, Roy D. Marangoni & John H. Lienhard. (1993). Mechanical Measurement, Addison-Wesley Publishing Company, pp.(515-520).

7

Cross-Flow-Induced-Vibrations in Heat Exchanger Tube Bundles: A Review

Shahab Khushnood et al.*

University of Engineering & Technology, Taxila
Pakistan

1. Introduction

Over the past few decades, the utility industry has suffered enormous financial losses because of vibration related problems in steam generators and heat exchangers. Cross-flow induced vibration due to shell side fluid flow around the tubes bundle of shell and tube heat exchanger results in tube vibration. This is a major concern of designers, process engineers and operators, leading to large amplitude motion or large eccentricities of the tubes in their loose supports, resulting in mechanical damage in the form of tube fretting wear, baffle damage, tube collision damage, tube joint leakage or fatigue and creep etc.

Most of the heat exchangers used in nuclear, petrochemical and power generation industries are shell and tube type. In these heat exchangers, tubes in a bundle are usually the most flexible components of the assembly. Because of cross-flow, tubes in a bundle vibrate. The general trend in heat exchanger design is towards larger exchangers with increased shell side velocities, to cater for the required heat transfer capacity, improve heat transfer and reduce fouling effects. Tube vibrations have resulted in failure due to mechanical wear, fretting and fatigue cracking. Costly plant shutdowns have lead to research efforts and analysis for flow- induced vibrations in cross-flow of shell side fluid. The risk of radiation exposure in steam generators used in pressurized water reactor (PWR) plants demand ultimate safety in designing and operating these exchangers.

(Erskine & Waddington, 1973) have carried out a parametric form of investigation on a total of nineteen exchanger failures, in addition to other exchangers containing no failures. They realized that these failures represent only a small sample of the many exchangers currently in service. The heat exchanger tube vibration workshop (Chenoweth, 1976) pointed out a critical problem i.e., the information on flow-induced vibration had mostly been withheld because of its proprietary nature.

---

* Zaffar Muhammad Khan[1], Muhammad Afzaal Malik[2], Zafarullah Koreshi[2],
Muhammad Akram Javaid[1], Mahmood Anwer Khan[3], Arshad Hussain Qureshi[4],
Luqman Ahmad Nizam[1], Khawaja Sajid Bashir[1], Syed Zahid Hussain[1]
[1]*University of Engineering & Technology, Taxila, Pakistan*
[2]*Air University, Islamabad Pakistan*
[3]*College of Electrical & Mechanical Engineering NUST, Rawalpindi, Pakistan*
[4]*University of Engineering & Technology, Lahore, Pakistan*

Failure of heat exchanger tubes in a bundle due to flow-induced vibrations is a deep concern, particularly in geometrically large and highly rated units. Excessive tube vibration may cause failure by fatigue or by fretting wear. Each tube in a bundle is loosely supported at baffles, forming multiple supports often with unequal support spacing. Reactor components like heat exchanger tubes, fuel rods and piping sections may be modeled as beams on multiple supports. It is important to determine whether any of the natural frequencies be within the operating range of frequencies. Considerable research efforts have been carried out, which highlight the importance of the problem.

Tube natural frequency is an important and primary consideration in flow-induced vibration design. A considerable research has been carried out to calculate the natural frequencies of straight and curved (U-tubes) by various models for single and multiple, continuous spans, in air and in liquids for varying end and intermediate support conditions. (Chenoweth, 1976), (Chen & Wambsganss, 1974), (Shin & Wambsganss,1975), (Wambsganss, et al., 1974), (Weaver, 1993), (Brothman, et al., 1974), (Lowery & Moretti, 1975), (Elliott & Pick, 1973), (Jones, 1970), (Ojalvo & Newman, 1964) and (Khushnood et al., 2002), to name some who have carried out research and highlighted the importance of the calculation of natural frequencies of heat exchanger tubes in a bundle.

The dimensionless parameters required for modeling a system may be determined as follows (Weaver, 1993):

- Through non-dimensionalizing the differential equations governing the system behavior.
- From application of Buckingham Pi-theorem.
- This theorem only gives the number of πs, and not a calculation procedure. So we rely on (i) essentially.

(Shin & Wambsganss, 1975), and (Khushnood et al., 2000) gave the basics of model testing via dimensional analysis. (Blevins, 1977) has described non-dimensional variables such as geometry, reduced velocity, dimensionless amplitude, mass ratio, Reynolds number and damping factor as being useful in describing the vibrations of an elastic structure in a subsonic steady flow. However, other non-dimensional variables such as Mach number, capillary number, Richardson number, Strouhal number and Euler number are also useful in case effects such as surface tension, gravity, supersonic flow or vortex shedding are also considered.

It is generally accepted that the tube bundle excitation mechanisms are (Weaver, 1993, Pettigrew et al., 1991)

- Turbulent buffeting
- Vorticity excitation
- Fluid-elastic excitation
- Acoustic resonance

Turbulent buffeting cannot be avoided in heat exchangers, as significant turbulence levels are always present. Vibration at or near shedding frequency has a strong organizing effect on the wake. Vorticity or vortex shedding or periodic wake shedding is a discrete, periodic, and a constant Strouhal number phenomenon. Strouhal number is the proportionality constant between the frequency of vortex shedding and free stream velocity divided by cylinder width. Fluid-elastic instability is by far the most dangerous excitation mechanism and the most common cause of tube failure. This instability is typical of self-excited vibration in that it results from the interaction of tube motion and flow. Acoustic resonance is caused by some flow excitation (possibly vortex shedding) having a frequency, which coincides with the natural frequency of the heat exchanger cavity.

With regard to dynamic parameters, including added mass and damping, the concept of added mass was first introduced by DuBuat in 1776 (Weaver, 1993). The fluid oscillating with the tube may have an appreciable affect on both natural frequency and mode shape. Added mass is a function of geometry, density of fluid and the size of the tube (Moretti & Lowry, 1976). Several studies including (Weaver, 1993, Lowery, 1995, Jones, 1970, Chen et al., 1994, Taylor et al., 1998, Rogers et al., 1984, Noghrehkar et al., 1995, Carlucci, 1980, Pettigrew et al., 1994, Pettigrew et al., 1986, Zhou et al., 1997) have targeted damping in heat exchanger tube bundles in single-phase and two-phase cross-flow. (Rogers et al., 1984) have given identification of seven separate sources of damping.

(Ojalvo & Newman, 1964) have presented design for out-of-plane and in-plane frequency factors for various modes. (Jones, 1970) carried out experimental and analytical analysis of a vibrating beam immersed in a fluid and carrying concentrated mass and rotary inertia. (Erskine & Waddington, 1973) conducted parametric form of investigation on a total of 19 exchanger failures along with other exchangers containing no failures, for comparative purpose, indicated the incompleteness of methods available till then and emphasized the need for a fully comprehensive design method. Finite element technique applied by (Elliott & Pick, 1973), concluded that the prediction of natural frequencies was possible with this method and that catastrophic vibrations might be prevented by avoiding matching of material and excitation frequencies. Lack of sufficient data to support comprehensive analytical description for several fundamentally different vibration excitation mechanisms for tube vibration have been indicated in Ontario Hydro Research Division Report (Simpson & Hartlen, 1974). The report also gives response in terms of mid-span amplitude to a uniformly distributed lift for a simply supported tube. A simple graphical method for predicting the in-plane and out-of-plane frequencies of continuous beams and curved beams on periodic, multiple supports with spans of equal length have been presented by (Chen & Wambsganss, 1974). They have given design guidelines for calculating natural frequencies of straight and curved beams. (Wambsganss, et al., 1974) have carried out an analytical and experimental study of cylindrical rod vibrating in a viscous fluid, enclosed by a rigid, concentric cylindrical shell, obtaining closed-form solution for added mass and damping coefficient. (Shin & Wambsganss, 1975) have given information for making the best possible evaluation of potential flow-induced vibration in LMFBR steam generator focusing on tube vibration. A simple computer program developed by (Lowery & Moretti, 1975), calculates frequencies of idealized support with multiple spans. (Chenoweth, 1976), in his final report on heat exchanger tube vibration, pointed out the slow progress and inadequacy of existing methods and a need for field data to test suitability of design procedures. It stressed the need for testing specially built and instrumented industrial- sized heat exchangers and wind tunnel based theories to demonstrate interaction of many parameters that contribute to flow-induced vibrations. (Rogers et al., 1984) have modeled mass and damping effects of surrounding fluid and also the effects of squeeze film damping. (Pettigrew et.al., 1986) have treated damping of multi-span heat exchanger tubes in air and gases in terms of different energy dissipation mechanisms, showing a strong relation of damping to tube support thickness.

(Price, 1995) has reviewed all known theoretical models of fluid-elastic instability for cylinder arrays subject to cross-flow with particular emphasis on the physics of instability mechanisms. Despite considerable difference in the theoretical models, there has been a general agreement in conclusions. (Masatoshi et al., 1997) have carried tests on an intermediate heat exchanger with helically coiled tube bundle using a partial model to investigate the complicated vibrational behavior induced by interaction through seismic stop between center pipe and tube bundle. They have indicated the effect of the size of gap between seismic stop and tube support of the bundle.(Botros & Price, 2000) have carried out a study of a large heat exchanger tube bundle of styrene monomer plant, which experienced severe fretting and leaking of tubes and considerable costs associated with operational shutdowns. Analysis through Computational Fluid Dynamics (CFD) and fluid-elastic instability study resulted in the replacement of a bundle with shorter span between baffles, and showed no signature of vibration over a wide range of frequencies. (Yang, 2000) has postulated that crossing-frequency can be used as a measure of heat exchanger support plate effectiveness. Crossing-frequency is the number of times per second the vibrational amplitude crosses the zero displacement line from negative displacement to positive displacement.

The wear of tube due to non-linear tube-to-tube support plate (TSP) interactions is caused by the gap clearances between the two interacting components. Tube wall thickness loss and normal work-rates for different TSP combination studies have been the target. Electric Power Research Institute (EPRI), launched an extensive program in early 1980's for analyses of fluid forcing functions, software development and studying linear and non-linear tube bundle dynamics. Other studies include (Rao et al., 1988), (Axisa & Izquierdo, 1992), (Payen et al., 1995), (Peterka, 1995), (Hassan et al., 2000), (Charpentier and Payen, 2000) and (Au-Yang, 1998).

Generally, there are three geometric configurations in which tubes are arranged in a bundle. These are triangular, normal square and rotated square. Relatively little information exists on two-phase cross-flow induced vibration. Not surprisingly as single-phase flow-induced vibration is not yet fully understood. Vibration in two-phase is much more complex because it depends upon two-phase flow regime and involves an important consideration, the *void fraction*, which is the ratio of volume of gas to the volume of the liquid gas mixture. Two-phase flow experimentation is much more expensive and difficult to carry out usually requiring pressurized loops with the ability to produce two-phase mixtures of desired void fractions.

Two-phase flow research includes the models, such as, Smith Correlation (Smith, 1968), drift-flux model developed by (Zuber and Findlay, 1965), Schrage correlation (Schrage, 1988), and Feenstra model (Feentra et al., 2000). (Frick et al., 1984) has given an overview of tube wear-rate in two-phase flow. (Pettigrew et al., 2000), (Mirza & Gorman, 1973), (Taylor et al.,1989), (Papp, 1988), (Wambsganss et al., 1992) and others have carried out potential research for vibration response. Earlier reviews on two-phase cross flow are provided by (Paidoussis, 1982), (Weaver & Fitzpatrik, 1988), (Price, 1995), and (Pettigrew & Taylor, 1994).

Two-phase cross-flow induced vibration in tube bundles of process heat exchangers and U-bend region of nuclear steam generators can cause serious tube failures by fatigue and fretting wear. Tube failures could force entire plant to shut down for costly repairs and suffering loss of production. Vibration problems may be avoided by thorough vibration analysis. However, this requires an understanding of vibration excitation and damping mechanism in two-phase flow. A number of flow regimes (Table 1) can occur for a given boundary configuration, depending upon the concentration and size of the gas bubbles and on the mass flow rates of the two-phases. Two-phase (khushnood, et al., 2004) flow characteristics greatly depend upon the type of flow occurring.

Flow Type	Average Void Fraction	Specification
Bubble	~0.3	Some bubbles are present in liquid flow and move with the same velocity.
Slug	0.3-0.5	Liquid slugs flow intermittently.
Froth	0.5-0.8	More violent intermittent flow.
Annular	0.8-0.9	Mainly gas flow. Liquid adheres to the tube surface.
Mist	~0.9	Almost gas flow. Mist sometimes causes energy dissipation.

Table 1. Types of Flow in Two-Phases (khushnood, et al., 2004)

Vibration of tube in two-phase flow displays different flow regimes i.e., gas and liquid phase distributions, depending upon the void fraction and mass flux. It is known that four mechanisms are responsible for the excitation of tube arrays in cross-flow (Pettigrew, et al., 1991). These mechanisms are: turbulence buffeting, vortex shedding or Strouhal periodicity, fluid-elastic instability and acoustic resonance. Table 2 presents a summary of these vibration mechanisms for single cylinder and tube bundles for liquid, gas and liquid-gas two-phase flow respectively. Of these four mechanisms, fluid-elastic instability is the most damaging in the short term, because it causes the tubes to vibrate excessively, leading to rapid wear at the tube supports. This mechanism occurs once the flow rate exceeds a threshold velocity at which tubes become self-excited and the vibration amplitude rises rapidly with an increase in flow velocity.

Flow Situation (Cross-Flow)	Fluid-Elastic Instability	Periodic Shedding	Turbulence Excitation	Acoustic Resonance
Single Cylinder				
Liquid	o	*	*	o
Gas	o	Δ	Δ	o
Two-phase	o	o	*	o
Tube Bundle				
Liquid	*	Δ	Δ	o
Gas	*	o	Δ	*
Two-phase	*	o	*	o

Unlikely	o
Possible	Δ
Most Important	*

Table 2. Vibration Excitation Mechanisms (Pettigrew, et al., 1991)

Typically, researchers have relied on the Homogeneous Equilibrium Model (HEM) (Feentra et al., 2000) to define important fluid parameters in two-phase flow, such as density, void fraction and velocity. This model treats the two-phase flow as a finely mixed and homogeneous in density and temperature, with no difference in velocity between the gas and liquid phases. This model has been used a great deal because it is easy to implement and is widely recognized which facilitated earlier data comparison. Other models include Smith correlation (Smith, 1968), drift-flux model developed by (Zuber and Findlay, 1965), Schrage Correlation (Schrage, 1988), which is based on empirical data, and Feenstra model (Feentra et al., 2000), which is given in terms of dimensionless numbers.

Dynamic parameters such as added or hydrodynamic mass and damping are very important considerations in two-phase cross-flow induced vibrations. Hydrodynamic mass depends upon pitch-to-diameter ratio and decrease with increase in void fraction. Damping is very complicated in two-phase flow and is highly void fraction dependent. Tube-to-restraint interaction at the baffles (loose supports) can lead to fretting wear because of out of plane impact force and in-plane rubbing force. (Frick et al., 1984) has given an overview of the development of relationship between work-rate and wear-rate. Another important consideration in two-phase flow is the random turbulence excitation. Vibration response below fluid-elastic instability is attributed to random turbulence excitation.

(Pettigrew et al., 2000), (Mirza & Gorman, 1973), (Taylor et al.,1989), (Papp, 1988), and (Wambsganss et al., 1992) to name some, have carried out research for Root Mean Square (RMS) vibration response, encompassing spatially correlated forces, Normalized Power Spectral Density (NPSD), two-phase flow pressure drop, two-phase friction multiplier, mass flux, and coefficient of interaction between fluid mixture and tubes. More recently researchers have expanded the study to two-phase flow which occur in nuclear steam generators and many other tubular heat exchangers, a review of which was last given by (Pettigrew & Taylor, 1994). A current review on this topic is given by (Khushnood et al., 2004)

The use of Finite Element Method (FEM), Computational Fluid Dynamics (CFD) and Large Eddy Simulation (LES) have proved quite useful in analyzing flow-induced vibrations in tube bundles in recent years. Earlier on, only pressure drop and heat transfer calculations were considered as the basis of heat exchanger design. Recently, flow-induced tube vibrations have also been included in the design criteria for process heat exchangers and steam generators.

1.1 Regimes

(Kim et al., 2009) have carried flow induced vibrations (Experimental study of two circular cylinders in tandem arrangement) and examined three different experimental conditions both cylinders allowed to vibrate, the upstream cylinder is allowed to vibrate with the downstream cylinder fixed and downstream cylinder allowed to vibrate with upstream cylinder fixed. The results include five regimes depending upon L/D, fluctuating lift forces and vibration characteristics of the cylinder as given in Table 3.

Regimes	I	II	III	IV	V
Range	$0.1 \leq L/D \geq 0.2$	$0.2 \leq L/D \geq 0.6$	$0.6 \leq L/D \geq 2.0$	$2.0 \leq L/D \geq 2.7$	$L/D \geq 2.7$
Response	Vibration absent	Violent vibrations of both For $U_r > 6$	Convergent vibrations at $U_r \approx 6.7$	Vibration absent	Each vibrating like isolated cylinder at $U_r \approx 6.7$
Characteristics		Vibration amplitude is strongly dependant on whether upstream cylinder is fixed or vibrating	Upstream cylinder vibration is completely suppressed when downstream cylinder is fixed but the downstream cylinder is independent of upstream cylinder.		Downstream vibration is strongly dependant on upstream cylinder but upstream cylinder vibrations is insensitive to downstream cylinder.

Table 3. Regimes of vibration for Circular cylinders tandem (Kim, et al., 2009)

2. Excitation mechanisms

2.1 Fluid-elastic instability

Fluid-elastic instability is by far the most dangerous excitation mechanism in heat exchanger tube bundles and the most common cause of tube failures. The forces associated with fluid- elastic instability exist only because of the motion of the body. (Price, 1995) has presented comprehensive review on fluid-elastic instability of cylinder arrays in cross-flow. According to Price, the nature of fluid-elastic instability can be illustrated as a feedback mechanism between structural motion and the resulting fluid forces. A small structural displacement due to turbulence alters the flow pattern, inducing a change in fluid forces. This in turn leads to a further displacement, and so on. If the displacement increases (positive feedback), then fluid-elastic instability occurs. Three mechanisms (Price, 1995), which enable the cylinder to extract energy from flow:

- require a phase difference between cylinder displacement and fluid force generated.
- relies on there being at least two-degrees of freedom with a phase difference between them.
- because of non-linearities, the fluid force is hysteretic and its magnitude depends on the direction of cylinder motion.

A considerable theoretical and experimental research has been undertaken in the past three decades to arrive at a safe and reliable design criteria against fluid-elastic instability. The topic has been reviewed on regular basis from time to time by various researchers including (Paidoussis, 1980, 1981, 1987, 1987), (Chen, 1984, 1987, 1987, 1989), (Zukauskas et al., 1987), (Weaver & FitzPatrick, 1988), (Moretti, 1993) and (Price, 1995).

2.2 Fluid-elastic instability models

2.2.1 Jet switch model

(Roberts, 1962, 1966) considered both a single and a double row of cylinders normal to flow. His analysis was limited to in-flow motion (experiments indicated that instability was purely in the in-flow direction). Roberts assumed that the flow downstream of two-adjacent cylinders could be represented by two wake regions, one large and one small, and a jet between them as shown in Figure 1.

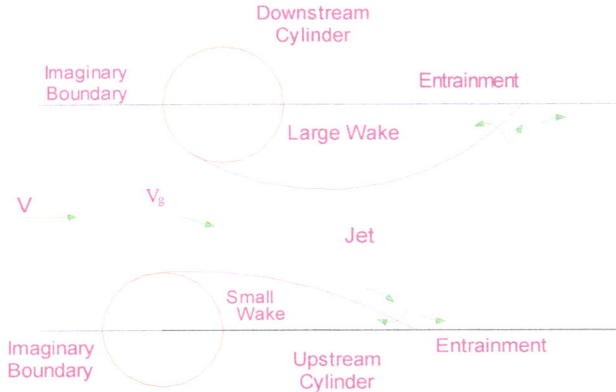

Fig. 1. Idealized model of the jet-flow between two cylinders in a staggered row of cylinders (Roberts, 1962).

Considering a downstream cylinder moving upstream; as the two cylinders cross, insufficient fluid flows in to the large wake region to maintain the entrainment, causing the wake to shrink and the jet to switch directions. Roberts has given the flow equation of motion for a cylinder or tube in a row.

$$\frac{d^2x}{d\tau^2} + 2\zeta\frac{dx}{d\tau} + x = \frac{\rho V^2}{2m\omega_n^2}\left\{0.717\left(1 - C_{pb}(x,\tau)\right) - 2\left(\frac{\omega_n D}{V}\right)\left(1 - C_{pb}\right)_{mean}\frac{dx}{d\tau}\right\} \tag{1}$$

where C_{pb} is the base pressure coefficient, τ is non-dimensional time $(t\omega_n)$, D is the tube diameter, ω_n is the natural frequency, x is the in-flow cylinder displacement, ζ is the damping factor or ratio, δ is the logarithmic decrement, and m is mass of the tube. Equation 1 was solved using the method of Krylov and Bogoliubov (Minorsky, 1947) giving V_c, the velocity just sufficient to initiate limit cycle motion for any $m\delta / \rho D^2$. Neglecting unsteady terms and fluid damping, the solution reduces to

$$\frac{V_c}{\omega_n \varepsilon D} = K\left(\frac{m\delta}{\rho D^2}\right)^{0.5} \tag{2}$$

where ε is the ratio of fluid-elastic frequency to structural frequency, which is approximately 1.0 and ρ is the fluid density. This has the same form as the classical Connors equation (Blevins, 1979). Figure 2 presents Robert's experimental data for pitch-to-dia. ratio ($P / D = 1.5$), showing a good agreement with this theoretical model.

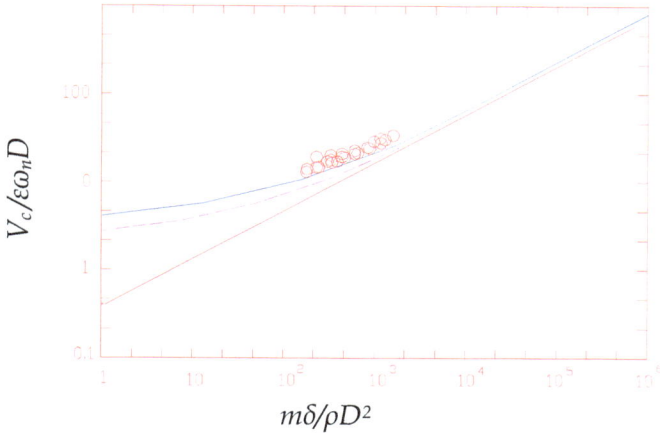

—— Solution including time for jet reversal and aerodynamic damping
- - - Solution assuming instantaneous jet reversal but still including aerodynamic damping
___ Solution assuming instantaneous jet reversal and neglecting aerodynamic damping
O Roberts' experimental results

Fig. 2. Theoretical stability boundary for fluid-elastic instability obtained by Roberts for a single flexible cylinder in a row of cylinders (Roberts, 1966).

2.2.2 Quasi-static models

Using a quasi-static analysis, (Connors & Parrondo, 1970) and later (Dalton & Helfinstine, 1971) developed the fluid-elastic instability prediction for cylinders (single row of cylinders) subjected to cross-flow. Connors measured the fluid forces instead of predicting these using pitch to dia. ratio of $P/D=1.41$. He observed many different model patterns at instability, but suggested that the most dominant was elliptical motion (whirling). Using the measured fluid stiffness, Connors obtained energy balances in the in-and cross-flow directions, which must be satisfied simultaneously giving

$$\frac{V_{pc}}{f_n D} = K \left(\frac{m\delta}{\rho D^2} \right)^{0.5} \tag{3}$$

where K is the so-called Connors constant, f_n is the frequency of oscillation. V_{pc} is the so-called pitch velocity given by

$$V_{pc} = \frac{VP}{P - D} \tag{4}$$

where P being the centre-to-centre inter cylinder pitch

(Blevins, 1974) has derived Equation 3 by assuming that the fluid forces on any cylinder are due to relative displacements between itself and its neighboring cylinders. Later, (Blevins, 1979) modified his original analysis to account for flow dependent fluid damping giving

$$\frac{V_{pc}}{f_n D} = K \left[\frac{m}{\rho D^2} 2\pi \, (\zeta_x \zeta_y)^{1/2} \right]^{1/2} \tag{5}$$

where ζ_x and ζ_y are total damping factors in the in-and cross-flow directions.

2.2.3 In viscid model

Despite the obviously viscous nature of the interstitial flow through arrays of cylinders, the compactness of some arrays suggests that the cylinder wake regions are small, especially for normal triangular arrays with small P/D (Price, 1995). Hence under this assumption wake regions are neglected and flow is treated as inviscid. Many solutions based upon potential flow theory have been given, including (Dalton & Helfinstine, 1971), (Dalton, 1980), (Balsa, 1977), (Paidoussis et al., 1984), (Vander Hoogt & Van Compen, 1984) and (Delaigue & Planchard, 1986). The results obtained from potential flow analyses are somewhat discouraging (Price, 1995). Recent flow visualizations suggest that even though the wake regions are small, the interstitial flow is more complex than that accounted for in these analyses.

2.2.4 Unsteady models

The unsteady models measure the unsteady forces on the oscillating cylinder directly. (Tanaka & Takahara, 1980, 1981) and (Chen, 1983) have given theoretical stability boundary for fluid-elastic instability as shown in Figures 3 and 4 respectively.

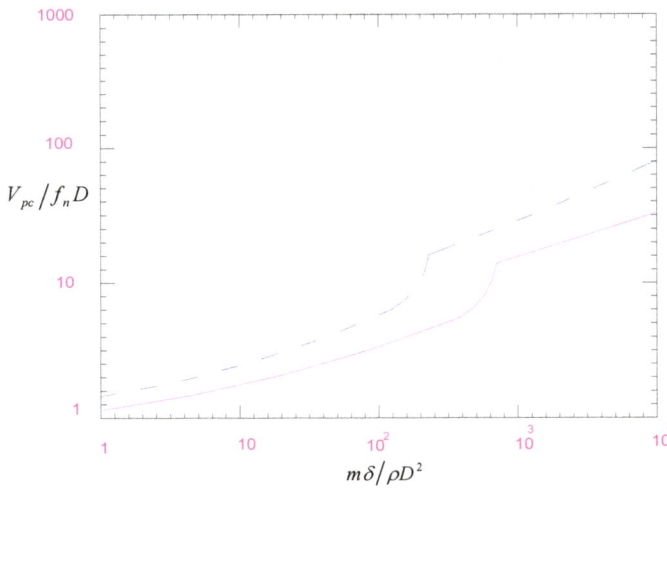

___ δ =0.01
- - - δ =0.03

Fig. 3. Theoretical stability boundary for fluid-elastic instability for an in-line square array, P/D=1.33, obtained by (Tanaka & Takahara, 1980, 1981).

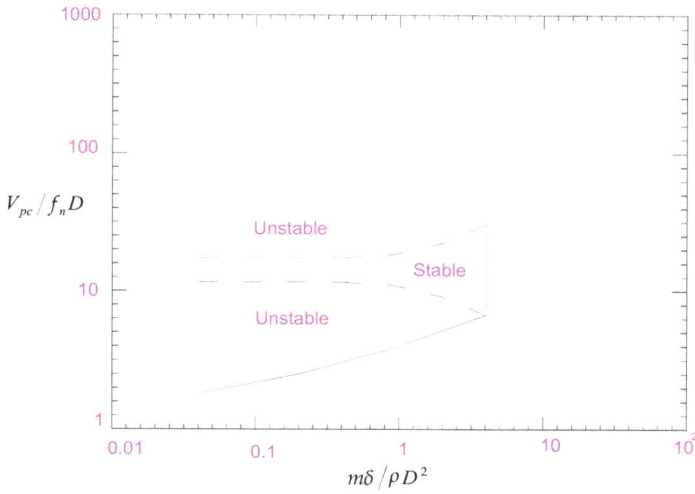

- - -Theoretical solution showing multiple instability boundaries
_____Practical stability boundary

Fig. 4. Theoretical stability for fluid-elastic instability predicted by (Chen, 1983), for a row of cylinders with P/D=1.33.

2.2.5 Semi-analytical model

Out of many semi analytical models, Figure 5 shows theoretical stability boundary for fluid-elastic instability obtained by (Lever & Weaver, 1986).

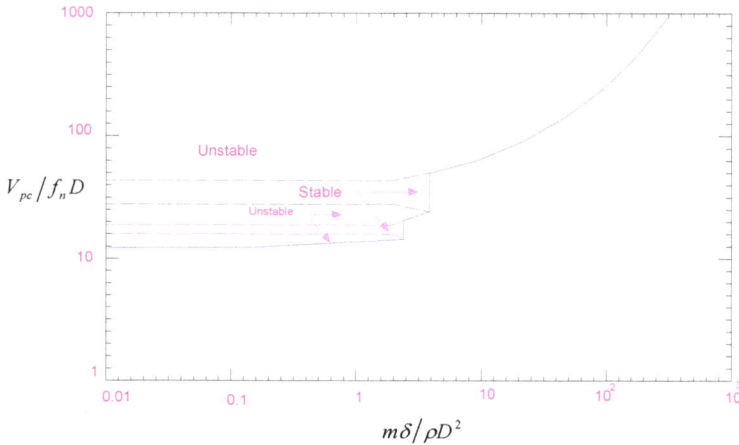

- - -Theoretical solution showing multiple instability boundaries
_____Practical stability boundary

Fig. 5. Theoretical stability boundary for fluid-elastic instability obtained by (Lever & Weaver, 1986) for single flexible cylinder in a parallel triangular array with P/D=1.375.

2.2.6 Quasi-steady model

(Price, 1995) remarks that Fung and Blevins have concluded that quasi-steady fluid dynamics is valid provided $\dfrac{V}{f_n D} \geq 10$; however, experiments by Price, Paidoussis and Sychterz and others suggest that for closely spaced bodies the restriction on the use of quasi-steady fluid dynamics is much more severe than that suggested by Fung or Blevins. (Gross, 1975) carried out first quasi-steady analysis of cylinder arrays subjected to cross-flow concluding that instability in cylinder arrays is due to two distinct mechanisms: negative damping and stiffness controlled instability.

2.2.7 Computational fluid dynamic (CFD) models

The CFD solutions applicable to fluid-elastic instability and other problem areas of flow-induced vibrations are increasing. These include (Marn and Catton, 1991) and (Planchard & Thomas, 1993).

2.2.8 Non-linear models

The first non-linear model was given by (Roberts, 1962, 1966), who employed Krylov and Bogoliubov method (Minorsky, 1947) of averaging to solve the non-linear equations. Two-motivating forces have been remarked by (Price, 1995) for non-linear analyses. Firstly because of manufacturing tolerances and thermal constraints, there are likely to be small clearances between heat exchanger tubes and intermediate supports. Hence, large lengths of unsupported tubes, having very low natural frequencies. These low-frequencies may suffer from fluid-elastic instability at relatively low V_{pc}. A second and more academic motivating force for these non-linear analysis has been to investigate the possibility of Choatic behavior of tube motion.

2.2.9 Recent researches in fluid-elastic instability

A summary of some recent efforts on the analysis of fluid-elastic instability in heat exchanger and steam generator tube bundles is given in Table 4.

Researchers	Flow (phase)	Analysis type	Frequency	Span type	Model type	Remarks
(Hassan et al., 2011)	Single phase	Simulation (linear/ non-linear)	Up to 90Hz approx.	Loosely supported multispan	Comparison with several time domains fluid force model	Tube supports interaction parameters Impact Force Contact rates Normal wave rate considered.

Researchers	Flow (phase)	Analysis type	Frequency	Span type	Model type	Remarks
(Sim & Park, 2010)	Two phase	Experimental test section consists of flexible and rigid cylinders	Frequency Range 8.25-12 Hz	Cantilevered flexible cylinders	Normal square tube bundles	Dimensionless flow velocity and mass-damping parameter considerations and fluid-elastic instability coefficients considerations
(Ishihara & Kitayama, 2009)	Single phase	Experimental		Tube banks such as boilers and heat exchangers in power plant	Experimental	Onset of fluid-elastic instability and geometry relationship considerations
(Mitra et al., 2009)	Single & two phase (Air-water & air-steam flow)	Experimental	Frequency range 7.6 - 13.74 Hz		Fully flexible tube arrays and single flexible tube (Normal square array)	Displacement and damping mechanisms Critical flow velocity was found proportional to tube arrays
(Mahon & Meskell, 2009)	Single phase	Experimental $P/D = 1.32$	Excitation frequency 6.62 Hz	Second array flexible tube with electro-magnetic damper	Normal Triangular	Time delay considerations
(Hassan & Hayder, 2008)	Single phase	Modeling and simulation (Linear/ Non-linear)	Up to 60 Hz		Time domain modeling of tube forces	Critical velocity predictions dependent upon i.e. sensitive to both gap size and turbulence level
(Chung & Chu, 2006)	Two phase Void Fraction 10-95%	Experimental $P/D = 1.633$ $100 m^3/hr$ 50 m Water Head	Strouhal number 0.15-0.19	Cantilevered straight tube bundles	Experimental	Hydro dynamic coupling effects consideration

Researchers	Flow (phase)	Analysis type	Frequency	Span type	Model type	Remarks
(Mureithi et al., 2005).	Single phase 0.44% damping	Experimental Wind tunnel partially fixed flexible array $P/D = 1.633$	18.74 Hz	Preferentially flexible	Rotated triangular array	Investigation of stability behavior and AVB's considerations

Table 4. A summary of recent fluid-elastic instability research

2.3 Vorticity induced instability

Flow across a tube produces a series of vortices in the downstream wake formed as the flow separates alternatively from the opposite sides of the tube. This shedding of vortices produces alternating forces, which occur more frequently as the velocity of flow increases. For a single cylinder, frequency of vortex shedding f_{vs} is given below by a dimensionless Strouhal number S.

$$f_{vs} = \frac{SV}{D} \qquad (6)$$

where V is the flow velocity and D is the tube diameter. For a single cylinder, the vortex shedding Strouhal number is a constant with a value of about 0.2 (Chenoweth, 1993). Vortex shedding occurs for the range of Reynolds number $100 < R_e < 5 \times 10^5$ and $> 2 \times 10^6$ whereas it dies out in-between. The gap is due to a shift of the flow separation point in vortices in the intermediate transcritical Reynolds number range. Vortex shedding can excite tube vibration when it matches with the natural frequency of the tubes. For tube banks with vortex shedding, Strouhal number is not constant, but varies with the arrangement and spacing of tubes, typical values for in-line and staggered tube bundle geometries are given in (Karaman, 1912, Lienhard, 1966). Strouhal numbers for in-line tube banks are given in Figure 6.

The vortex shedding frequency can become locked-in to the natural frequency of a vibrating tube even when flow velocity is increased (Blevins, 1977). Earlier on, the mechanism of vortex shedding has been investigated by a number of researchers. These include Sipvack (Sipvack, 1946) and, Thomas and Kraus (Thomas & Kraus, 1964) who investigated the vortex shedding of two cylinders arranged parallel and perpendicular to flow direction respectively. Grotz and Arnold (Groth & Arnold, 1956) measured for the first time systematically the vortex shedding frequencies in in-line tube bank for various tube spacing ratios.

The cause of vorticity excitation has been disputed in literature (Owen, 1965), but recent studies of (Weaver, 1993) and, (Oengoren & Ziada, 1993) have confirmed its cause of existence as periodic vortex formation. Vorticity shedding can cause tube resonance in liquid flow or acoustic resonance of the tube bundles or acoustic resonance of the tube bundles' containers in gas flows (Oengoren & Ziada, 1995). (Chen, 1990), (Zaida & Oengoren, 1992) and (Weaver, 1993) have summarized the recent research efforts targeted at improvement in Strouhal number charts for vortex shedding and acoustic resonance for in-line tube bundles.

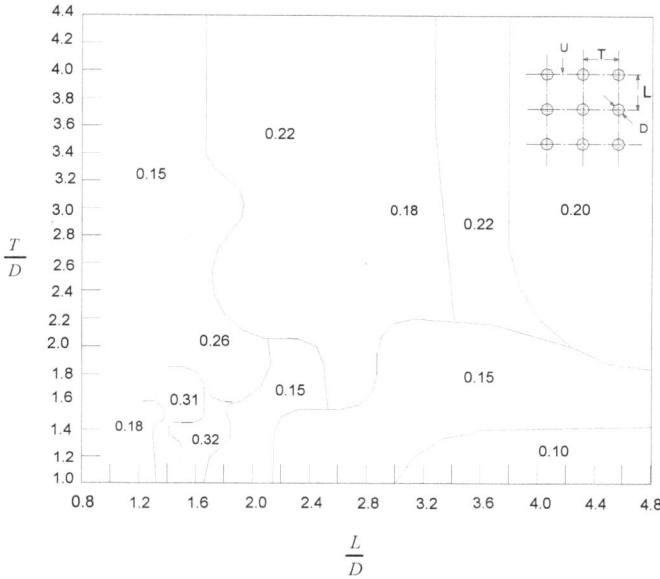

Fig. 6. Strouhal numbers for in-line tube banks (Karaman, 1912).

(Oengoren and Ziada, 1992) have investigated the coupling between the acoustic mode and vortex shedding, which may occur near the condition of frequency coincidence. They have investigated the system response both in the absence and in the presence of a splitter plate, installed at the mid-height of the bundle to double the acoustic resonance frequencies and therefore double the Reynolds number at which frequency coincidence occurs. They have also investigated the effect of row number on vortex shedding and have carried out flow visualization in Reynolds number range of ≤ 355000. Figure 7 is a typical example of the mechanism of vortex shedding from the tubes of the first two rows displaying a time series of symmetric and anti-symmetric patterns (Oengoren & Ziada, 1993).

(Liang et al., 2009) has addressed numerically the effect of tube spacing on vortex shedding characteristics of laminar flow past an inline tube arrays. The study employs a six row in-line tube bank for eight pitch to diameter $(P/_D)$ ratios with Navier-Strokes continuity equation based unstructured code (validated for the case of flow past two tandem cylinders) (Axisa & Izquierdo, 1992) . A critical spacing range between 3.0 and 3.6 is identified at which mean drag as well as rms lift and drag coefficients for last three cylinders attain maximum values. Also at critical spacing, there is 180° phase difference in the shedding cycle between successive cylinders and the vortices travel a distance twice the tube spacing within one period of shedding.

(Williamson & Govardhan, 2008) have reviewed and summarized the fundamental results and discoveries related to vortex induced vibrations with particular emphasis to vortex dynamics and energy transfer which give rise to the mode of vibrations. The importance of mass and damping and the concept of "critical mass", "effective elasticity" and the relationship between force and vorticity. With reference to critical mass, it is concluded that as the structural mass decreases, so the regime of velocity (non-dimensional) over which there is large amplitude of vibrations increases. The synchronizing regime become infinitely wide not simply when mass become zero but when a mass falls below special critical value when the numerical value depends upon the vibrating body shape.

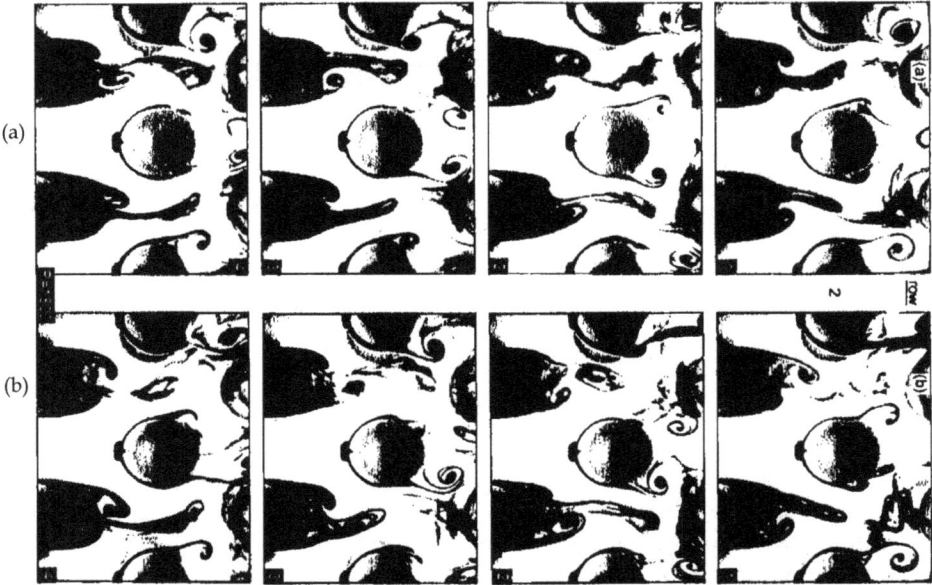

Fig. 7. Time sequence of the two transient modes of vortex shedding, (a) symmetric and (b) anti-symmetric, behind the first two rows of the intermediate spacing array (Leinhard, 1966).

(Williamson & Govardhan, 2000) present a large data set for the low branch frequency f^*_{lower} plotted versus m^* (mass ratio) yielding a good collapse of data on to single curve base equation 7.

$$f_{lower} = \sqrt{\frac{m^*+1}{m^*-0.54}} \tag{7}$$

This equation provides a practical and simple means to calculate the frequency attained by vortex induced vibrations. The critical mass ratio is given by

$$m^*_{crit} = 0.54 \pm 0.002 \tag{8}$$

Below which the lower branch of response can never be attained. With respect to combine mass-damping parameter's capability to reasonably collapse peak amplitude data in Griffins plot, a number of parameters like stability parameter, Scrutom number and combined response parameter termed as Skop-Griffins parameter given by (S_G):

$$S_G = 2\pi^3 S^2 (m^*) \tag{9}$$

Where S stands for single vortices and S_c is the Scruton number.

(Hamakawa & Matsue, 2008) focused on relation between vortex shedding and acoustic resonance in a model (boiler plant) for tube banks to clarify the interactive characteristics of vortex shedding and acoustic resonance. Periodic velocity fluctuation due to vortex shedding was noticed inside the tube banks at the Reynolds number (1100-10000) without acoustic resonance and natural vortex shedding frequency of low gap velocities. Kumar et al., 2008 in their review stated that controlling or suppressing vortex induced vibrations is of importance in practical applications where active or passive control could be applied.

(Paidoussis, 2006) specially addressed real life experiences in vortex induced vibrations and concludes with this mechanism in addition of other already clarified mechanisms of flow induced vibrations. Vortex induced vibrations of ICI nozzles and guide tubes in PWR for ICI thimble guiding into the core of the reactor to monitor reactivity may witness breakage of ICI nozzles resulting in strange noises experience in the reactor. Analysis of shedding frequencies confirmed the vortex induced vibrations to be the culprit partially due to the large values of varying lift coefficients and partially due to lock-in.

(Hamakawa & Fukano, 2006) also focused vortex shedding in relation with the acoustic resonance in staggered tube banks and observe three Strouhal number (0.29, 0.22 and 0.19). In cases with no resonance inside tube banks, the last rows of tube banks and in both regimes respectively. The vortices of 0.29 and 0.22 components alternatively irregularly originated.

(Pettigrew & Taylor, 2003) discussed and overviewed procedures and recommended design guidelines for periodic wave shedding in addition to other flow induced vibration considerations for shell and tube heat exchangers. It concludes that the fluctuating forces due to periodic wave shedding depends on the number of considerations like geometric configuration of tube bundles, its location, Reynolds number, turbulence, density of fluid and pitch to diameter P/D ratio.

2.4 Turbulence excitation

Extremely turbulent flow of the shell-side fluid contains a wide spectrum of frequencies distributed around a central dominant frequency, which increases as the cross-flow velocity increases. This turbulence buffets the tubes, which extract energy from the turbulence at their natural frequency from the spectrum of frequencies present. When the dominant frequency for the turbulent buffeting matches the natural frequency, a considerable transfer of energy is possible leading to significant vibration amplitudes (Chenoweth, 1993). Turbulent flow is characterized by random fluctuation in the fluid velocity and by intense mixing of the fluid. Nuclear fuel bundles and pressurized water reactor (PWR) steam generators are existing examples (Hassan & Ibrahim, 1997).

Turbulence is by nature three-dimensional (Au-Yang, 2000). Large-Eddy Simulation, (LES) incorporated in three-dimensional computer codes has become one of the promising techniques to estimate flow turbulence. (Hassan & Ibrahim, 1997) & (Davis & Hassan, 1993) have carried out Large Eddy Simulation for turbulence prediction in two-and three-dimensional flows. The primary concern in turbulence measurements is how the energy spectrum or the power spectral density (PSD) of the eddies are distributed. The PSD of the velocity profile E(n) is numerically equal to the square of the Fourier Transform of $U'(t)$, and is defined to be (Hassan & Ibrahim, 1997).

$$\overline{U'^2} = \int\limits_{-\infty}^{+\infty} E(n)dn \tag{10}$$

where $E(n)$ is the sum of power at positive and negative frequency n.

$$E(n) = \frac{4\pi^2 |a(n)|^2}{T} \tag{11}$$

where T is the time period over which integration is performed, and $a(n)$ is the Fourier Transform coefficient.

An important parameter of flow turbulence is the correlation function. The Lagrangian (temporal) auto-correlation over a time T gives the length of time (past history) that is related to a given event (Hassan & Ibrahim, 1997).

$$(\text{Non-dimensional}) \quad R(\tau) = \frac{\overline{U'(t)U'(t+\tau)}}{\overline{U'(t)U'(t)}} \tag{12a}$$

$$(\text{Dimensional}) \quad R(\tau) = \lim_{\tau \longrightarrow \infty} \frac{1}{T} \int\limits_{t=0}^{t=T} U'(t)U'(t+\tau)dt \tag{12b}$$

Physically $R(\tau)$ represents the average of the product of fluctuating velocity U' values at a given time and at a time τ later. $R(\tau)$ gives information about whether and for how long the instantaneous value of U' depends on its previous values. Cross-correlation curves can also be obtained as a function of the time delay to give the correlation between the velocities at consecutive separated location points (Owen, 1965).

$$R_{12}(\tau) = \frac{1}{T} \int\limits_{t=0}^{t=T} U_1'(t)U_2'(t+\tau)dt \tag{13}$$

where R_{12} gives the cross-correlation of the U-velocity component at 1- and 2- point locations.

Recently (Au-Yang, 2000) has reviewed the acceptance integral method to estimate the random vibration, Root Mean Square (RMS) of structures subjected to turbulent flow (random forcing function). The acceptance integral is given by:

$$J_{\alpha\beta}(\omega) = \frac{1}{\ell} \int_0^\ell \int_0^\ell \phi_\alpha(x') \left[S_p(x',x'',\omega) / S_p(x',\omega) \right] \phi_\beta(x'')dx'dx'' \tag{14}$$

When $\alpha = \beta$, $J_{\alpha\alpha}$ is known as joint acceptance

where

$J_{\alpha\alpha}$	=	Joint acceptance for α_{th} mode
$J_{\alpha\beta}$	=	Cross-acceptance
ℓ	=	Surface of 2-D structure of length of 1-D structure
x	=	Position vector
S_p	=	Double sided pressure power spectral density.
ϕ_α	=	Mode shape function
ϕ_β	=	Mode shape function
ω	=	Frequency
α, β	=	Modal indices

Yang obtained closed form solutions for the joint acceptances for two special cases of spring-supported and simply-supported beams. A review of turbulence in two-phase is presented by (Khushnood et al., 2003).

(Endres & Moller, 2009) present the experimental analysis of disturbance propagation with a fixed frequency against cross flow and its effect on velocity fluctuations inside the bank. It is concluded that continuous wavelet transforms of the signals. Figure 8 indicates the disturbance frequency to be showing steady behavior. Generally designing for enhanced heat exchange ratios in thermal equipments ignores the structural effects caused by turbulent flow.

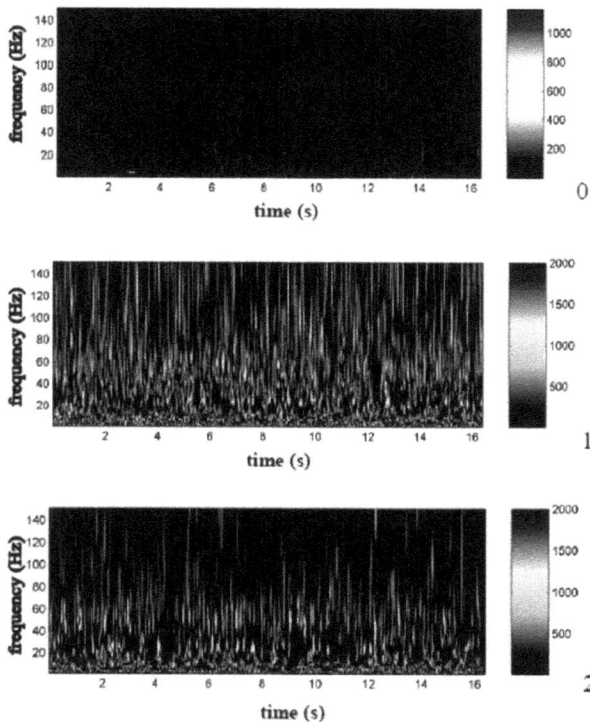

Fig. 8. Continuous wavelet transforms of the signals at locations 0, 1 and 2. Tube bank with P/D = 1.26, vortex generator #2, Re_G = 6.46X10⁴ (Endres & Möller, 2009).

(Pascal-Ribot and Blanchet, 2007) proposed a formulation to collapse the dimensionless spectra of buffeting forces in a single characteristic curve and gives edge to the formulation over previously normalized models in terms of collapse of data.

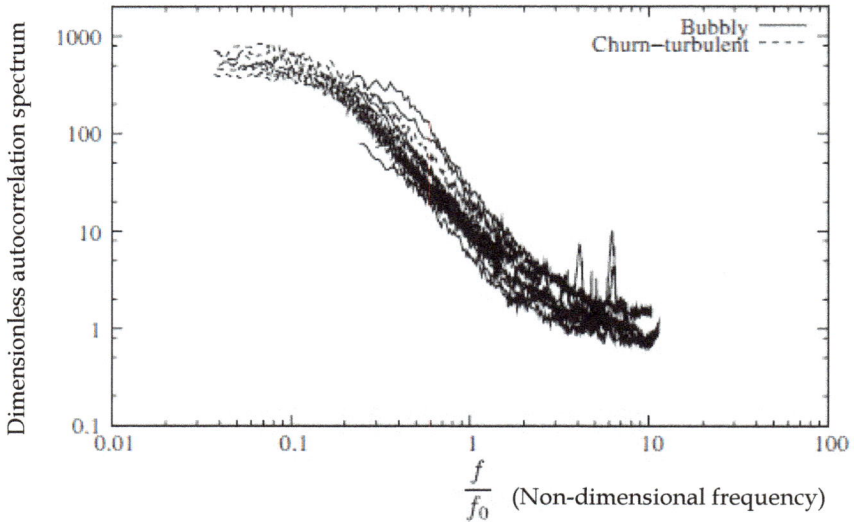

Fig. 9. Dimensionless reference equivalent spectra (Barrington, 1973)

Figure 9 shows the dimensionless spectra calculated with equations 15 & 16 respectively.

$$P_o = k\rho_1 g \sqrt{\frac{\sigma}{\Delta\rho g}} \left[\alpha(1-\alpha)\right]^2 \tag{15}$$

$$P_0 = k\rho_1 g \sqrt{\frac{\sigma}{\Delta\rho g}} \left[\alpha_{ct}(1-\alpha_{ct})\right]^2 : \alpha_{ct} = 0.4 \tag{16}$$

Where α is the void fraction.

(Wang et al., 2006) concludes the physically realistic solutions for turbulent flow in a staggered tube banks can be realized by FLUENT (with 2-D Reynolds stress model).

Figure 10 shows the consistency of turbulence intensity contours obtained through standard wall function approach and non-equilibrium wall function approach whereas near-wall treatment model and near -wall turbulence model predicts much higher results (Wang, et al., 2006).

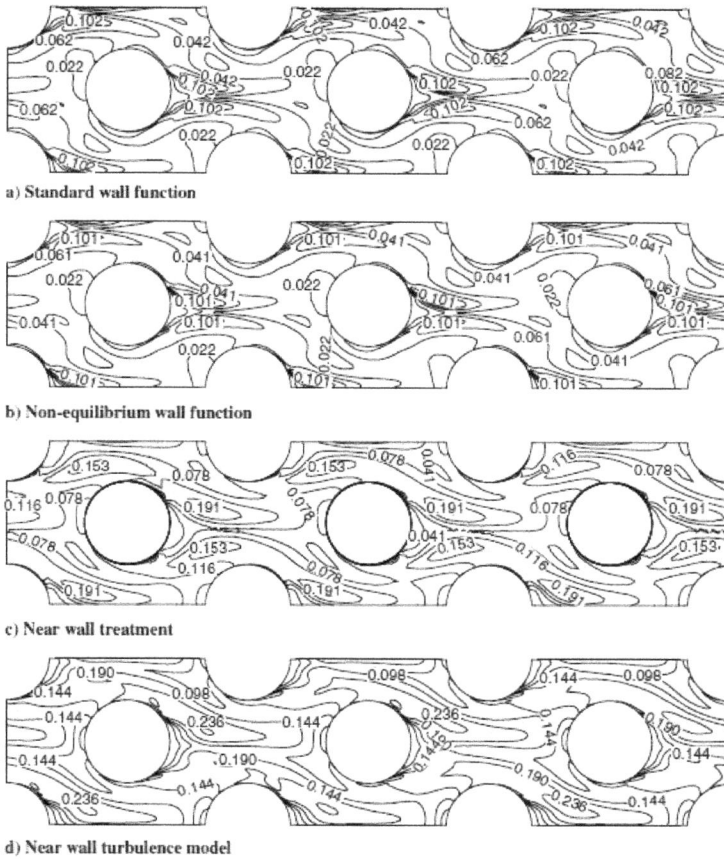

a) Standard wall function

b) Non-equilibrium wall function

c) Near wall treatment

d) Near wall turbulence model

Fig. 10. Turbulence intensity contours (Barrington, 1973)

2.5 Acoustic resonance

Acoustic vibration occurs only when the shell-side fluid is a vapor or a gas. The characteristic frequency of acoustic vibration in a heat exchanger depends on some characteristic length, usually the shell diameter and the velocity of sound in shell-side fluid, U_{sound}. The acoustic frequency (Chenoweth, 1993) can be predicted by the following equation.

$$f_a = \frac{m\ U_{sound}}{2d} \tag{17}$$

where m is the mode number (a dimensionless integer), and d is the shell diameter. The lowest acoustic frequency is achieved when $m = 1$ and the characteristic length is the shell diameter. The acoustic frequencies of an exchanger can be excited by either vortex shedding or turbulent buffeting (Chenoweth, 1993). (Barrington, 1973) indicated that so long as the exciting frequencies are within 20% of an acoustic frequency, a loud sound may be produced. Acoustic vibration becomes destructive when it is in resonance with some component of exchanger. The acoustic frequencies of shell can be changed by inserting a detuning plate parallel to the direction of cross-flow to alter the characteristic length (Chenoweth, 1993). There are a number of published acoustic vibration criteria to predict strong acoustic vibration within a tube bank, including (Eisinger et al., 1994), (Groth & Arnold, 1956), (Chen, 1968), (Fitzpatrick, 1986), (Ziada et al.,) and Blevins (Blevins, 1990).

(Hanson & Ziada, 2011) have investigated the effects of acoustic resonance on the dynamic lift force acting on the central tube. Two effects of resonant sound field includes generation of "sound induced" dynamic lift because of resonant acoustic pressure distribution on the tube surface and synchronization of vorticity shedding. Sound enhancements coefficients and sound induced lift force development is carried through numerical solution. (Hanson et al., 2009) investigated aeroacoustic response of two side-by-side cylinders against cross flow. It is concluded that acoustic resonance synchronizes vortex shedding and eliminates bistable flow phenomenon. Vortex shedding is noticed a particular strouhal number which excites acoustic resonance. Figure 11 and figure 12 gives the pressure spectra for two side-by-side cylinders and aeroacoustic response of two side-by-side cylinders.

(Eisinger & Sullivan , 2007) considers strong acoustic resonance with acoustic pressure reading 165 dB for package boiler at near full load, suppression of resonance (lower frequency) through baffle covering with downstream section and the development of another resonance (higher frequency) in the unbaffled upstream section.

(Feenstra et al., 2006) carried out experimental investigation of the effects of width of test section for measuring the acoustic resonance with a small pitch rates staggered tubes. The conclusion was that the maximum acoustic pressure versus input energy parameter of Blevins and Bressler is not a reliable preditor and it over predicts.

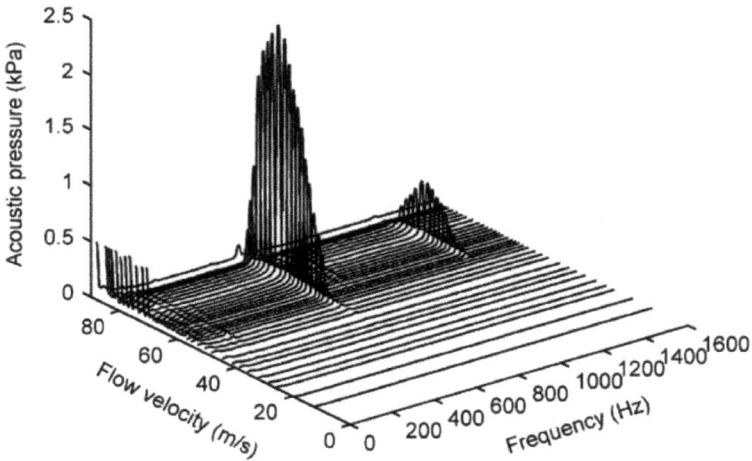

Fig. 11. Pressure spectra for the two side-by-side cylinders for T/D=1.25 (Hanson, et al., 2011)

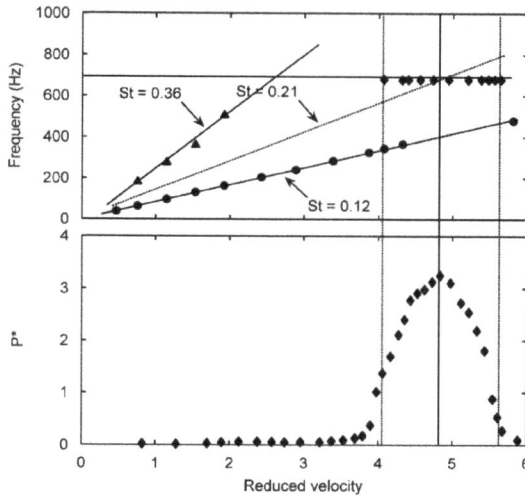

Fig. 12. Aeroacoustic response of the two side-by-side cylinders with T/D =1.25 and D =21.8 mm (Hanson, et al., 2011) (Eisinger & Sullivan, 2005) test results concluded that wide test sects gives the maximum acoustic pressure (lower acoustic mode) at P=53.4 MPa which is 4.27 times greater than predicted by Blevins and Bressler.

3. Natural frequencies of tube vibration

In flow-induced vibration design of heat exchanger tube bundles, resonant conditions must be suppressed by ensuring separation of natural frequencies of the tubes and exciting frequencies (Shin & Wambsganss,1975). A number of techniques are available for computing natural frequencies of straight, curved, single and multiple span tubes. These tubes may be subjected to varying end conditions. Loose baffles act like "kinfe-edged rings" supports (Timoshenko, 1955) . Tubes are rigidly fastened to the tube sheet and supported at intermediate points along their lengths by baffles or support plates. Some tubes in the centre may be supported by every baffle, whereas tubes that pass thorough baffle window may be supported by every second and third baffle. Table 5 (MacDuff & Feglar, 1957, Kissel, 1977) gives some of the formulas/techniques for estimating the natural frequencies of straight-and curved-or U-tubes.

3.1 Variables affecting tube natural frequencies

The tube natural frequencies are affected by tube-to-baffle hole clearance, axial stress, fins, span length, span shape (straight, U-tube), support type, temperature and tube vibration (Chenoweth, 1976, Elliot & Park, 1973, Pettigrew et al., 1986, Simpson et al., 1974, Tanaka & Takahara, 1981, TEMA Standards, 8th edition). At about one tenth of TEMA allowable clearance, the frequency is about 7% higher than that predicted for simple supports. For most exchangers, tube-to-hole clearance is not a significant parameter for controlling natural frequencies, but it may be important in damping and tube wear (Chenoweth, 1993). Due to manufacturing procedures, the tubes may be under a tension or compression axial loading. (Kissel, 1972) found a variation in natural frequency due to axial loading in a typical exchanger as much as ± 40%. The natural frequency varies as reciprocal of the span length squared (unsupported span). For finned tubes, effective diameter for the outside diameter should be used to find the area moment of inertia for natural frequency calculation (Chenoweth, 1993). Currently, software by TEMA (FIV) (TEMA Standards, 8th edition) is capable of predicting the natural frequencies.

Formula/ Procedure	Conditions
$f_n = \left(\dfrac{1}{2\pi}\right)\dfrac{\lambda_n}{l^2}\left(\dfrac{EI}{m}\right)^{1/2}$ (Jones, 1970)	Straight beams / single span n is the mode number and λ_n is a frequency factor which depends upon the end conditions
$f_n = \dfrac{1}{2\pi}\lambda_n\left(\dfrac{1}{R\alpha}\right)\left(\dfrac{EI}{m}\right)^{1/2}$ (Archer, 1960)	Curved beams/ single span λ_n is a frequency factor R is the radius of curvature and α is the subtended angle
$f_n = 59.55\dfrac{C_u}{L^2}\left(\dfrac{EI}{M_e}\right)^{0.5}$ (TEMA, 6th Edition, 1978)	U-tube curved C_u is the first mode U-tube constant
Experimental/ computer program $f_n = \dfrac{(\beta_n L)^2}{2\pi L^2}\sqrt{\dfrac{EIg}{W}}$ (Lowery & Moretti, 1975)	Straight/multiple, free-free spans (1-5 span tests); idealized support conditions, $(\beta_n L)^2$ is eigen value
FEM in-plane and out of plane Experimental/ analytical (Elliott & Pick, 1973)	Straight/curved
Beams immersed in liquids, air, kerosene, and oil (Jones, 1970)	Straight/simply supported/clamped
Out of plane: $f_n = 3.13\dfrac{\lambda_n}{R^2}\sqrt{\dfrac{C}{\gamma A}}$ $\lambda_n = \dfrac{n(n^2-1)}{\sqrt{1+kn^2}}$ (Ojalvo & Newman, 1964)	Clamped ring segments n is mode number; k is bending stiffness γ is specific weight; C is the torsional stiffness. A is cross-sectional area
Graphical in-plane and out of plane (Chen & Wambsganss, 1974)	Straight/curved, single span / multiple span

Formula/ Procedure	Conditions
Analytical/experimental (Khushnood et al., 2000)	Straight tubes single/multiple spans with damped/ fixed boundaries, Experimentation on refinery research exchanger (in-service)
Plucking and transient decay (Simpson & Hartlen, 1974)	Tubes were not fully straightened. (Slight residual wiggleness) Wind tunnel determination of fluid-elastic thresholds Tubes were found sensitive to temperature

Table 5. Tube natural frequencies (MacDuff et al., 1957, Kissel, 1977)

4. Dynamic parameters

Added mass and damping are known to be dependent on fluid properties (in particular, fluid density and viscosity) as well as functions of component geometry and adjacent boundaries, whether rigid or elastic. Nuclear reactor components are typically immersed in a liquid coolant and are often closely spaced (Wambsganss, et al., 1974).

4.1 Added mass

(TEMA, 7th Edition, 1988) defines hydrodynamic mass as an effect which increases the apparent weight of the vibrating body due to the displacement of the shell-side flow resulting from motion of vibrating tubes, the proximity of other tubes within the bundles and relative location of shell-wall. The so-called "virtual mass" for a tube is composed of the mass of the tube, mass of the fluid contained in the tube and the inertia M' imposed by the surrounding fluid. This hydrodynamic mass M' is a function of the geometry, the density of the fluid, and the size of the tube. In an ideal fluid, it is proportional to the fluid density and to the volume of the tube (Moretti & Lowry, 1976), and hence may be expressed, per unit length as:

$$\frac{M'}{L} = C_m \rho \pi r^2 \tag{18}$$

where L is the tube length, r the radius of the tube and ρ is the mass per unit volume of the surrounding fluid, C_m is called the inertia coefficient which is a function of the geometry, and is discussed by Lamb (Lamb, 1932, 1945). If the moving tube is not infinitely long, the flow is three-dimensional and leads to smaller values of C_m (Moretti & Lowry, 1976). For a vibrating tube in a fluid region bounded by a circular cylinder, Stokes (Endres & Moller, 2009) has determined hydrodynamic mass per unit length as given by:

$$m_h = C_m m_a \tag{19}$$

where $C_m = \dfrac{R^2 + r^2}{R^2 - r^2}$, R is the outer radius of annulus and, $m_a = \rho \pi r^2$, where ρ is fluid density, and r is the tube radius.

(Wambsganss, et al., 1974) have published a study on the effect of viscosity on C_m. Hydrodynamic mass M' for a tube submerged in water was determined by measuring its natural frequency, f_a, in air and, f_w, in water. Neglecting the density of air compared to water, the following equation may be obtained from beam equation (Moretti & Lowry, 1976).

$$\frac{M'}{L} = \mu \left[\left(\frac{f_a}{f_w} \right)^2 - 1 \right] \tag{20}$$

where μ is the tube mass per unit length. The inertia coefficient, C_m, can be obtained from Equation 18. Figure 13 gives the results showing the variation of C_m with pitch-to-diameter ratios (Wambsganss, et al., 1974).

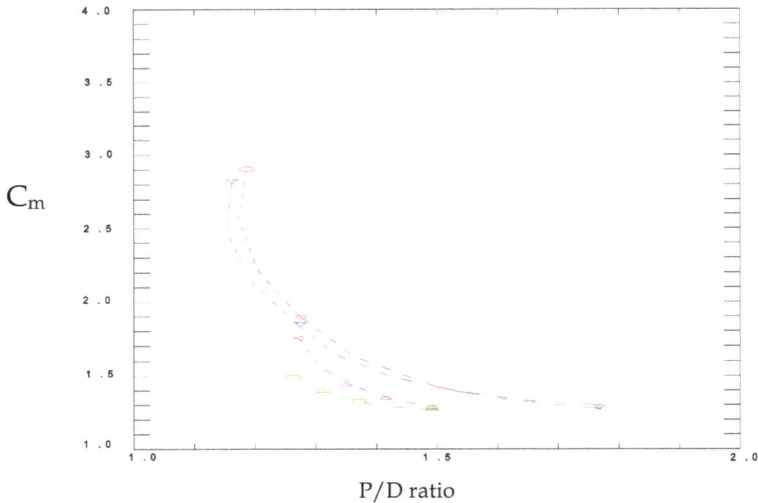

Fig. 13. Experimental C_m-values in tube bundles (Wambsganss et al., 1974, Moretti et al., 1976).

4.2 Damping

System damping has a strong influence on the amplitude of vibration. Damping depends upon the mechanical properties of the tube material, the geometry of intermediate supports and the physical properties of the shell-side fluid. Tight tube-to-baffle clearances and thick baffles increase damping, as does very viscous shell-side fluid (Chenoweth, 1993). (Coit et al., 1974) measured log decrements for copper-nickel finned tubes of 0.032 in still air. The range of most of the values probably lies between 0.01 and 0.17 for tubes in heat exchangers (Chenoweth, 1993). From (Wambsganss, et al., 1974), damping can be readily obtained from the transfer function or frequency response curve as

$$\zeta = \frac{1}{2\sqrt{N^2 - 1}} \frac{\Delta f_n}{f_n} \tag{21}$$

with $\Delta f_n = f_N^{(1)} - f_N^{(2)}$,

where f_n is the resonant frequency and $f_N^{(1)}$ and $f_N^{(2)}$ are the frequencies at which the response is a factor $\frac{1}{N^{th}}$ of resonant response.

(Lowery & Moretti, 1975), have concluded that damping is almost entirely a function of the supports. More complex support conditions (non-ideal end supports or intermediate supports with a slight amount of clearance) lead to values around 0.04. From analytical point of view, (Jones, 1970) has remarked that in most cases, the addition of damping to the beam equation re-couples its modes. Only a beam, which has, as its damping function, a restricted class of functions can be uncoupled. (Chen et al., 1994) have found the fluid damping coefficients from measured motion-dependent fluid forces. (Pettigrew et al., 1986, 1991) outlines the energy dissipation mechanisms that contribute to tube damping as given in Table 6:

Type of damping	Sources
Structural	Internal to tube material
Viscous	Between fluid forces and forces transferred to fluid
Flow-dependent	Varies with flow regime.
Squeeze film	Between tube and fluid as tube approaches support
Friction	Coulomb damping at support
Tube support	Internal to support material
Two-phase	Due to liquid gas mixture
Thermal damping	Due to thermal load

Table 6. Energy dissipation mechanisms (Pettigrew et al., 1986, 1991)

4.3 Parameters influencing damping

(Pettigrew et al., 1986) further outlines the parameters that influence damping as given below:

The Type of tube motion. There are two principal types of tube motion at the supports, rocking motion and lateral motion. Damping due to rocking is likely to be less. Rocking motion is pre-dominant in lower modes. Dynamic interaction between tube and supports may be categorized in three main types, namely: sliding, impacting, and scuffing, which is impacting at an angle followed by sliding:

Effect of number of supports. The trend available in damping data referenced in (Pettigrew et al., 1986), when normalized give

$$\zeta_n = \zeta N / (N - 1) \qquad (22)$$

where ζ_n is the normalized damping ratio, N is the number of spans, and ζ is the damping ratio.

Effect of tube Frequency. Frequency does not appear to be significant parameter (Pettigrew et al., 1986).

Effect of vibration amplitude. There is no conclusive trend of damping as a function of amplitude. Very often, amplitude is not given is damping measurements (Pettigrew et al., 1986).

Effect of diameter or mass. Large and massive tubes should experience large friction forces and the energy dissipated should be large. However, the potential energy in the tube would also be proportionally large in more massive tubes. Thus, the damping ratio, which is related to the ratio of energy dissipated per cycle to the potential energy in the tube should be independent of tube size or mass (Pettigrew et al., 1986).

Effect of side loads. In real exchangers, side loads are possible due to misalignment of tube-supports or due to fluid drag forces. Side loads may increase or reduce damping. Small side loads may prevent impacting, and thus reduce damping, whereas large side loads may increase damping by increasing friction (Pettigrew et al., 1986).

Effect of higher modes. Damping appear to decrease with mode order, for mode order higher than the number of spans, since these higher modes involve relatively less interaction between tube and tube-support (Pettigrew et al., 1986).

Effect of tube support thickness. Referenced data in (Pettigrew et al., 1986) clearly indicates that support thickness is a dominant parameter. Damping is roughly proportional to support thickness. (Pettigrew et al., 1986) corrected the damping data line for support width less than 12.7mm such that

$$\zeta_{nc} = \zeta_n \left(\frac{12.7}{L} \right) \qquad (23)$$

where L is support thickness in mm and ζ_{nc} is the corrected normalized damping ratio.

Effect of clearance. For the normal range of tube-to-support diametral clearances (0.40mm-0.80mm), there is no conclusive trend in the damping data reviewed (Pettigrew et al., 1986).

Design Recommendations (Pettigrew et al., 1986, 1991, Taylor et al., 1998)

Friction damping ratio in a multi-span tube (percentage)

$$\text{(For liquid) } \zeta_F = 0.5 \left(\frac{N-1}{N} \right) \left(\frac{L}{l_m} \right)^{1/2} \qquad (24)$$

$$\text{(For gas) } \zeta_F = 5.0 \left(\frac{N-1}{N} \right) \left(\frac{L}{l_m} \right)^{1/2} \qquad (25)$$

where N is the number of tube spans, L is the support thickness, l_m is the characteristic span length usually taken as average of three longest spans.

Viscous damping ratio

Rogers simplified version of Chens' cylinder viscous damping ratio (percentage) of a tube in liquid.

$$\zeta_F = \frac{100\pi}{\sqrt{8}} \left(\frac{\rho D^2}{m} \right) \left(\frac{2\upsilon}{\pi f D^2} \right)^{1/2} \left[\frac{1+(D/D_e)^3}{(1-(D/D_e)^2)^2} \right]$$ (26)

where ρ is the fluid density, m is the mass per unit length of tube (interior fluid and hydrodynamic mass), D_e is the equivalent diameter to model confinement due to surrounding tubes, D is the tube diameter, f is the frequency of tube vibration and υ is the fluid kinematic viscosity. The term $S = \frac{\pi f D^2}{2\upsilon}$ is the Stoke number.

Squeeze film damping ratio

$$\text{(For multi-span tube) } \zeta = \left(\frac{N-1}{N} \right) \left(\frac{1460}{f} \right) \left(\frac{\rho D^2}{m} \right) \left(\frac{L}{l_m} \right)^{1/2}$$ (27)

Support damping

(Pettigrew, et al., 1991) has developed a semi-empirical expression to formulate support damping, using Mulcahys' theory (Mulcahy, 1980).

$$\zeta_s = \left(\frac{N-1}{N} \right) \left(\frac{2200}{f} \right) \left(\frac{\rho D^2}{m} \right) \left(\frac{L}{l_m} \right)^{0.6}$$ (28)

(TEMA, 6th Edition, 1978, TEMA, 7th Edition, 1988)

According to TEMA standards, ζ is equal to greater of ζ_1, and ζ_2

(For shell-side liquids)

$$\zeta_1 = \frac{3.41 d_o}{W_o f_n}$$ (29)

$$\zeta_2 = \frac{0.012 d_o}{W_o} \left[\frac{\rho_o v}{f_n} \right]^{1/2}$$ (30)

where v is the shell fluid velocity, d_o is the outside tube diameter, ρ_o is the density of shell-side fluid, f_n is the fundamental frequency of tube span, and W_o is the effective tube weight.

(For shell-side vapors)

$$\zeta = 0.314 \frac{N-1}{N}\left(\frac{t_b}{l}\right)^{1/2}$$

(31)

where N is the number of spans, t_b is the baffle or support plate thickness, and l is the tube unsupported span. A review of two-phase flow damping is presented by (Khushnood et al., 2003).

5. Damage numbers for collision and baffle damage (Chenoweth, 1976, Shin et al., 1975, Brothman et al., 1974)

Two types of vibration damage are prevalent in cross-flow regions of steam generators (Shin & Wambsganss, 1975).

- Tube-to-baffle impact.
- Tube-to-tube collision.

(Thorngren, 1970) deduced "damage numbers" for the two types of damage, based on the assumption that tube is supported by baffles and deflected by a uniformly distributed lift force. These damage numbers are given by following equations:

$$\text{(Baffle damage number)} \quad N_{BD} = \frac{d\rho V^2 l^2}{\beta_1 S_m g_c A_m B_t}$$

(32)

$$\text{(Collision damage number)} \quad N_{CD} = \frac{0.625\, d\rho V^2 l^4}{\beta_1^4 g_c A_m (d^2 + d_i^2) C_T E}$$

(33)

where

$N_{BD} < 1$ for safe design.

C_T = Maximum gap between tubes.

d_i = Tube inner diameter.

E = Modulus of elasticity.

B_t = Baffle thickness.

l = Length of tube between supports.

A_m = Tube cross-sectional area $\left[\frac{\pi}{4}(d^2 - d_i^2)\right]$ and

S_m = Maximum allowable fatigue stress [ASME Pressure Vessel code Sec. III].

$N_{CD} < 1$ for safe design.

d = Tube outer diameter.

β_1 = Tube-to-baffle-hole clearance factor.

g_c = Gravitational constant.

ρ = Mass density of shell side fluid.

V = Free stream velocity.

Collision damage is usually predicted together with baffle damage, whereas the latter can be predicted without collision damage being indicated, i.e., baffle damage is important factor when appraising design (Erskine et al., 1973). (Burgreen et al., 1958) were the first to conduct an experiment to investigate vibration of tube for fluid flowing parallel to tube axis. (Quinn, 1962) and (Paidoussis, 1965) have developed analytical and empirical expressions respectively for peak amplitude. Paidoussis give the following expression:

$$\frac{\Delta}{d} = \alpha_1^{-4} \frac{U^{1.6} \varepsilon^{1.8} R_e^{0.25}}{1+U^2} \left(\frac{d_h}{d}\right)^{0.4} \frac{\beta^{2/3}(5 \times 10^{-4} \times K_p)}{1+4\beta} \tag{34}$$

where

K_p	=	Flow condition constant
Δ	=	Maximum vibration amplitude at mid-span
d	=	Outer diameter of tube
α_1	=	First mode beam eigen value of the tube
U	=	Dimensionless flow velocity
$\varepsilon = \dfrac{\ell}{d}$		
ℓ	=	Tube length
d_h	=	Hydraulic diameter
R_e	=	Reynolds number
I	=	Moment of inertia and
β	=	Added mass fraction

Later on a number of expressions for peak and RMS amplitudes have been developed (Shin et al., 1975, Blevins, 1977).

6. Wear work-rates

In fretting wear, work-rate is defined as the rate of energy dissipation when a tube is in contact with its support. Energy is being dissipated through friction as the tube moves around in contact with its supports. A force (the contact force between tube and support) multiplied by a displacement (as the tube slides) results in work or dissipated energy required to move the tube (Taylor et al., 1998, Au-Yang, 1998). Normal work-rate W_n for different tube and tube support plate material combinations and different geometries (Au-Yang, 1998) is defined.

$$W_n = \frac{1}{T} \int_0^T |F_n| dS \tag{35}$$

where T is the total time, F_n is the normal contact force, and S is the sliding distance.

(Au-Yang, 1998) has assessed the cumulative tube wall wear after 5, 10, and 15, effective full power years of operation of a typical commercial nuclear steam generator, using different wear models.

The EPRI data reproduced from (Hofmann & Schettlet, 1989) in Figure 14 shows the wear volume against normal work-rate for the combination of Inconel 600 tube (discrepancy as plot shows J 600 whereas text indicates Inconel 600) and carbon tube support plate, a condition that applies to many commercial nuclear steam generators (Hofmann & Schettlet, 1989). Figure 15 shows the tube wall thickness loss against volumetric wear for different support conditions (Hofmann & Schettlet, 1989).

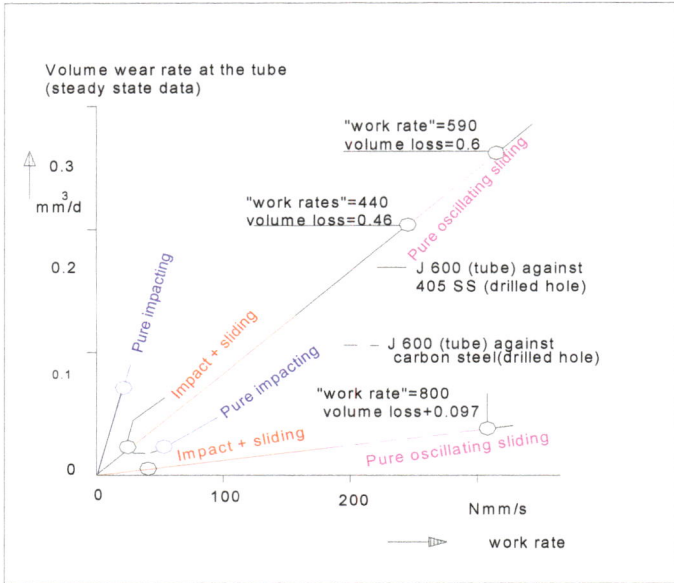

Fig. 14. Volumetric wear rate versus normal work-rate for different material combinations (Hofmann & Schettlet, 1989).

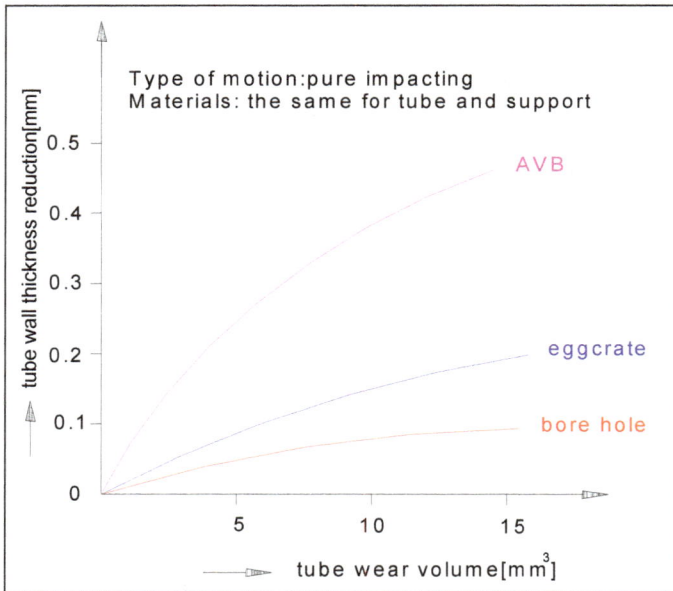

Fig. 15. Tube wall thickness loss versus volumetric wear for different support conditions, from Hofmann and Schettler (Hofmann & Schettlet, 1989).

(Payen et al., 1995) have carried predictive analysis of loosely supported tubes vibration induced by cross-flow turbulence for non-linear computations of tube dynamics. They have analyzed the gap effect and have concluded that wear work-rate decreases when the gap value increase at low velocities. (Peterka, 1995) has carried out numerical simulation of the tubes impact motion with generally assumed oblique impacts. (Charpentier and Payen, 2000) have carried out prediction of wear work-rate and thickness loss in tube bundles under cross-flow by a probabilistic approach. They have used Archard's Law and wear correlation depending on the contact geometry, and have concluded that most sensitive parameters that affect the wear work-rate are the coefficient of friction, the radial gap and the spectral level of turbulent forces.

(Paidoussis & Li, 1992) and (Chen et al., 1995) have studied the chaotic dynamics of heat exchanger tubes impacting on the generally loose baffle plates, using an analytical model that involves delay differential equations. They have developed a Lyapunov exponent technique for delay differential equations and have shown that chaotic motions do occur. They have performed analysis by finding periodic solutions and determining their stability and bifurcations with the Poincare map technique. Hopf bifurcation is defined as the loss of stable equilibrium and onset of amplified oscillation (Paidoussis & Li, 1992).

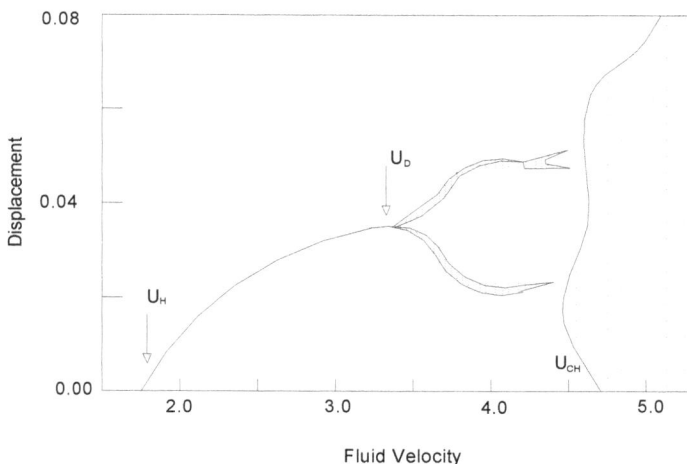

Fig. 16. The bifurcation diagram (Paidoussis & Li, 1992, Chen et al., 1995)

A typical bifurcation diagram for the symmetric cubical model with $P/D = 1.5$, is given in Figure 16 showing dimensionless mid-point displacement amplitude in terms of dimensionless fluid velocity. Where U_H denotes critical U for Hopf bifurcation, U_D, is the first post Hopf bifurcation, and U_{CH} denotes the onset of chaos. Total wear work rates against pitch velocity and mass flux have been given by (Taylor et al., 1995) and (Khushnood et al., 2003).

Researchers	Salient fretting - wear features
(Rubiolo & Young, 2009)	• The evaluation of turbulence excitation is very challenging. • Identification of key wear factors that can be correlated to assembly operating conditions. • Functional dependence of wear damage against identified factors. • Grid cell clearance size and turbulence forces as key risk factors for PWR fuel assemblies. • Grid misalignment and cell tilts are less important. • Minimization of wear risk through modification in core loading.
(Jong & Jung, 2008)	• Fretting wear in helical coil tubes steam generator. • Thermal-Hydraulic prediction through FEM. • Emphasis on the effects of number of supports, coil diameter and helix pitch on free vibration modes. • Design guidelines for designers and regulatory reviewers.
(Attia, 2006$_a$)	• Investigation of fretting wear of Zr-2.5% Nb alloy. • Experimental setup includes special design fretting wear tribometers. • Fretting wear is initially dominated by adhesion and abrasion and then delamination and surface fatigue. • Volumetric wear loss decreased with number of cycles.
(Attia, 2006$_b$)	• Fitness for service and life management against fretting fatigue. • Examples of fretting problems encountered in nuclear power plants. • Methodology to determine root cause. • Non-linearity of the problem and risk management. • Critical role of validation experimentally (long term) under realistic conditions and to qualify in-situ measurements of fretting damage non-destructive testing.
(Rubiolo, 2006)	• Probabilistic method of fretting wears predictions in fuel rods. • Non-linear vibration model VITRAN (Vibration Transient Analysis). • Numerical calculations of grid work and wear rates. • Monte carlo method applied for transient simulations (due to large variability of fuel assemble parameters). • Design preference of fuel rods.
(Kim et al., 2006)	• A way toward efficient of restraining wear. • Increase in contact area through two different contours of spacer grid spacing. • Consideration of contact forces, slip displacement and wear scars on rods to explore mechanical damage phenomenon. • It concludes that the contact shape affects the feature and behavior of length, width and volumetric shape of wear. • A new parameter "equivalent depth" is introduced to represent wear severity.

Table 7. Salient features of some recent researches on fretting wear Damage in Tube Bundles.

A generalized procedure to analyze fretting - wear process and its self - induced changes in properties of the system and flow chart for fretting fatigue damage prediction with the aid of the principles of fracture mechanics is presented in figure 17 & 18 respectively.

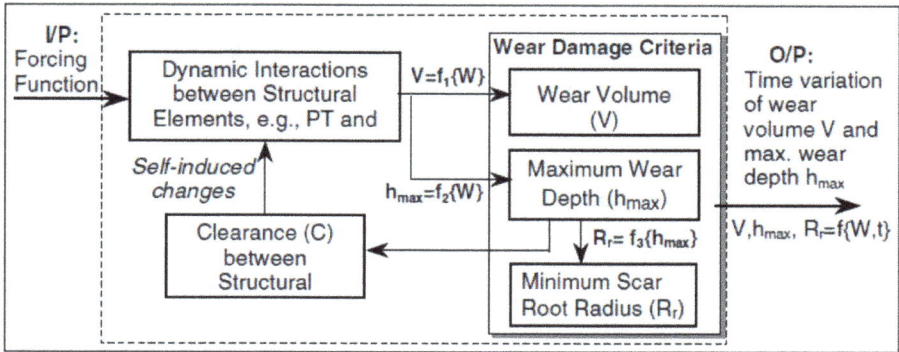

Fig. 17. System approach to the fretting wear process and its self-induced changes in the system properties (Attia, 2006$_a$).

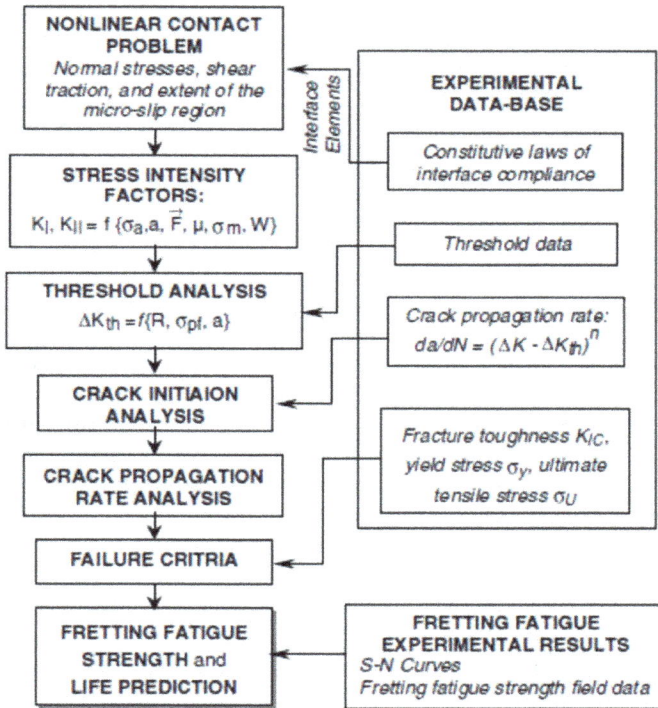

Fig. 18. Flow chart for the prediction of fretting fatigue damage, using fracture mechanics principles (Attia, 2006$_a$).

7. Tube bundle vibrations in two-phase cross-flow

7.1 Modeling two-phase flow

Most of the early experimental research in this field relied on sectional models of tube arrays subjected to single-phase fluids such as air or water, using relatively inexpensive flow loops and wind tunnels. The cheapest and simplest approach to model two-phase flow is by mixing air and water at atmospheric pressure. However, air-water flows have a much different density ratio between phases than steam-water flow and this will affect the difference in the flow velocity between the phases. The liquid surface tension, which controls the bubble size, is also not accurately modeled in air-water mixtures. Table 8 gives the comparison of liquid and gas phase of refrigerants R-11, R-22 and air-water mixtures at representative laboratory conditions with actual steam-water mixture properties at typical power plant conditions (Feentra et al., 2000). This comparison reveals that the refrigerants approximate the liquid surface tension and liquid dynamic viscosity of steam-water mixtures more accurately than air-water mixtures.

Property	R-11	Air-water	R-22	Steam-water
Temperature (^{0}C)	40	22	23.3	260
Pressure (kPa)	175	101	1000	4690
Liquid Density (kg/m^3)	1440	998	1197	784
Gas Density (kg/m^3)	9.7	1.18	42.3	23.7
Liquid kinematic viscosity (μm^2/sec)	0.25	1.0	0.14	0.13
Gas kinematic Viscosity (μm^2/sec)	1.2	1.47	0.30	0.75
Liquid Surface Tension (N/m)	0.016	0.073	0.0074	0.0238
Density Ratio	148	845	28.3	33
Viscosity Ratio	0.20	0.70	0.47	0.17

Table 8. Comparison of properties of air-water, R-22, and R-11 with steam-water at plant conditions (Feentra et al., 2000)

Typical nuclear steam generators such as those used in the CANDU design utilize more than 3000 tubes, 13mm in diameter, formed into an inverted U-shape. In the outer U-bend region, these tubes are subject to two-phase cross-flow of steam-water which is estimated to be of 20% quality. It is highly impractical and costly to perform flow- induced vibration experiments on a full-scale prototype of such a device so that small-scale sectional modeling is most often adopted. R-11 simulates the density ratio, viscosity ratio and surface tension of actual steam-water mixtures better than air-water mixtures and it also allows for localized phase change which air-water mixture does not permit. While more costly and difficult to use than air-water mixture, R-11 is a much cheaper fluid to model than steam-water because it requires 8% of the energy compared with water to evaporate the liquid and operating pressure is much lower, thereby reducing the size and cost of the flow loop (Feentra et al., 2000).

7.2 Representative published tests on two-phase flow across tube arrays

Table 6, an extension of period beyond 1993 (Pettigrew et al., 1973) presents a summary of salient features of the experimental tests performed on the three possible tube arrangements (triangular, normal square, and rotated square).

Researchers	Fluid	Tube Array	Void Fraction	Tube Length (mm)	Natural Frequency (Hz)	Damping Ratio (%)
(Pettigrew et al., 1973)	Air-Water	Triangular/ Parallel. Square/Rotated Square	10-20% (quality)	50.8	17- 30	2.5-2.7
(Heilker & Vincent, 1981)	Air-Water	Triangular/ Rotated Square	0.5 – 0.87	910	56-62	0.8-4
(Hara et al., 1981)	Air-Water	Single Tube	0.02 – 0.61	60	Rigid	–
(Remy, 1982)	Air-Water	Square	0.65 – 0.85	1000	56.6	0.6-1.75
(Nakamura et al., 1982)	Air-Water	Square/ Rotated Square	0.2 – 0.94	190	142	1.3-1.7
(Pettigrew et al., 1985)	Air-Water	Triangular/ Square	0.05-0.98	600	26-32	0.9-8.0
(Axisa et al., 1984)	Steam-Water	Square	0.52-0.98	1190	74	0.2-3.0
(Nakamura et al., 1986)	Steam-Water	Square	0.75-0.95	174	15.2 –16	4.0-8.0
(Hara, 1987)	Air-Water	Single/Row	0.01-0.5	58	6.0-8.4	2.9-15.6
(Goyder, 1988)	Air-Water	Triangular	0.5-0.8	360	175	–
(Gay et al., 1988)	Freon	Triangular	0.58-0.84	1000	39.8	0.89-1.7
(Nakamura & Fujita, 1988)	Air-Water	Square	0.02-0.95	600	52	2.1
(Funakawa et al., 1989)	Air-Water	Square/ Triangular	0.0-0.6	100	12.8	3.3
(Nakamura et al., 1990)	Steam-water	Square	0.33-0.91	174	94-137	–
(Axisa et al., 1990)	Steam-water	Square/ Triangular Parallel/ Triangular	0.52-0.99	1190	72	0.2-4
(Papp & Chen, 1994)	–	Normal triangular / Normal Square/ Parallel Triangular	25-98%	–	–	–
(Pettigrew, Taylor, Jong & Currie, 1995)	Freon	Rotated triangular	40-90% 10-90%	609	28-150	0.15
(Noghrehkar et al., 1995)	Air-Water	Square fifth & sixth row	0-90%	200	15-25	0.3-3.9
(Taylor et al., 1995)	Air-Water	U-bend tube bundle with 180° U-tubes parallel triangular configuration	0-90%	U-Tube radii 0.6-0.7m	23-114	1.5-2
Marn & Catton (Marn & Catton, 1996)	Air-Water	Normal triangular, Parallel triangular, & rotated square	5-99%	–	–	0.7-21
(Taylor & Pettigrew, 2000)	Freon	Rotated triangular & Rotated Square	50-98%	609	–	0.2-5
(Pettigrew et al., 2000)	Air-Water	Normal 30° & rotated 60° triangular, Normal 90° & rotated 45° square	0-100%	600	30-160	1-5

Researchers	Fluid	Tube Array	Void Fraction	Tube Length (mm)	Natural Frequency (Hz)	Damping Ratio (%)
(Inada et al., 2000)	Air-Water	Square	0-70%	198	15	-2.5 - +1.6% Eq. added damping coefficient
(Nakamera et al., 2000)	Freon	46x5 U-bend tubes, specification of actual westinghouse type-51 series steam generator.	(Not considered) based on Connors single-phase relation.	--	16-26	Damping ratio <1%
(Feentra et al., 2000)	R-11	Parallel/Triangular	0-0.99	--	0-100	1.1-2.9
(Pettigrew & Taylor, 2003)	Steam-water general overview	General overview	General overview	General overview	General overview	General overview
(Chung & Chu, 2005)	Air-water	Normal square/rotated square	Void fraction 10-95%	--	18.65-20.7 Hz	0.01-0.05
(Parsad et al., 2007)	Steam-water general overview	General overview	General overview	General overview	General overview	General overview
(Pascal-Ribot & Blanchet, 2007)	Air-water	Rigid cylinder	10-80%	Dia=100 mm	0-25 Hz	--
(Kakac & Bon, 2008)	Steam-water general overview	General overview	General overview	General overview	General overview	General overview
(Mitra et al., 2009)	Air-water / Steam-water	Normal square tube array suspended from piano wires	Void fraction 0-45%	Dia=190 mm	7.84-13.9 Hz	3.3-5.2
(Sim & Park, 2010)	--	Cantilevered flexible cylinder	3-38%	Dia=123 mm	12Hz	0.1128-0.1154
(Chu et al., 2011)	Air-water	U-tube rotated square	70-95%	Dia=19.05mm	4-12 Hz	0.0038

* Results 1973-1993 (Pettigrew et al., 1973)

Table 9. Representative Published Tests on Two- Phase Flow

7.3 Thermal hydraulic models

Considering two-phase flow, homogenous flow assumes that the gas and liquid phases are flowing at the same velocity, while other models for two-phase flow, such as drift-flux assume a separated flow model with the phases allowed to flow at different velocities. Generally the vapor flow is faster in upward flow because of the density difference.

7.3.1 The homogenous equilibrium model

Homogenous Equilibrium Model (HEM) treats the two-phase flow as finely mixed and homogeneous in density and temperature with no difference in velocity between the gas and liquid phases.

A general expression for void fraction α, is given in (Feentra et al., 2000).

$$\alpha = \left[1 + S \frac{\rho_G}{\rho_L} (\frac{1}{x} - 1) \right]^{-1} \qquad (36)$$

where ρ_G and ρ_L are the gas and liquid densities respectively and S is the velocity ratio of the gas and liquid phase (i.e. $S = U_G / U_L$). The quality of the flow x is calculated from energy balance, which requires measurement of the mass flow rate, the temperature of the liquid entering the heater, the heater power, and the fluid temperature in the test section. The HEM void fraction α_H is the simplest of the two-phase fluid modeling, whereby the gas and liquid phases are assumed to be well mixed and velocity ratio S in Equation 36 is assumed to be unity. The average two-phase fluid density ρ is determined by Equation 37.

$$\rho = \alpha \rho_G + (1 - \alpha)\rho_L \qquad (37)$$

The HEM fluid density ρ_H is determined using Equation 32 by substituting α_H in place of α. The HEM pitch flow velocity V_P is determined by

$$V_P = G_P / \rho_H \qquad (38)$$

Where G_p=Pitch mass flux

7.3.2 Homogenous flow (Taylor & Pettigrew, 2000)

This model assumes no relative velocity between the liquid velocity U_1 and the gas velocity U_g. Slip S between the two-phases is:

$$S = 1 : U_h = U_g = U_1;$$

$$\varepsilon_g = \frac{j_g}{j_g + j_l} \qquad (39)$$

where U_h is the homogeneous velocity, U_g is the gas phase velocity, U_l is the liquid phase velocity, ε_g is the homogenous void fraction, j_g is superficial gas velocity and j_l is the superficial liquid velocity.

7.3.3 Smith correlation

(Smith, 1968) assumes that kinetic energy of the liquid is equivalent to that of the two-phase mixture and a constant fraction k of liquid phase is entrained with the gas phase. The value $k = 0.4$ was chosen to correspond with the best agreement to experimental data for flow in a vertical tube. Using the Smith correlation, the slip is defined as follows.

$$S = k + (1-k)\left[\frac{x\rho_l / \rho_g + k(1-x)}{x + k(1-x)}\right]^{1/2} \tag{40}$$

where x is the mass quality, ρ_g is the density of the gas phase and ρ_l is the density of the liquid phase.

7.3.4 Drift-flux model

The main formulation of drift-flux model was developed by (Zuber and Findlay, 1965). This model takes into account both the two-phase flow non-uniformity and local differences of velocity between the two phases. The slip is defined as follows.

$$S = \frac{(1-\varepsilon_g)}{\dfrac{1}{C_0 + \dfrac{\overline{U_{gj}}}{j}} - \varepsilon_g} = \frac{C_0 + \dfrac{\overline{U_{gj}}}{j} - \dfrac{x}{x(1-\dfrac{\rho_g}{\rho_l}) + \dfrac{\rho_g}{\rho_l}}}{1 - \dfrac{x}{x(1-\dfrac{\rho_g}{\rho_l}) + \dfrac{\rho_g}{\rho_l}}} \tag{41}$$

where $\overline{U_{gj}}$ is averaged gas phase drift velocity.

$$j = j_g + j_l = \dot{m}\left(\frac{x}{\rho_g} + \frac{(1-x)}{\rho_l}\right) \tag{42}$$

Where $\dot{m}$ is the mass flux

The remaining two unknowns are empirical and (Lellouche et al., 1982) is used to estimate these.

$$C_0 = \frac{L}{K_1 + (1-K_1)\varepsilon_g^r} \tag{43}$$

7.3.5 Schrage correlation

The correlation by (Schrage, 1988) is based on empirical data from an experimental test section, which measures void fraction directly. This test section has two valves capable of isolating a part of the flow almost instantaneously.

The correlation is based on physical considerations and assumes two different hypotheses:

The Schrage correlation is as follows:

$$\varepsilon_g / \varepsilon_{gh} = 1+0.123 \, F_r^{-0.191} \ln x \tag{44}$$

with

$$F_r = \frac{\dot{m}}{\rho_l \sqrt{gD}} \tag{45}$$

This correlation was established with an air-water mixture, but it remains valid for any other phase flow.

7.3.6 Feenstra model

In this model (Feentra et al., 2000), predicted velocity ratio of the phases is given by

$$S = 1+25.7(R_i \times Cap)^{0.5} \, (P/D)^{-1} \tag{46}$$

Where Cap is the capillary number and R_i is the Richardson number

7.3.7 Comparison of void fraction models

The HEM greatly over-predicts the actual gamma densitometer void fraction measurement and the prediction of void fraction model by Feenstra et al., is superior to that of other models. It also agrees with data in literature for air-water over a wide range of mass flux and array geometry (Feentra et al., 2000). The main problem with using the HEM is that it assumes zero velocity ratios between the gas and liquid phases. This assumption is not valid in the case of vertical upward flow, because of significant buoyancy effects.

7.4 Dynamic parameters

7.4.1 Hydrodynamic mass

Hydrodynamic mass m_h is defined as the equivalent external mass of fluid vibrating with the tube. It is related to the tube natural frequency f in two-phase mixture as discussed in (Carlucci & Brown, 1997) and is given below:

$$m_h = m_t[(f_g / f)^2 - 1] \tag{47}$$

where m_t is the mass of tube alone and f_g is the natural frequency in air.

Hydrodynamic mass depends on the pitch-to-diameter ratio of the tube, and is given by (Pettigrew et al., 1989)

$$m_h = \left(\frac{\rho \pi d^2}{4} \right) \left[\frac{(D_e / d)^2 + 1}{(D_e / d)^2 - 1} \right] \tag{48}$$

where ρ is the two-phase mixture density.

$$D_e / d = (0.96 + 0.5P / d)P / d \text{ , for a triangular bundle.} \qquad (49a)$$

$$D_e / d = (1.07 + 0.56P / d)P / d \text{ , for a square bundle.} \qquad (49b)$$

where D_e is equivalent diameter to model confinement due to the surrounding tubes as given by (Rogers et al., 1984).

Early air-water studies (Carlucci, 1980) showed that added mass decreases with the void fraction as shown in Figure 19. It is also less than $(1-\alpha)$, where α is the void fraction. This deviation from expected $(1-\alpha)$ line is caused by the air bubble concentrate at the flow passage center. Surprisingly added mass has attracted very little attention of researchers which is a potential avenue for future researches.

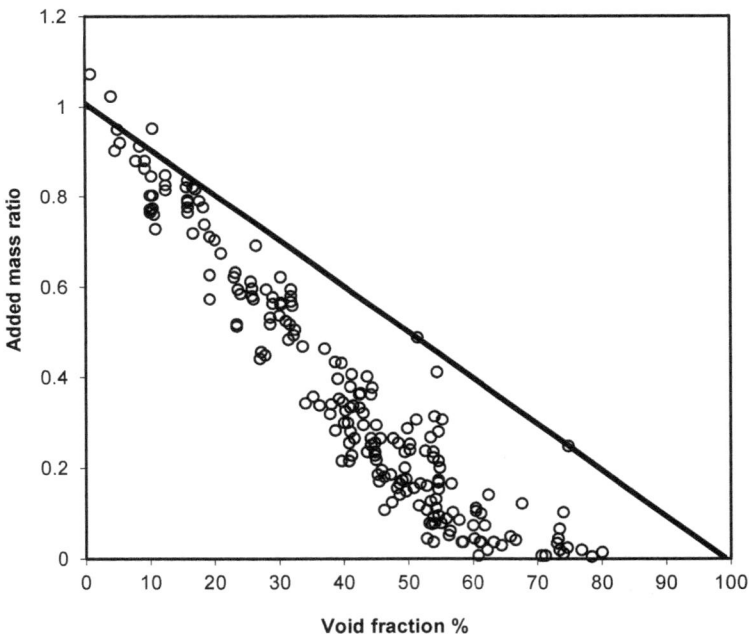

Fig. 19. Added mass as a function of void fraction (Carlucci, 1980).

7.4.2 Damping in two-phase

Subtracting the structural damping ratio from the total yields the two-phase fluid-damping ratio (Noghrehkar et al., 1995). Total damping includes structural damping, viscous damping and a two-phase component of damping as explained by (Pettigrew et al. 1992). The damping ratio increases as the void fraction increases and peaks at 60% (Carlucci, 1980), then the ratio decrease with α (Figure 20). Damping also decreases as the vibration frequency increases (Pettigrew et al., 1985).

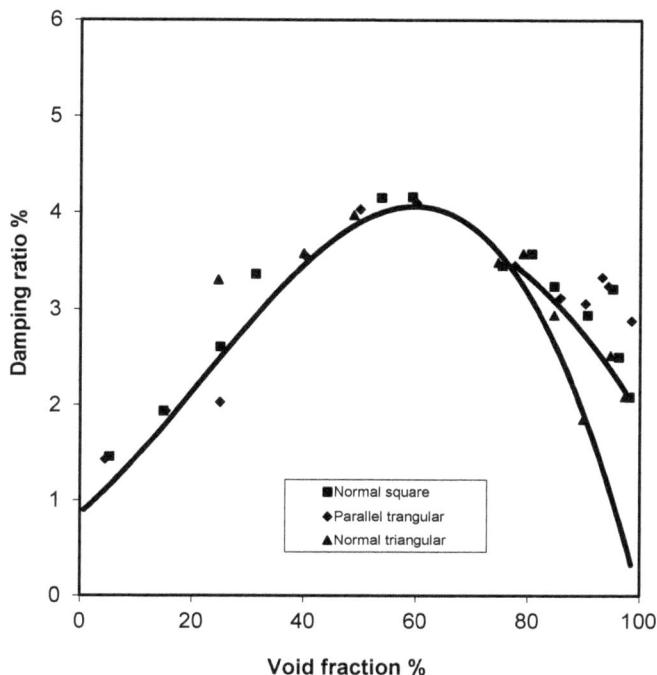

Fi3. 20. Damping ratio as a function of void fraction (Carlucci, 1983).

Demping in two-phase is very complicated. It is highly dependent upon void fraction and flcw regime. The results for the two-phase component of damping can be normalized to taFe into account the effect of confinement due to surrounding tubes by using the zonfinement factor C (Pettigrew et al., 2000). This factor is a reasonable formulation of the conrinement due to P/D. As expected, greater confinement due to smaller P/D increase damping. The confinement factor is given by equation below:

$$C = \frac{[1+(D/D_e)^3]}{[1-(D/D_e)^2]^2} \tag{50}$$

7.5 Flow regimes

Many researchers have attempted the prediction of flow regimes in two-phase vertical flow. As yet, a much smaller group has examined flow regimes in cross-flow over tube bundles. Some of the first experiments were carried out by I.D.R. Grant (Collier, 1979) as it was the only available map at the time. Early studies in two-phase cross-flow used the Grant map to assist in identifying tube bundle flow regimes (Pettigrew et al., 1989) and (Taylor et al., 1989). More recently, Ulbrich & Mewes [180] performed a comprehensive analysis of available flow regime data resulting in a flow regime boundaries that cover a much larger range of flow rates. They found that their new transition lines had an 86% agreement with available data. Their flow map is shown in Figure 21 by (Feenstra et al., 1990) with the flow regime boundary transitions in solid lines and the flow regimes identified with upper-case text. The dotted lines outline a previous flow regime map based on Freon-11 flow in a vertical tube from (Taitel et al., 1980).

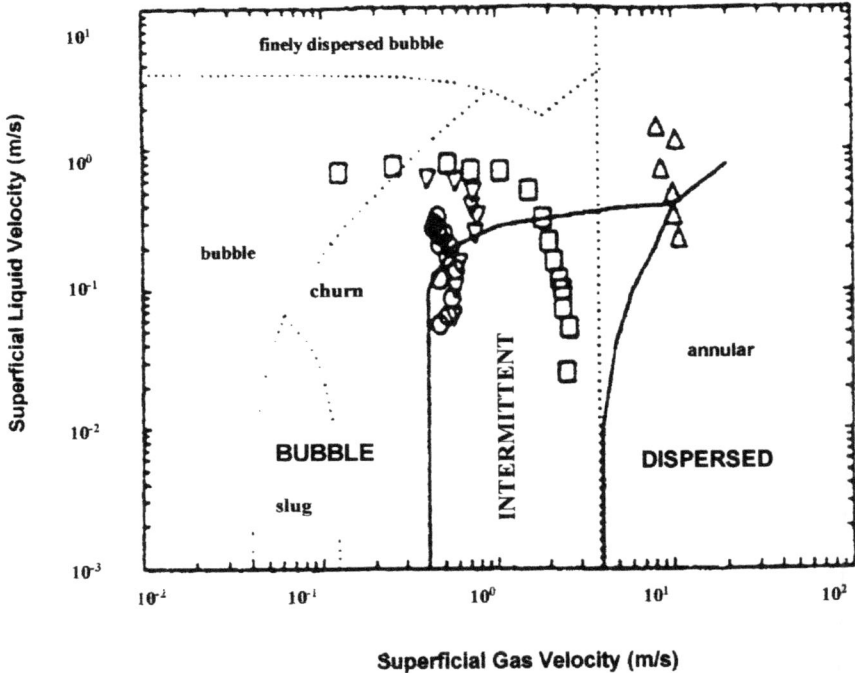

Superficial Gas Velocity (m/s)

Fig. 21. Flow regime map for vertically upward two-phase flow: From (Feenstra et al., 1986, Taitel et al., 1980). □ (Pettigrew et al., 1989), Δ (Axisa, 1985), ▽ (Pettigrew et al., 1995), O (Feenstra et al., 1995).

Almost every study of flow regimes in tube bundles has concluded that three distinct flow regimes exist. In fact, several studies have shown that these regimes can easily be identified by measuring the probability density function (PDF) of the gas component of the flow (Ulbrich & Mewes, 1997), (Noghrehkar et al., 1995) and (Lian et al., 1997).

7.6 Tube to restraint interaction (wear work-rate)

Significant tube-to-restraint interaction can lead to fretting wear. Large amplitude out-of-plane motion will result in large impact forces and in-plane motion will contribute to rubbing action. Impact force and tube-to-restraint relative motion can be combined to determine work-rate. Work-rate is calculated using the magnitude of the impact force and the effective sliding distance during line contact between the tube and restraint (Chen et al., 1995). The work-rate is given below in Equations 54 and 55.

$$W = \frac{1}{T_s} \int_{i=0}^{n} F_i dS_i \tag{51}$$

$$W = \frac{1}{T_s} \sum_{i=0}^{n} F_i \Delta S_i = \frac{1}{T_s} \sum_{i=0}^{n} \frac{F_i + F_{i+1}}{2} \Delta S_i \tag{52}$$

where F_i is the instantaneous normal force, ΔS_i is the sliding distance during line contact and n is the number of points discretized over the sample duration T_s. As the work-rate increases, the effective wear rate increases and the operational life of the U-bend tube decreases. Implementation of the technology is described in detail by (Fisher et al., 1991). Measured values of wear work-rate for pitch velocity and mass flux (Chen et al., 1995) are presented in Figures 22a and 22b respectively. The effect of fluid-elastic forces is very evident in the measured work-rates.

It is interesting to note that at higher pitch velocities and/or mass fluxes, the wear work-rate does not increase. Further study is required to understand why the flow-rates do not affect the work-rates. This may be related to the fact that at high void fractions and high flow rates the random excitation forces are constant with increasing flow rate (Taylor, 1992).

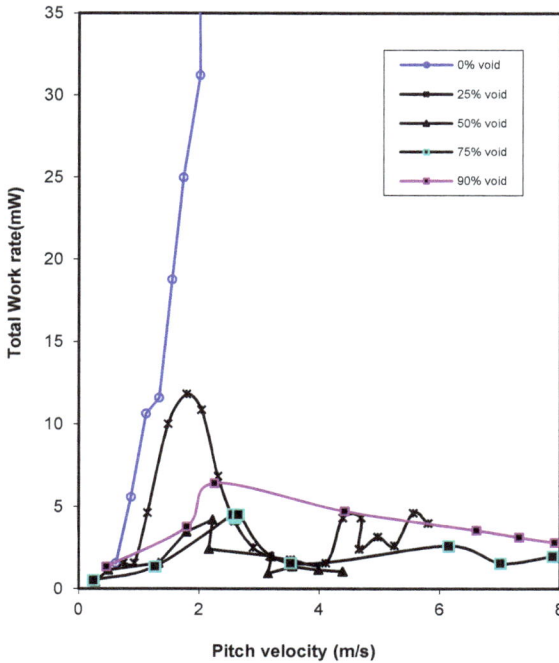

Fig. 22(a). Measured work-rate versus pitch velocity (Chen et al., 1995)

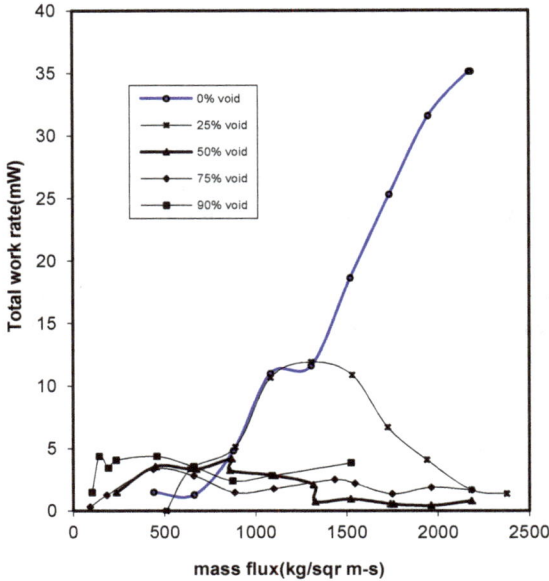

Fig. 22(b). Measured work-rate versus mass flux (Chen et al., 1995).

7.7 Measurement of void fraction

In general, the surveyed research indicates two types of void fraction measurements (Feentra et al., 2000). The HEM void fraction and RAD void fraction. HEM refers to Homogeneous Equilibrium Model and RAD refers to Radiation Attenuation Method. The determination of fluid parameters (fluid density and flow velocity) are quite different when these two methods are used (Feentra et al., 2000). In RAD method (Feenstra et al., 2000, Wright & Bannister, 1970) gamma flux from radiation source which penetrates the test section will be attenuated by different amounts depending upon the average density of the two-phase flow. Void fraction α can be determined by interpolating the average density of the fluid between the benchmark measurements for one hundred percent liquid and gas according to the following equation.

$$\alpha = \ln(N / N_L) / \ln(N_G / N_L) \tag{53}$$

where N represents the gamma counts obtained during an experimental trial, N_L and N_G are the reference counts obtained prior to the experiment for 100% liquid and 100% gas respectively. Gas phase velocity, U_G, and liquid phase velocity U_L can be calculated by Equations below:

$$U_G = \frac{xG_P}{\alpha \rho_G} \tag{54}$$

$$U_L = \frac{(1-x)G_P}{(1-\alpha)\rho_L} \tag{55}$$

where G_p is the pitch mass flux.

A logical measure of an equivalent two-phase velocity, V_{eq} is determined from averaging the dynamic head of the gas and liquid phases as given by equation below:

$$V_{eq} = \sqrt{[\alpha\rho_G U_G^2 + (1-\alpha)\rho_L U_L^2]/\rho} \tag{56}$$

8. Conclusions

Loss of Millions of Dollars through Cross-Flow-Induced-Vibrations related problems in steam generators and heat exchangers excitations has been a cause of major concern in process, power generation and nuclear industries. Flow-Induced Vibration pose a potential problem to designers, process engineers and plant operating and maintenance personnel. Such vibrations lead to motion of tubes in loose supports of baffles of tube bundles, resulting in mechanical damage, fretting wear, leaking and fatigue etc. Heat exchanger tubes are the most flexible components of the assembly. The risk of radiation exposure is always present in case of leakage in steam generator of PWR plants due to vibration related tube failures.

A number of design consideration have been reviewed in this chapter in order to achieve design improvements to support large scale heat exchangers with increased shell-side cross-flow-velocities. The prime consideration is the natural frequency of tubes in a bundle against cross-flow-induced-vibrations. Various analytical, experimental and computational techniques for straight & curved tubes have been discussed with reference to single and multiple spans and varying end and intermediate support conditions. Earlier, Flow-Induced-Vibration analysis was based upon the concept of two types of damage numbers (Collision damage and baffle damage). Discussion on these damage numbers and on the parameters that influence damping has been included.

Next consideration is the generally accepted following four tube bundle vibration excitation mechanisms (various models have been discussed & reviewed) including steady, unsteady, analytical, FEM based, CFD based, experimental, empirical correlation based, large eddy simulation (LES) based, linear and non-linear etc.

Turbulent Buffeting	-	It can not be avoided in Heat Exchangers and is caused due to turbulence.
Vorticity Excitation	-	Vortex shedding or periodic wake Shedding
Fluid-Elastic Instability	-	Self excited vibration resulting from interaction of tube motion and flow is the most dangerous excitation mechanism.
Acoustic Resonance	-	Caused by some flow excitation having frequency which coincides with natural frequency.

Dynamic parameters like added mass and damping which are function of geometry, density of fluid and tube size have been targeted by a number of researches in single-phase and two-phase flow. These researches have identified seven separate sources of damping which have been highlighted.

Tube wear due to non-linear tube-to-tube support plate interactions caused by gap clearances between interacting components resulting in thickness loss and normal wear work-rates have been reviewed. Chaotic dynamics of tubes impacting generally on loose baffle plates with consideration of stability and bifurcation have been discussed.

Two-phase Cross-Flow-Induced-Vibrations in tube bundles of process heat exchanger and U-bend region of Nuclear Steam generators can cause serious tube factures by fatigue and fretting wear. Solution to such problems require understanding of vibration excitation and damping mechanism in two-phase flow. This further requires consideration of different flow regimes which characterize two-phase flow. The discussion includes the most important parameter which is void fraction, various thermal-hydraulic models, dynamic parameters, wear work-rates, void fraction measurement and application of TEMA/ASME and other codes have been reviewed. In conclusion the objective of this chapter is to suggest improvements in the design guidelines from the available researches to use the related equipment at optimal performance level.

9. Acknowledgements

We are deeply indebted to University of Engineering & Technology, Taxila – Pakistan, PASTIC, Islamabad – Pakistan and College of EME NUST, Rawalpindi – Pakistan for providing financial, administrative and technical support. We sincerely appreciate the support provided by Mr. Zahid Iqbal, Mr. Riffat Iqbal and Mr. Muhammad Shafique in finalizing the manuscript.

10. References

Antunes, J., Villard, B., Axisa, F., 1985," Cross-flow induced vibration of U-bend tubes of a steam-generator ", Proceedings of SMiRT, F-16/9, pp. 229-239.

Archer, R. R., 1960, "Small vibrations of thin incomplete circular rings", International Journal of Mech. Science, Vol. 1, pp. 45-66.

Attia, M. H. 2006$_a$, "Fretting fatigue and wear damage of structural components in nuclear power stations-Fitness for service and life management perspective," Tribolgy International, 39, 1294-1304.

Attia, M. H. 2006$_b$, "On the fretting wear mechanism of Zr-alloys," Tribolgy International, 39, 1320-1326.

Au-Yang, M. K., 1998, "Flow-induced wear in steam generator tubes-prediction versus operational experience", ASME Journal of Pressure Vessel Technology, Vol. 120, pp. 138-143.

Au-Yang, M. K., 2000, "Joint and cross-acceptances for cross-flow induced Vibration-Part I Theoretical and Finite Element Formulations", Journal of Pressure Vessel Technology Vol. 122, pp.349 -354.

Au-Yang, M. K., 2000, "The crossing frequency as a measure of heat exchanger support plate effectiveness", Proceedings of the 7th International Conference on Flow-

Induced Vibrations, FIV-2000, Lucerne, Switzerland, June 19-22, ISBN 90 5809, 1295, pp. 497-504.

Axisa, F., and Izquierdo, P., 1992, "Experiments on vibro-Impact-Dynamics of loosely supported tubes under harmonic excitation", Proceedings of International Symposium on Flow-Induced Vibration and Noise, Vol. 2, ASME, PVP. 242, eds., Paidoussis, M. J., Chen, S.S., and Steininger, D.A.

Axisa, F., Antunes, J. and Villard, B., 1990, "Random excitation of heat exchanger tubes by cross-flows", Journal of Fluid and Structures, Vol. 4, p. 321.

Axisa, F., Villard, B., Gibert, R. J., Hostroni, G., and Sundheimer, P., 1984, "Vibration of tube bundles subjected to air-water and steam-water cross-flow", ASME Winter Annual Meeting, p. 269.

Balsa, T. F., 1977, "Potential flow interactions in an array of cylinders in cross-flow", Journal of Sound and Vibration, Vol. 50, pp. 285-303.

Barrington, E. A., 1973, "Experience With acoustic Vibrations in tubular exchangers", Chem. Eng. Prog. Vol. 69. No. 7, pp. 62-68.

Blevins, R. D., 1974, "Fluid-elastic whirling of a tube row", ASME Journal of Pressure Vessel Technology, Vol. 96, pp. 263-267.

Blevins, R. D., 1977, "Flow-induced vibration", Van Nostrand Reinhold Company.

Blevins, R. D., 1979, "Fluid damping and the whirling instability of tube arrays", In flow-Induced Vibrations (eds. S. S., Chen and M.D. Bernstein), New York: ASME, pp. 35-39.

Blevins, R. D., 1990, "Flow-Induced Vibration", Second Edition, Van Nostrand Reinhold Co., New York, N.Y.

Botros K. K., and Price G., 2000, "A case study of fluid -elastic instability of a large heat exchanger in a petrochemical process plant", Proceedings of the 7th International Conference on Flow-Induced Vibrations, FIV-2000, Lucerne, Switzerland, June 19-22, ISBN 90 5809, 1295, pp. 489-496.

Brothman, A., Devore, A., Hollar, G.B., Horowitz, A., and Lee, H.T., 1974, "A tube vibration analysis method", AIChE Symposium Series No. 138, Vol. 70, pp. 190-204.

Burgreen, D., Byrnes, J. J., and Benforado, 1958, "Vibration of rods induced by water in parallel flow", Trans. ASME, Vol. 80, pp. 991-1003.

Carlucci, L. N., 1980, "Damping and hydrodynamic mass of a cylinder in simulated two-phase flow", Journal of Mech. Design, Vol. 102, pp. 597-602.

Carlucci, L. N., and Brown, J.D., 1983, "Experimental studies of damping and hydrodynamic mass of a cylinder in confined two-phase flow", ASME Journal of Vibration, Acoustics, Stress Reliability in Design, Vol. 105, pp.83-89.

Charpentier, J., and Payen, Th., 2000, "Prediction of wear work-rate and thickness loss in tube bundles under cross-flow by a probabilistic approach", Proceedings of the 7th International Conference on Flow-Induced Vibrations, FIV-2000, Lucerne, Switzerland, June 19-22, ISBN 90 5809, 1295, eds., Ziada, S., and Staubli, T, pp.521-528.

Chen, S. S., 1983, "Instability mechanisms and stability criteria of a group of circular cylinders subject to cross-flow", Part-II: Numerical Results and Discussion, Journal of Vibration, Acoustics, Stress and Reliability in Design, Vol. 105, pp. 253-260.

Chen, S. S., 1984, "Guidelines for the instability flow velocity of tube arrays in cross-flow", Journal of Sound and Vibration, Vol. 93, pp. 439-455.

Chen, S. S., 1987, "A general theory for the dynamic instability of tube arrays in cross-flow", Journal of Fluids and Structures, Vol. 1, pp.35-53.

Chen, S. S., 1989, "Some issues concerning fluid-elastic instability of a group of cylinders in cross-flow", ASME Journal of Pressure Vessel Technology, Vol. 111, pp. 507-518.

Chen, S. S., 1990, "Unsteady fluid forces and fluid-elastic vibration of a group of circular cylinders", Proceedings of ASME, Forum on unsteady flow (ed. Rathe, P.M.) ASME Publication, PVP, Vol. 204, pp. 1-6.

Chen, S. S., Zhu, S., and Cai, Y., 1995, "Experiment of chaotic vibration of loosely supported tube in cross-flow", ASME Journal of Pressure Vessel Technology, Vol. 117, pp. 204-211.

Chen, S.S., 1987, "Flow-induced vibration of circular cylindrical structures", Washington: Hemisphere Publishing.

Chen, S.S., and Wambsganss, M.W., 1974, "Design guide for calculating natural frequencies of straight and curved beams on multiple supports", ANL-CT-74-06, Components Technology Division, Argonne National Laboratory, Argonne, Illinois.

Chen, S.S., Zhu, S., and Jendrzejczyk, J.A., 1994, "Fluid damping and fluid stuffiness of a tube row in cross-flow", ASME Journal of Pressure Vessel Technology, Vol. 116, pp. 370-383.

Chen, Y. N., 1968, "Flow-induced vibration and noise in tube bank of heat exchangers due to von Karaman Streets", ASME Journal of Engineering Industry, Vol. 90, pp. 134-146.

Chenoweth, J. M., Chisholm, D., Cowie, R. C., Harris, D., Illingworth, A., Loncaster, J. F., Morris, M., Murray, I., North, C., Ruiz, C., Saunders, E. A. D., Shipes, K.V., Dennis Usher, and Webb, R. L., 1993, Heat Exchanger Design Handbook HEDH, Hemisphere Publishing Corporation.

Chenoweth, J.M., 1976, "Flow-Induced vibrations in shell-and-tube heat exchangers", Final report: Research on Heat Exchanger Tube Vibration, Contract No. EY-76-C-03-1273, Workshop, June 27-28, Pasadena, California.

Chu, I. C., Chung, H. J., & Lee, S. 2011, "Flow-induced vibration of nuclear steam generator U-tubes in two-phase flow," Nuclear Engineering and Design, 241, 1508–1515.

Chung, H. J., & Chu, I. C. 2006, "Fluid-Elastic Instability Of Rotated Square Tube Array In An Air-Water Two-Phase Cross-Flow," Nuclear Engineering And Technology, Vol.38, No.1.

Coit, R. L., Ritland, P. D., Rabas, T. J., and Viscovich, P. W., 1974, "Moisture Separator Reheater: Entering the Second Decade, Two-phase flow in turbines", Short course, Rhode-Saint-Genese, Belguim.

Collier, J., 1979, "Convective boiling and condensation", Oxford Science Publication.

Connors, H. J., and Parrondo, J., 1970, "Fluid-elastic vibration of tube arrays excited by cross-flow", In Flow-Induced Vibration in Heat Exchangers (ed. D. D., Reiff), New York: ASME, pp. 42-56.

Dalton, C., 1980, "Inertia coefficients for riser configurations", In Energy Technology Conference and Exhibition, New Orleans, paper 80-Pet-21, York: ASME.

Dalton, C., and Helfinstein, R. A., 1971, "Potential flow past a group of circular cylinders", ASME Journal of Basic Engineering, Vol. 93, pp. 636-642.

Davis (Jr), F.J., and Hassan, Y. A., 1993, "A two-dimensional finite element method large eddy Simulation for application to turbulent steam generator flow".

Delaigue, D., and Planchard, J., 1986, "Homogenization of potential flow models for the dynamics of cylinder arrays in transient cross-flow", In ASME Symposium on Flow-Induced Vibrations (ed S. S. Chen, J. C. Simonis and Y. S. Shin), PVP-Vol. 104, New York: ASME, pp. 139-145.

Eisinger, F. L., & Sullivan, R. E. 2005, "Acoustic Vibration Behavior of Full Size Steam Generator and Tubular Heat Exchanger In-line Tube Banks," Proceedings of ASME Pressure Vessels & Piping Division Conference, July 17- 21, Denver USA.

Radiochemical Separation of Nickel for ^{59}Ni and ^{63}Ni Activity Determination in Nuclear Waste Samples

Aluísio Sousa Reis, Júnior, Eliane S. C. Temba,
Geraldo F. Kastner and Roberto P. G. Monteiro
Centro de Desenvolvimento da Tecnologia Nuclear – (CDTN)
Brazil

1. Introduction

For legal and regulatory purposes, the International Atomic Energy Agency (IAEA, 1994) defines radioactive waste as "waste that contains or is contaminated with radionuclides at concentrations or radioactivity levels greater than clearance levels as established by the regulatory body". The radioactive wastes are residues that have been produced by human nuclear activity and for which no future use is foreseen. Besides the nuclear power plants, the nuclear weapons testing, medical uses and various research studies involve a large number of radionuclides. In particular the nuclear accidents such as Three Mile Island Nuclear Power Station, where some gas and water were vented to the environment around the reactor, Chernobyl Nuclear Power Plant, the effects of the disaster were very widespread and Fukushima II Nuclear Power Plant have also released a large amount of radionuclides to environment.

In the case of radioactive wastes each country has its own classification, in general we can identify three types of wastes, that are, Low Level Waste (LLW), the LLW wastes contain primarily short lived radionuclides which refer to half-lives shorter than or equal to 30-year half-life, Intermediate Level Waste (ILW), radioactive non-fuel waste, containing sufficient quantities of long-lived radionuclides which refer to half-lives greater than 30 years. And a third one that is High Level Waste (HLW), arise from the reprocessing of spent fuel from nuclear power reactors to recover uranium and plutonium, containing fission products that are high radioactive, heat generating and long-lived. We would like to call attention to the fact that the waste classification LLW, ILW, HLW used here is only one of several alternative schemes; we adopted the simplest one.

Identification and characterization of radioactive wastes is a technical challenge because of their importance in choosing the appropriate permanent storage mode or further processing. Characterization definition of nuclear waste by IAEA (IAEA, 2003) is "the determination of the physical, chemical and radiological properties of the waste to establish the need for further adjustment, treatment, conditioning, or its suitability for further handling, processing, storage or disposal. Thus, it involves a collection of data that pertains to specific waste properties as well as processing parameters and quality assurance, some of which include the following: thermal, mechanical, physical, biological, chemical and radioactivity properties (IAEA, 2007).

Testing and analyzes to demonstrate the radioactive content and the quality of final waste forms and waste packages are key components of this knowledge and control and are essential to accurate characterization of the waste. Physical characterization involves inspection of the waste to determine its physical state (solid, liquid or gaseous), size and weight, compactability, volatility and solubility, including closed waste packages which can be done using a variety of techniques, such as radiography (X-ray). Chemical waste characterization involves the determination of the chemical components and properties of the waste that is, potential chemical hazard, corrosion resistance, organic content, reactivity. This is most often done by chemical analysis of a waste sample. The radioactive inventory of various materials needs to be assessed for the classification of the nuclear waste. Radiological waste characterization involves detecting the presence of individual radionuclides and its properties such as half-life, intensity of penetrating radiation, activity and concentration and quantifying their inventories in the waste. This can be done by a variety of techniques, such as radiometric methods, mass spectrometric methods depending on the waste form, radionuclides involved and level of detail/accuracy required.

Furthermore, for developing a scaling factor (IAEA, 2009) to be applicable to the assessment of the radioactive inventory of the wastes with various matrices, it is indispensable to prepare a database compiled with a large numbers of information related to the radioactive inventory of long lived alpha and beta emitting nuclides which are difficult to measure (DTM) and gamma emitting nuclides which are easy to measure (ETM). It is necessary to develop analytical techniques for the DTM nuclides.

The aim of this work was to develop a sensitive analytical procedure for simultaneous determination of radionuclides difficult to measure. Between them is the ^{59}Ni and ^{63}Ni determination in low and intermediate level wastes from Brazilian Nuclear Power Plants – Eletrobrás Termonuclear according to an analytical protocol developed based on sequential separation of different radionuclides presents in the waste matrices (Reis et al, 2011). Sources for ^{59}Ni are austenitic steel in the reactor and activation of nickel dissolved in the coolant and in corrosion particles deposited on the core. The content of nickel in stainless steel is around to 10% and in Inconel in the range of 50–75%. Furthermore, nickel is found as an impurity in Zircaloy, ~ 40 ppm, and in reactor fuel, ~ 20 ppm (Lingren et al, 2007).

2. The radiometric detection and techniques for ^{59}Ni and ^{63}Ni

Radioactive wastes are residues with different radionuclide compositions, placing, therefore considerable demands by measurement techniques used in their characterization. All radioisotopes, at some stage, require quantitation of the isotope, which is done by measuring the intensity of radiation emitted for the three main types of ionizing radiation. Radioactive isotopes of elements are normally determined by their characteristic radiation, i. e., by radiometric methods. Radiometric determination is performed by instrumental analysis using sophisticated methods such as liquid scintillation counters that allow beta spectrometry, alpha spectrometry with semiconductor detectors and high resolution gamma spectrometry for high and low energy gamma emitting nuclides. Besides, mass spectrometric methods can be also used for the determination of radionuclides once they are normally used for determination of isotopes of elements.

There are several types of detectors that can be used for the measurement of ionizing radiation. In the specific case of ^{59}Ni and ^{63}Ni, the more common radiation detection systems are ultra low energy gamma detection and liquid scintillation detection on the basis of charge carriers (holes and electrons) and liquid scintillation phenomena, respectively. Furthermore, for sequential analysis of these radionuclides alpha spectrometry can be applied to alpha emitters associated and presents in ILW and LLW samples.

2.1 The semiconductors detectors

The most recent class of detector developed is the *solid-state semiconductor detector*. In these detectors, radiation is measured by means of the number of charge carriers set free in the detector, which is arranged between two electrodes. Ionizing radiation produces free electrons and holes. The number of electron-hole pairs is proportional to the energy transmitted by the radiation to the semiconductor. As a result, a number of electrons are transferred from the valence band to the conduction band, and an equal number of holes are created in the valence band. Under the influence of an electric field, electrons and holes travel to the electrodes, where they result in a pulse that can be measured in an outer circuit. Solid state detectors are fabricated from a variety of materials including: germanium, silicon, cadmium telluride, mercuric iodide, and cadmium zinc telluride.

Germanium detectors are mostly used for spectrometry in nuclear physics and chemistry. The Ultra Low Energy Germanium (Ultra-LEGe) detectors extends the performance range of germanium detectors down to a few hundred electron volts, providing resolution, peak shape, and peak-to-background ratios once thought to be unattainable with semiconductor detectors. According to it specification this detector offers excellent performance over a wide range of detector sizes. The resolution, for example, of a 100 mm^2 Ultra-LEGe is less than 150 eV in terms of full-width-half-maximum (FWHM) at 5.9 keV.

Radionuclides commonly emit gamma rays in the energy range from a few keV to ~10 MeV, corresponding to the typical energy levels in nuclei with reasonably long lifetimes. The boundary between gamma rays and X rays is somewhat blurred, as X rays typically refer to the high energy electromagnetic emission of atoms, which may extend to over 100 keV, whereas the lowest energy emissions of nuclei are typically termed gamma rays, even though their energies may be below 20 keV. Therefore, ^{59}Ni that decays by electron capture with emission of 6.9 keV X-rays is suitable to be detected by low energy gamma spectroscopy using Ultra Low Energy Germanium detectors.

2.2 The liquid scintillation counting

Beta emitting radionuclides are normally measured by a gas ionization detector or liquid scintillation counting (LSC). In LSC the scintillation takes place in a solution, the cocktails contain two basic components, the solvent and the scintillator(s). This allows close contact between the isotope atoms and the scintillator what becomes an advantage in measuring low-energy electron emitters due to the absence of attenuation. Once the solvent must act as an efficient collector of energy, and it must conduct that energy to the scintillator molecules instead of dissipating the energy by some other mechanism (National Diagnostics, 2004). Liquid scintillation cocktails absorb the energy emitted by radioisotopes and re-emit it as flashes of light. A β particle, passing through a scintillation cocktail, leaves a trail of energized solvent molecules. These excited solvent molecules transfer their energy to scintillator molecules, which give off light. With LSC the short path length of soft β emissions is not an obstacle to detection. LSC can thus be used for the measurement of both high and low energy emitters.

A pulse height spectrum is a representation of the average kinetic energy associated with the decay of a particular isotope. When an isotope decays it liberates an electron or beta particle and a neutrino that have the energy associated shared between the two particles. As a result of that the resulting beta particles have a continuous distribution of energies from 0 to maximum decay energy (E_{max}). The amount of light energy given off is proportional to the amount of energy associated with the beta particle. Therefore, the beta decay shows a continuous energy distribution and beta particle spectrometry becomes an analytical thecnique in which it is difficult to identify individual contributions in the spectrum beta. The determination of various beta emitters such as ^{3}H, ^{14}C, ^{63}Ni, ^{55}Fe, ^{90}Sr requires chemical separation of the individual radionuclides from the matrix and from the other radionuclides before couting.

The isotope ^{63}Ni is an artificial radionuclide. It is a pure β emitter with a half-life of 100 years. The maximum energy of the emitted β-radiation is 67 keV. No γ radiation is observed. Except ^{59}Ni with a half-life of 7.6 x 10^4 years all nickel radionuclides have very short half-lifes. They range between 18 seconds and 54.6 hours. Therefore they don´t disturb a measurement of ^{63}Ni. Besides, LSC has a high couting efficiency for ^{63}Ni, about 70%., i. e., the ratio cpm/dpm, counts per minute to disintegration per minute expressed as a percentage, in other words, the percentage of emission events that produce a detectable pulse of photons, making the technique widely used for the determination of ^{63}Ni.

2.3 The alpha spectrometry

The sequential analyses determine in addition to ^{59}Ni and ^{63}Ni others DTM´s present in the nuclear waste including alpha emitters. Therefore, alpha spectrometry is one complementary technique for the nuclear waste characterization either ILW or LLW.

In this technique to achieve results with good quality, the sample must be converted into a chemically isolated, thin layered and uniform source. The preparation of an alpha sample contains three basic steps: preliminary treatment, chemical separation and source preparation

Alpha-emitting radioisotopes spontaneously produce alpha particles at characteristic energies usually between about 4 and 6 MeV. Alpha particles (or ^{4}He nuclei) are heavy charged, large and slow particles and loses some of its energy each time it produces an ion (its positive charge pulls electrons away from atoms in its path), finally acquiring two electrons from an atom at the end of its path to become a complete helium atom. These attenuation characteristics, which manifest themselves both within the sample and with any materials between the sample and the active detector volume, cause a characteristic tailing in the alpha peak. When tailing occurs (it is also called "spill down"), the accuracy with which the peak areas can be determined is compromised because the peaks tend to have an asymmetric shape rather than the Gaussian shape.

The alpha particle energies of many isotopes differ by as little as 10 to 20 keV (Canberra, n.d.). The relatively small difference in alpha particle energy between some alpha emitters makes it difficult to spectrometrically separate the peaks once this is near the resolution of the silicon detectors used in alpha spectrometers. If two of these alpha particle energies are so close, they cannot be spectrometrically separated and if they are chemically the same, they cannot be chemically separated and analyzed.

Resolution is the ability of the spectrometry system to differentiate between two different alpha particles and its quantitative measure is the FWHM. Besides, a FWHM of about 15 keV can be achieved with electroplated sources because they have very little mass to slow down the alpha particles. For this reason it is essential that a thin source to be prepared in alpha spectrometry.

3. Radiochemical for radionuclides difficult to measure

3.1 Radiochemical separation

The methods for separating, collecting, and detecting radionuclides are similar to ordinary analytical procedures and employ many of the chemical and physical principles that apply to their nonradioactive isotopes. One of the differences is interesting from the viewpoint of methodology. Substance separation in analytical chemistry in the majority of cases is not an end in itself. In radiochemistry, separation is most often an end in itself, for example, when a radionuclide is purified of other radioactive elements (Zolotov, 2005). Techniques used for separation include co-precipitation, liquid-liquid extraction, ion exchange and extraction chromatography. In some cases, two or more of these techniques are combined.

In order to account for the inevitable loss of the sample during separation, a specific isotope or tracer is added to the sample. A tracer represents the addition to an aliquot of sample a known quantity of a radioactive isotope that is different from that of the isotope of interest but expected to behave in the same way. Sample results are normally corrected based on tracer recovery. The percent of tracer lost in the chemical processes is equal to the percent of sample lost, assuming the tracer is homogeneously mixed with the sample and is brought into chemical equilibrium with the sample. Radiochemical analysis frequently requires the radiochemist to separate and determine radionuclides that are present at extremely small quantities. The amount can be in the picomole range or less, at concentrations in the order of 10^{-15} to 10^{-11} molar (United States Environmental Protection Agency, 2004). The use of a material that is different in isotopic make-up to the analyte and that raises the effective concentration of the material to the macro level is referred to as a carrier, a substance that has a similar crystalline structure that can incorporate the desired element.

Radiochemical waste characterization is the identification of radionuclides contained in a package of nuclear waste and the determination of their concentration. The problem the waste producers have to cope with comes from the fact that those nuclides which are mainly (pure) β- or α-emitters cannot be measured by direct methods such as γ-scanning. In the waste packages produced by a nuclear power reactor the radionuclides may be originated as fission products from the nuclear fuel, activation products and transmutation nuclides, Table 1.

Products	Radionuclides	Decay mode
Fission products from the nuclear fuel	^{90}Sr, ^{99}Tc, , ^{137}Cs, ^{129}I ^{134}Cs	β γ
activation	^{3}H, ^{14}C, ^{94}Nb,^{60}Co, ^{63}Ni, ^{54}Mn ^{55}Fe, ^{59}Ni	β EC
Transmutation nuclides	^{241}Am, ^{242}Cm, ^{244}Cm, ^{235}U , ^{238}U and ^{239}Pu, ^{240}Pu, ^{242}PU ^{241}Pu	α β

Table 1. Radionuclides obtained as products of nuclear power plants and their origin

Identification of these nuclides requires methods that, in general, involve analyses of waste samples using complex chemical analysis to separate the various radionuclides for measurement. Among the various proposed methods there are those who seek the identification of a radionuclide isolated or those seeking to identify by simultaneous determination two or more radionuclides in the same analysis.

The main constraint for a new protocol is to obtain a high recovery yield, a high-energy resolution and low interferences of other radionuclides. Thus, it is necessary to develop accurate and reliable methods for the determination of radionuclides in the low and intermediate radioactive samples. A simultaneous determination procedure was developed for the separation of Pu isotopes, ^{241}Am, ^{242}Cm, ^{244}Cm, ^{89}Sr and ^{90}Sr using precipitation by oxalate, ion exchange resin, extraction of plutonium by TTA (thenoyltrifluoro acetone/benzene) and Sr by precipitation techniques. This method was applied for determination of these radionuclides in the grass, collected near Munich after the fallout from the nuclear accident at Chernobyl (Bunzl & Kracke, 1990). In another case, Pu, Am and Cm were determined by extraction chromatography using an organophosphorus compound immobilized on an inert support commercially available under the name TRU Resin (for Transuranium specific) from Eichrom Technologies, Inc. This method was used in samples from nuclear power plants such as spent ion exchange resins and evaporator concentrates (Rodriguez et al., 1997). Besides, combined procedure was used for the determination of ^{90}Sr, ^{241}Am and Pu isotopes by anion exchange for Pu isotopes analysis, the selective method for Sr isolation based on extraction chromatography using Sr Resin and the TRU Resin for separation of Am (Moreno et al., 1997). In the radiological characterization of low- and intermediate-level radioactive wastes the separation of Pu isotopes, ^{241}Am, ^{237}Np and ^{90}Sr was performed by anion-exchange chromatography, extraction chromatography, using TRU and Sr Resin, and precipitation techniques (Tavcar et al., 2007).

3.2 Combined procedure for Ni radionuclides separation

An analytical procedure for radiochemical characterization of radioactive waste material containing some of the radionuclides cited in Table 1 was developed. Radionuclides ^{242}Pu, ^{238}Pu, $^{239 + 240}$Pu, ^{241}Am, ^{235}U and ^{238}U were determined by alpha spectrometry whilst ^{241}Pu, ^{90}Sr, ^{55}Fe and ^{63}Ni were determined by LSC and ^{59}Ni by low energy gamma spectrometry. ^{242}Pu, ^{238}Pu, ^{243}Am and ^{232}U were used as tracers and Sr (2 mg/mL), Fe (3 mg/mL) and Ni (2 mg/mL) were used as carriers. In this work was developed a sensitive method for sequential analyses of the radionuclides in samples of radioactive waste. The samples analyzed were evaporator concentrate, resin and filter originated from Brazilian Nuclear Power Plants located at Angra dos Reis city (Reis et al, 2011).

The radiochemical procedure consists of three steps performed by anion-exchange chromatography, precipitation techniques and extraction chromatography, using TRU, Sr and Ni resins. In the first step, it was made the separation of ^{242}Pu, ^{238}Pu, $^{239+240}$Pu and ^{241}Pu of the matrix by ion exchange chromatography using an anion exchange column (Dowex 1X8, Cl-form, 100-200 mesh, Sigma Chemical Co., USA). The separation is based on the formation of anionic complexes of Pu (IV) with NO_3^- or Cl^- in concentrated HNO_3 or HCl. In the second one, the effluent from the exchange column was used to separate Am and Sr by co-precipitation with oxalic acid of Fe, U and Ni that are retained in the filtrate. Americium and Sr isolation was done using commercially available resins, TRU resin and Sr Resin, respectively. These resins can be used for a number of analytical purposes, including the separation of actinides as a group from the matrix, separation of Sr from the matrix and sequential separation of individual actinides and Sr. In the third step Ni was separated by co-precipitation of Fe and U. And after that, Fe and U were separated by ion exchange chromatography using the anion exchange column (Dowex 1X8, Cl form, and 100-200 mesh) and Ni was isolated by Ni Resin extraction chromatography column from Eichrom Technologies, Inc. This work represents a fundamental step in establishing an analytical protocol for radioactive waste management system.

The safety planning for disposal of LLW and ILW radioactive waste takes account in special long half-life radionuclides. Both ^{59}Ni and ^{63}Ni are activation products of stable nickel, which was present as an impurity in fuel cladding materials or the uranium fuel of reactors (Kaye et al., 1994). ^{59}Ni (half-life 7.6×10^4 years) is produced by neutron irradiation of ^{58}Ni and decays by electron capture to stable ^{59}Co with emission of 6.9 keV x-rays. ^{63}Ni (half-life 100 years) emits only low-energy beta rays with a maximum energy of 67 keV, and is produced through neutron irradiation of ^{62}Ni. Counting requirements dictated that prior the measurement these isotopes should be separated and purified with the purpose of removing the radiometric and chemical interferent elements so that they are essentially free of significant radioactive contamination.

Hou (Hou et al., 2005) proposed an analytical method for the determination of ^{63}Ni and ^{55}Fe in nuclear waste samples. Hydroxide precipitation was used to separate ^{63}Ni and ^{55}Fe from the interfering radionuclides as well as from each other. The separated ^{63}Ni was further purified by extraction chromatography. According to him the recovery of Fe and Ni by hydroxide precipitation using NH_4OH, was about 99, 9% and 21, 9%, respectively. Lee (Lee et al., 2007) proposed a sequential separation procedure developed for determination of ^{99}Tc, ^{94}Nb, ^{55}Fe, ^{90}Sr and $^{59/63}$Ni in various radioactive wastes. Ion exchange and extraction chromatography were adopted for the individual separation of the radionuclides. According to him Ni separation on the cation-exchange resin column was not selective enough therefore a further purification of Ni was performed by precipitation with dimetylglyoxime.

The aim of this work is the sequential analysis of nuclear waste containing several radionuclides (Pu, U, Am, Sr, Fe e Ni) where the last step consists in the separation of U, Fe and Ni. Thus we established the procedure for sequential separation of Pu, Am, Sr (Reis et al., 2011) in which we also included one step that is the hydroxide precipitation to separate U and Fe from Ni because Ni remains in solution in the co-precipitation of U and Fe.

3.3 Experimental

3.3.1 Reagents and apparatus

All reagents used were analytical grade. The detection of radioactive ^{63}Ni was carried out by Liquid Scintillation Counting (LSC), using the Quantulus 1220 spectrometer, the vials used were the 20 mL polyethylene and the scintillation cocktail was the Optiphase Hisafe 3, all from PerKinElmer Inc. (PerkinElmer Inc., Finland). The column materials used in the analysis were Ni Resin in pre-packed 2 mL columns, 100-150 μ particle size, an extraction chromatographic material available from Eichrom Technologies (USA) and the anion exchange resin Dowex 1x8, Cl- form, from Sigma-Aldrich Chemical Co., (USA). ^{59}Ni was analyzed using Ultra-LEGe Detector (GUL) with a cryostat window of beryllium low energy γ-detector containing an active area of 100 mm², efficiency 5.9 keV for ^{55}Fe with a resolution of 160 eV in terms of FWHM, from Canberra (USA). The recovery was obtained analyzing stable nickel by ICP-AES.

3.3.2 Separation and purification of nickel

The sequential determination is based on radiochemical procedure that consists of three steps performed by anion-exchange chromatography, extraction chromatography, using Eichrom resins, and precipitation techniques. For each aliquot was added 2 mL of Ni (0.01 mol L^{-1}), 1 mL of Sr (0.02 mol L^{-1}) and 2 mL of Fe (0.01 mol L^{-1}) as carriers and yield monitor. In the first step, the separation of Pu by ion exchange chromatography, anion exchange column (Dowex 1x8, Cl-form. 100-200 mesh, Sigma Chemical Co. USA), is based on the formation of anionic complexes of the Pu(IV) with NO$_3$- or Cl- in concentrated HNO$_3$ or HCl. In the second one the effluent from the anion exchange column was used to separate Am and Sr by co-precipitation with oxalic acid of U, Fe and Ni that remains in the filtrate.

In the third step we use the filtrate to separate Ni from U and Fe. The filtrate was heated to dryness and the solid obtained was dissolved in 30 mL of concentrate nitric acid and heated to dryness in order to destroy the excess of oxalic acid. The solid obtained was hot dissolved in 30 mL of 3:2 nitric acid and was diluted to 200 mL with deionized water. The pH of the solution was corrected to 9.0 with ammonia hydroxide for co-precipitation of iron hydroxide and uranium while Ni forms a soluble [Ni(NH$_3$)$_4$]$^{2+}$complex.

After filtration, the filtrate was heated to dryness and retaken with 20 mL of HCl concentrate and again heated to dryness. The solid obtained was dissolved in 25 mL of 1 mol L^{-1} HCl and was added 1 mL of ammonium citrate to the sample being the pH adjusted to 8-9 with ammonium hydroxide (Eichrom Technologies, 2003). A nickel resin extraction chromatography column (Eichrom Industries Inc. USA) was pre-conditioned with 5 mL of solution 0.2 mol L^{-1} ammonium citrate that has been adjusted to pH 8-9 with ammonium hydroxide. The column was loaded with the sample and rinsed with 20 mL of solution 0.2 mol L^{-1} ammonium citrate. Nickel was eluted with 10 mL of solution 3 mol L^{-1} HNO$_3$. Figure 1 represents the flowchart for sequential separation of radionuclides in a sample of radioactive waste.

3.3.3 Determination of ^{59}Ni by ultra low energy germanium detection

It was taken an aliquot of 3 mL from the 10 mL solution 3 mol L^{-1} HNO$_3$ eluted of the column. Measurements of ^{59}Ni were performed with Ultra-LEGe Detector (GUL) with a cryostat window of Beryllium low energy γ-detector containing an active area of 100 mm², and resolution less than 150 eV (FWHM) at 5.9 keV, from Canberra Industries (USA).

3.3.4 Determination of ⁶³Ni by LSC

It was taken an aliquot of 3 mL collected in a scintillation vial from the 10 mL solution 3 mol L⁻¹ HNO₃ eluted of the column,. It was added 17 mL of the scintillation cocktail and the vial was shaken vigorously. Before counting, in order of minimizing luminescence interferences, the vial was stored in the dark for 24 hours.

In order to calibrate the counter and to determine the counting conditions, it was prepared a ⁶³Ni standard solution and a blank solution, in the same conditions of the sample. The counting conditions set up were a time of counting of 60 minutes and a channel interval of 50-400.

3.4 Results and discussion

In the ⁶³Ni analysis by LSC the following parameters were determined. The counting efficiency was obtained by the Equation 1.

$$Eff = \frac{R_{st} - R_b}{A_{st} \cdot 60 \cdot Y} \tag{1}$$

where R_{st} is the count rate in counts per minute (cpm) of the ⁶³Ni standard, R_b is the cpm of the blank, Y is the chemical yield and A_{st} is the activity of the standard (in Bq).

The counting efficiency obtained was 71.5 %, with a background of 12.5 ± 1.76 cpm. If we compare with the values determined by Hou (Hou et al., 2005), that is, a counting efficiency of 71.2 % and a background of 1,30 cpm to 30 minutes of counting, in samples of graphite and concrete, it is observed the same efficiency, however, with a background increased. The sample activity was obtained by the Equation 2.

$$A = \frac{R_s - R_b}{Eff \cdot Y \cdot Q \cdot 60} \tag{2}$$

where R_s is the count rate (cpm) of the sample and Q is the quantity of sample.

The detection limit was calculated using the equation proposed by Currie (Currie, 1968) and according to Standard Methods (Standard Methods, 2005), Equation 3, where it is the total counting time for the blank and L_d is the limit of detection with 95 % at confidence level. The 95% confidence level means that, for a large number of observations, 95% of the observations indicate the presence of the analyte, whereas 5% of these observations reflect only random fluctuations in background intensity.

$$L_d = \frac{2.71 + 3.29\sqrt{t \cdot R_b}}{60 \cdot t \cdot Eff \cdot Q} \tag{3}$$

In the ⁵⁹Ni analysis by ultra low energy gamma spectrometry the corresponding gamma peak area (6.3 keV) was correlated with the gamma peak area (5.9 keV) of ⁵⁵Fe, taking account the efficiency curve of the detector. The activities were related with the nominal activity of ^{55}Fe standard solution. The Figures 2 and 3 show typical ^{59}Ni spectra of evaporator concentrate and resin samples.

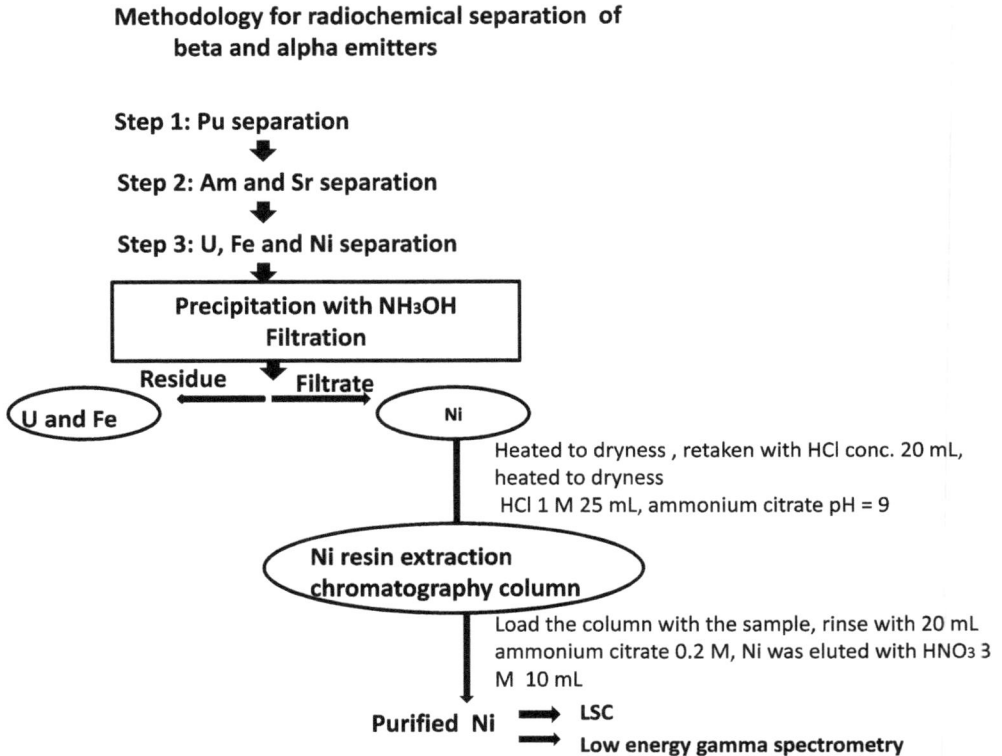

Methodology for radiochemical separation of beta and alpha emitters

Step 1: Pu separation
↓
Step 2: Am and Sr separation
↓
Step 3: U, Fe and Ni separation
↓

Precipitation with NH₃OH Filtration

Residue ↓ **Filtrate**

(**U and Fe**)   (Ni)

Heated to dryness , retaken with HCl conc. 20 mL, heated to dryness
HCl 1 M 25 mL, ammonium citrate pH = 9

(**Ni resin extraction chromatography column**)

Load the column with the sample, rinse with 20 mL ammonium citrate 0.2 M, Ni was eluted with HNO₃ 3 M 10 mL

Purified Ni ➡ **LSC**
➡ **Low energy gamma spectrometry**

Fig. 1. Flowchart for radiochemical separation and purification of Nickel

Hou (Hou & Ross, 2008) related background count rates from 3-10 cpm to LSC mean while the background obtained by us was 12.5 cpm. The L_d obtained was 12.0 Bq/L, and this relatively high value results from the high background count rate. The Figure 4 shows a LSC spectrum for an evaporator concentrate sample according to the parameters as were established.

The chemical yield was 58 % determined by measuring the stable Ni added as carrier using ICP-AES. This value is also used for the calculations of activities of ^{59}Ni.

The results obtained by LSC, using the parameters as set up, and that obtained by low energy γ detection for activities of ^{63}Ni and ^{59}Ni, respectively, are shown in the Table 2. Every measurement presented in Table 2 was considered along with a confidence interval, the uncertainty to the measurement.

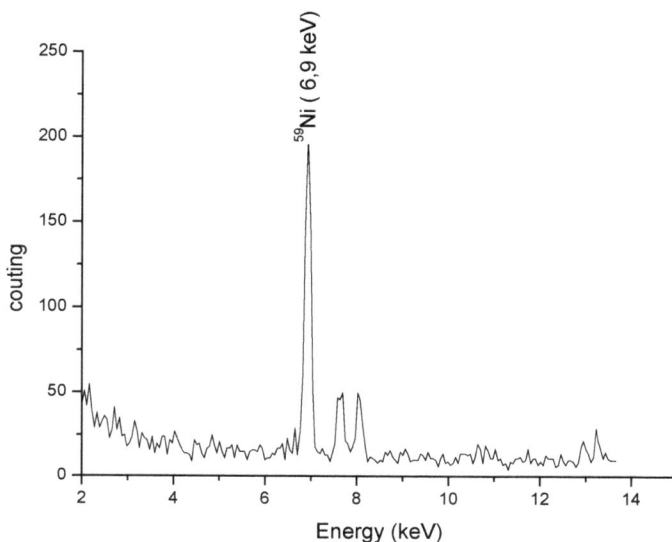

Fig. 2. ^{59}Ni spectrum for a radioactive waste concentrate evaporator sample

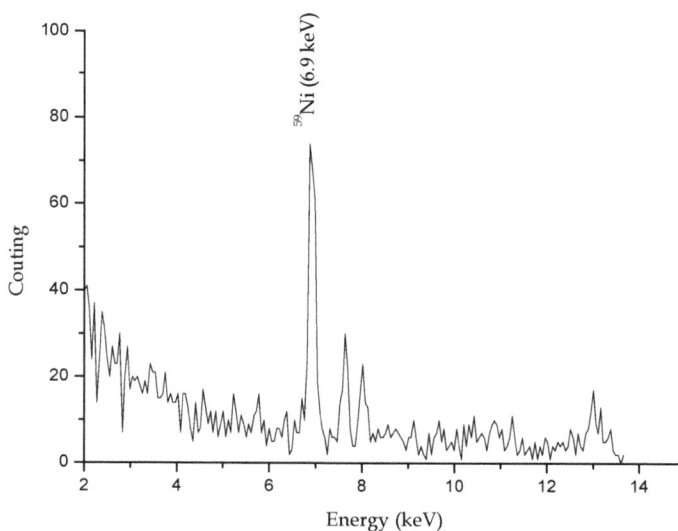

Fig. 3. ^{59}Ni spectrum for a radioactive waste resin sample

Due to the very long half-life of ^{59}Ni its radioactivity in radioactive waste samples is normally much lower than ^{63}Ni (Hou et al., 2005). This was verified by the values obtained for both radioisotopes to the same sample, in the Table 2. According to Scheuerer (Scheuerer et al., 1995) the ratio of activity concentrations of ^{59}Ni to ^{63}Ni is about 0.008 for environmental samples, steel and concrete. For the samples of nuclear waste analyzed by our laboratory the ratio found to vary from 0.03 to 0.14, these values indicate that the concentration activities for ^{63}Ni are yet bigger than that for ^{59}Ni. We can, therefore, consider that these ratios are in accordance with that ratio waited for ^{59}Ni/^{63}Ni. The Figure 4 shows a typical ^{63}Ni spectrum of evaporator concentrate sample.

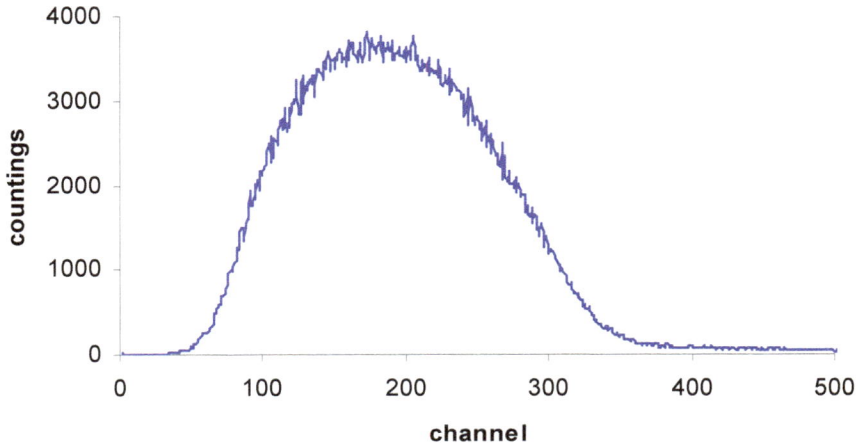

Fig. 4. ^{63}Ni spectrum for an evaporator concentrate sample

Sample	^{63}Ni (Bq g^{-1})	^{59}Ni (Bq g^{-1})
Evaporator concentrate	31 ± 2	0.86 ± 0.04
Ion-exchange Resin	1.48 ± 0.07 x 10^4	2.2 ± 0.1 x 10^3
Filter	5.4 ± 0.3 X 10^3	2.1 ± 0.2 x 10^2

Table 2. ^{63}Ni and ^{59}Ni activities obtained for different types of radioactive waste

4. Conclusions

An analytical procedure for the determination of nickel in nuclear waste samples was developed. The separation of various radionuclides was performed and the use of ammonium hydroxide for separation of nickel from uranium and iron occurred according was proposed. An extraction chromatography was used to purify Ni from the interfering radionuclides. Radionuclides ^{63}Ni and ^{59}Ni were determined by LSC and low gamma energy spectrometry after to be purified by Ni resin. It is possible to indicate that when Ni radioisotopes are analyzed by extraction chromatography there are no interferences in the measurement by the techniques utilized.

The chemical yield for Ni was 58% and the detection limit for LSC was 12 Bq/L and the couting efficiency was 75.1%, thus, experiences indicate so far that the method can be used for the analyses of radionuclides in the waste samples.

The ratio ^{59}Ni/^{63}Ni fell within a range that indicates a higher activity of ^{63}Ni, which was to be expected in view of the difference between the half-lives of radionuclides.

5. Acknowledgements

The authors are very grateful to Eletrobrás Termonuclear for its collaboration and samples supply and to Instituto de Radioproteção e Dosimetria (IRD)-CNEN for its radiotracer standards supply.

6. References

Bunzil, K., Kracke, W., (1990), Simultaneous determination of ^{238}Pu, $^{239+240}$Pu, ^{241}Pu, ^{241}Am, ^{242}Cm, $_{244}$Cm, ^{89}Sr and ^{90}Sr in vegetation samples, and application to Chernobyl-fallout contaminated grass, *Journal of Radioanalytical and Nuclear Chemistry*, 138(1), 83-91.

Canberra, n.d., A Practical Guide to Successful Alpha Spectroscopy, *Thechnical Literature*, 26/08/2011, http://www.canberra.com/literature/953.asp

Currie, A. L., (1968), Limits for Qualitative Detection and Quantitative Determination, *Analytical Chemistry*, 40(3), 586-593.

Eichrom Technologies Inc, (2003). Nickel 63/59 in Water, *Analytical Procedures, Rev. 1.2.* 06/09/2011, http://www.eichrom.com/docs/methods/pdf/niw01-12_ni-water.pdf

Hou, X., and Roos, P., (2008) Critical comparison of radiometric and mass spectrometric methods for the determination of radionuclides in environmental, biological and nuclear waste samples, *Analytica Chimica Acta*, 608, 105-139.

Hou, X., Ostergaard, L. F. and Nielsen, S. P., (2005) Determination of ^{63}Ni and ^{55}Fe in Nuclear Waste Samples using radiochemical separation and liquid scintillation counting, *Analytica Chimica Acta*, 535, 297-307.

INTERNATIONAL ATOMIC ENERGY AGENCY (1994) ,Classification of Radioactive Waste, *SAFETY SERIES* No. 111-G-1.1

INTERNATIONAL ATOMIC ENERGY AGENCY (2003), *Radioactive waste management glossary*, Viena, ISBN 92-0-105303-7

INTERNATIONAL ATOMIC ENERGY AGENCY (2007), Strategy and Methodology for Radioactive Waste Characterization, *IAEA Tecdoc-1537*

INTERNATIONAL ATOMIC ENERGY AGENCY (2009), Determination and use of scaling factors for waste characterization in nuclear power plants, *Nuclear Energy Series*, No. NW-T-1.18, Vienna.

Kaye, J. H., Strebin, R. S. and Nevissi, A. E., (1994) Measurement of ^{63}Ni in Highly Radioactive Hanford Waste by Liquid Scintillation Counting, *Journal of Radioanalytical and Nuclear Chemistry*,180(2), 197-200.

Lee, C. H., Suh, M. Y., Jee, K. Y. and Kim, W. H., (2007) Sequential separation of ^{99}Tc,^{94}Nb, ^{55}Fe, ^{90}Sr and $^{59/63}$Ni from radioactive wastes, *Journal of Radioanalytical and Nuclear Chemistry*, 272(1), 187-194.

Lindgren, M., Petersson M., Wiborgh, M. (2007) Correlation factors for C-14, Cl-36, Ni-59, Ni-63, Mo-93, Tc-99, I-129 and Cs-135, SFR1 SAR-08, Stockholm.

Moreno, J., Vajda, N., Danesi, P. R., Larosa, J. J., Zeiller, F., Sinojimeri, M., (1997), Combined Procedure for the determination of ^{90}Sr, ^{241}Am and Pu radionuclides in soil samples, *Journal of Radioanalytical and Nuclear Chemistry*, 226(1-2), 279-284.

National Diagnostics laboratory staff, (2004), Principles and applications of Liquid Scintillation Couting, 31/08/2011, http://www2.fpm.wisc.edu/safety/radiation/docs/lsc_guide.pdf

Reis, jr A. S., Temba, E. S. C., Kastner, G. F. and Monteiro, R. P. G., (2011) Combined procedure using radiochemical separation of plutonium, americium and uranium radionuclides for alpha-spectrometry, *Journal of Radioanalytical and Nuclear Chemistry*, 287(2), 567-572.

Rodriguez, M., Gascón, J. L., Suárez, J. A., (1997), Study of the interferences in the determination of Pu, Am and Cm in radioactive waste by extraction chromatography, *Talanta*, 45, 181-187.

Scheuerer C., Schupfner, R. and Schüttelkopf, H., (1995), A very sensitive LSC procedure to determine Ni-63 in environmental samples, steel and concrete. *Journal of Radioanalytical and Nuclear Chemistry*, 193(1), 127-131

Standard Methods on Line (2005), 7020 Quality Assurance/Quality Control, 21th edition, 11/11/2011, http:www.standardmethods.org/store/ProductView.cfm?ProductID=358

Tavcar, P., Smodis, B., Benedik, L., (2007) Radiologial characterization of low- and intermediate-level radioactive wastes, *Journal of Radioanalytical and Nuclear Chemistry*, 273(3), 393-396.

United States Environmental Protection Agency, 2004, Separation Techiniques, 29/08/2011, http://www.epa.gov/safetewater/radionulidees/training/resources/MARLAP_1 4_9.pdf

Zolotov, A. Yu., 2005, Seaparation in analytical chemistry and radiochemistry: similarities and differences, *Journal of Analytical Chemistry*, 60(11), 999.

Radiobiological Characterization Environment Around Object "Shelter"

Rashydov Namik et al.[*]
Institute Cell Biology & Genetic Engineering of NAS of Ukraine
Institute for Safety Problems of Nuclear Power Plants NAS of Ukraine
National University "Kyievo-Mogiljanskaja Academy"
Ukraine

1. Introduction

The quarter century away pass after Chornobyl catastrophe. As result there surrounding lands object "Shelter" remain heavily contaminated by long-living radioactivity isotopes for many years to come. Nuclear danger dividing materials of object "Shelter" represent open sources of ionizing radiation (IR) and define not potential only, but direct danger to the personnel and environment. Today in object "Shelter" is about 95 % of highly active fuel loading of a reactor (an order 180 - 190 T of uranium and over 400 kg of plutonium). These danger materials are in different updating - in the form of active zone fragments, warm carried assemblages with the fulfilled nuclear fuel, lava-like fuel-containing materials (LFCM); in dispersion condition (a dust and aerosols); in water solutions of salts of uranium. Fuel containing materials represent congestions glassy state masses in the form of black, brown both polychromic ceramics and pumice state pieces of grey-brown colors. Virtually all materials of "Shelter" are sources of ^{90}Sr, ^{137}Cs, ^{241}Pu, appreciable amounts of plutonium isotopes (^{238,239,240}Pu), as well as ^{241}Am accumulated after the radoactivy decay of isotope ^{241}Pu. Special attention found out and progressively developing intensive destruction of LFCM with formation highly dispersed "hot" particles (HP) on a surface.

It is shown, that dust generation ability of the fulfilled nuclear fuel of object "Shelter" from LFCM various type has high enough level both in the air environment, and in high vacuum. Annual total dust generation in object "Shelter" only at the expense of this mechanism it appears at level of several tens kg's of the irradiated fuel. In HP activity averages β - and γ-activity radionuclide decrease according to a half-life period about 30 years, and α-particles, on the contrary, increase at the expense of accumulation ^{241}Am as disintegration product ^{241}Pu additionally. HP from object "Shelter" are considered as the most radiation dangerous for biota because of the big factor of adjournment in breathe bodies (more than 40 %) and

[*] Kliuchnikov Olexander, Seniuk Olga, Gorovyy Leontiy, Zhidkov Alexander, Ribalka Valeriy, Berezhna Valentyna, Bilko Nadiya, Sakada Volodimir, Bilko Denis, Borbuliak Irina, Kovalev Vasiliy, Krul Mikola and Georgy Petelin
Institute Cell Biology & Genetic Engineering of NAS of Ukraine, Ukraine
Institute for Safety Problems of Nuclear Power Plants NAS of Ukraine, Ukraine
National University "Kyievo-Mogiljanskaja Academy", Ukraine

the high radiating weighing factor (more than 20). Long track α-emitters in a living tissue (33 to 40 microns) significantly increases the radiation dosemore than 50 times due to appearance of local foci of exposure and increased risk of subsequent development of cancer as well as non-neoplastic diseases.

Dissolution of fuel containing materials in water inside "Shelter" as the result of ability to living microorganisms, results in occurrence of new bindings radionuclide – with organic substance are potentially dangereos and more mobile. On the basis of the received experimental data the biotic factor connection with radioactive aerosols in object "Shelter" is investigated. Existence of nano- and micro- size dust as radioactive aerosols with a number of specific properties inside the object "Shelter" and absence strong barriers against entered an environment of catching filters represents high potential risk of occurrence of adverse biomedical consequences for the personnel and for ecology of a 30-km Chornobyl zone.

The complex studies living mammalian organisms and of Chornobyl zone grown plants using post genomic methodologies such as genomics, transcriptomics and proteomics might provide detailed insight into the biochemistry of living plant cells influence chronic ionizing radiation were investigated. For plant of flax the main objective of this experimental research is to elucidate molecular compare changes between plants grown during flowering and embryogenesis in contaminated and control fields in Chornobyl area.

It is shown, that cells of different types (splenocytes, hepatocytes, bone marrow and astroglia cells) obtained from irradiated mice irrespective of a mode of influence of IR (total one-time external exposure γ-radiation by 5.0 Sv, total external exposure γ-radiation by 0.290 Sv for 231 days, long-term (over 74 days) incorporation of ^{137}Cs with drink accumulation in the bodies of mice radioactivity (near 18 kBq) produce the factors not identified in this research in addition raising levels of single strand breaks (SSB) inside DNA for non-irradiated cells. In the conditions of a single exposition in γ-fields with achievement of a dose of an external irradiation nearby 5.0 Sv intensity of production of «bystander» signals above at mice with the raised level of genetically determined sensitivity to response chronic irradiation. Under the same conditions of influence γ-fields sheds light an induction of additional levels of SSB in DNA not irradiated cells on an extent at least one month after IR influence. The positive dynamics of serum levels of alanine aminotransferase and antibodies to the liver-specific lipoprotein - specific poly antigen for the liver, in the accumulation of dose 0.100 Sv is determined in animals that are under chronic exposure. Registered data correlated with pathomorphological changes in liver tissue.

An intra peritoneum injection of melanin with melanin-glucan complex from fungus *Fomes fomentarius* before irradiation procedure promotes essential decrease in production of «bystander» signals and normalization of hemopoietic progenitor cells, testifying in favor of free radical nature of their certain part. It is discovered also in our experimental research would be approved in Chornobyl zone the character of morphology changes develop, flowering, mature soybean seeds and flax of plant depend of chronic irradiation, changes in signal system and epigenetic changes as appear two peaks in curve depend of flowering rate during term of vegetation for flax which treatment with melanin content solution.

The researchers could use working models on base received experimental data in order to develop approach for living mammalian cells and plants which grown under influence chronic irradiation that were withstand consequences chronic irradiation of the radionuclide contamination environment around object "Shelter".

2. Consequences of the Chornobyl catastrophe for the biota

2.1 Structure of the radiation factor in object "Shelter"

The unique biota inside object "Shelter" during its existence was appeared. It is one more physical carrier of ionizing radiation. High humidity and positive temperatures inside "Shelter" even in winter are responsible for the formation of homogenecity thermostatic conditions favorable for the development of microorganism biocenoses. Authors [Pazuhin E.M., Krasnov V.A. & Lagunenko A.S. 2004] have shown that the fuel-containing materials are dissolved under the action of microorganisms. As a result, new compounds of radionuclides with organic matter appeared and they are potentially more mobile and more ecologically dangerous. It was found out the microorganisms on the irradiated nuclear fuel from "Shelter" by the method of electronic microscopy. They are capability to develop on the irradiated nuclear fuel (dioxide of uranium) as a source of mineral substances and to cause the characteristic damage of a surface of nuclear fuel. In Figures 1 and 2 the characteristic damages of a surface of the irradiated nuclear fuel is found.

Fig. 1. Clean the surface of irradiated nuclear fuel

Fig. 2. Microorganisms on the surface of irradiated nuclear fuel

The size of fuel particles in the sample is in the range of 1 to 100 mcm, which corresponds to the grain size of the irradiated fuel is not oxidized. The main source of radioactivity was radionuclide ^{137}Cs which moved through the aerosol pathway. Practically the entire aerosol ^{137}Cs is detected in the material of organic origin fuel particles coated this organic material. Presence of radiating fields of different intensity promotes strengthening of mutational process and selection of radio-resistant microorganisms with new properties was investigated.

Dust. Using special equipment as called impactor received gas-air mixture probe samples inside object "Shelter" was determinated. This instrument allows you to separately collect particles of different aerodynamic diameters (from 9.4 microns up to 0.1 microns) of the aerosol complex structure with an uneven distribution of radioactivity. The rate of pumping of air was 70 ± 4 l/sec. The aerosol particles in the electron microscope REM-100U were evaluated. The most typical kinds of aerosol particles are represented in the Figures 3 – 6.

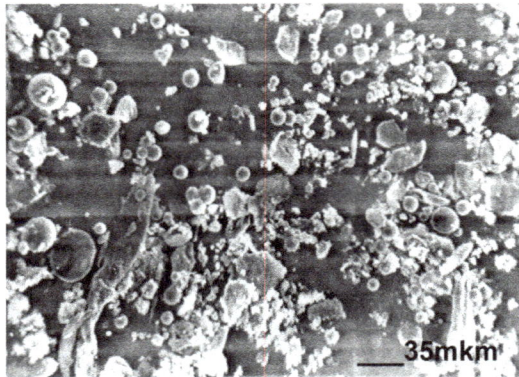

Fig. 3. The appearance of particles washed off from the first cascade impactor. The photograph shows that the aerosol is contained a large number of spherical and oval-elongated particles - bacterial spores and cocci.

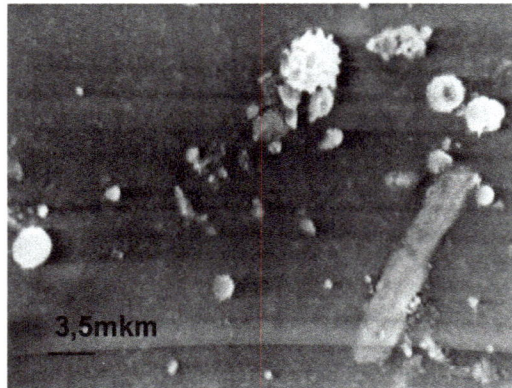

Fig. 4. Fragment the surface form the third cascade. The photograph identified particles of bacterial polysaccharide slime, as well as particles which can be produced during the destruction of microbial cells in the vacuum treatment

The particles of the correct spherical or oval-oblong, which are easily identified (spores, cocci, etc.) [Gusev M.V. & Mineeva L.A. 1992] and particles of irregular forms seen in presented figures. It is shown by electron microprobe elemental analysis that the major part of these particles consists of organic matter and it is of biological origined [V.B. Rybalka, G.F. Smirnova & G.I.Petelign, 2005].

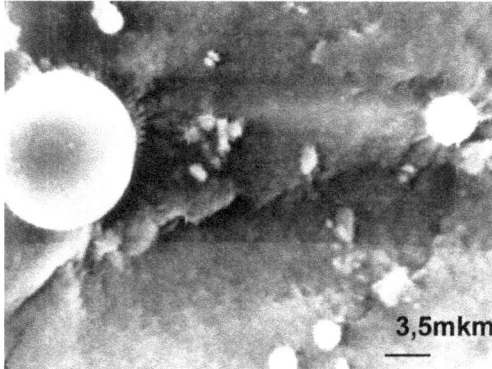

Fig. 5. The appearance of the surface of 4th level impactor. Cocci from 10 to 2 micron and submicron particles of indefinite form are visible

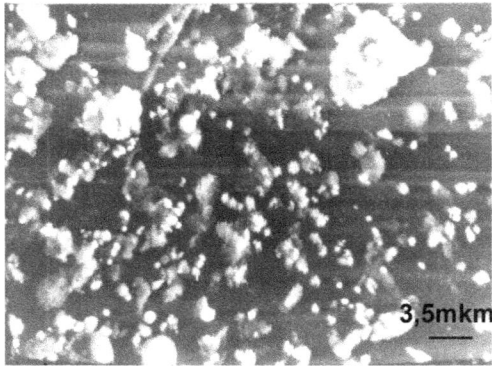

Fig. 6. The appearance of the surface of 4th level impactor. It is seen a large number of particles that are aggregates of loosely coupled small particles

Thus, there are a large number of particles consisting of organic matter in the investigated samples of aerosols originated from "Shelter" premises. A significant number of these particles are identified as the cells of microorganisms and spores.

Sub-micron hot particles (smHP) There is high-dispersive hot particles, which needs a special attention. Such kind of particles is a product of spontaneous dust productivity phenomenon which means dust generation from surfaces of irradiated nuclear fuel and LFCM surfaces. Such a phenomenon was discovered experimentally [Baryakhtar V.G., Gonchar V.V. & Zhidkov A.V. 1997] and for irradiated fuel was confirmed in [Walker C. 2000] later. The smHP grade distribution and possible physical mechanisms responsible for it was later identified in [Baryakhtar V., V.Gonchar & Zhidkov A. 2002]. Typical particle grade is 150 nm for fuel HP and near 50 nm for LFCM particles; all the particles have a complicated internal structure [Baryakhtar V., Gonchar V. & Zhidkov A. 2002]. Their radionuclide composition and specific radioactivity does correspond to those for FCM [Baryakhtar V.G., Gonchar V.V. & Zhidkov A.V. 1997]. Such kind of HP practically cannot be trapped by standard respirators, which usually at personnel's disposal. Annual estimated activity generated in a form of such kind aerosol does equal to a few tens kilograms of irradiated fuel [Baryakhtar V.G., Gonchar V.V. & Zhidkov A.V. 1997].

Such sub-micron HP (Figures 7, 8) to be considered as the most dangerous radiation-hazardous agent regarding to a few reasons: their behaviour in biological liquids is similar to those for the particles in true liquid solutions, smHP does not need solubility of fuel matrix for penetrating all natural biological barriers originated from cell membranes. Sub-micron HP aerosol provides near 80% of total inhalation dose and to be an agent determining effective dose formation for "Shelter" object personnel [Bondarenko O.A., Aryasov P.B. & Melnichuk D.V. 2001]. Existing national regulatory documents on radiation safety does not establish a tolerable concentration of such kind aerosols in the "Shelter" object atmosphere because any attempt to classify them (in accordance to accepted classification) turned out to be doubtful. There are, however, explanations in regulatory document СПРБ-ОУ (in Ukrainian) (Appendix 1), which prescripts what should be done when real aerosols characteristics differs from the typical ones. According to that document, when planned activity stipulates thermal or chemical impact on FCM congestions or heavily contaminated "Shelter" object elements and when revealing in aerosols of non-oxide uranium or TUE chemical compositions one should establish the tolerable concentration for α- and β-emitters in air basing on results of additional special investigation.

Water of the lower marks and technological waters Shelter. It was established that concentration of γ-emitting radionuclides reaches $3.8.10^{11}$ Bq/l, while the concentration of uranium measured up to 0.3 g/l in several thousand cubic meters of "block" and the technological water in premises of lower marks. These water accumulations influence the state of nuclear security of the object "Shelter". They may lead to a change in breeding properties of system "FBM + fragments of the core + water" and the emergence of emissions of short-lived radioisotopes such as iodine. This radionuclide is known to cause the spectrum of thyroid diseases, including cancer.

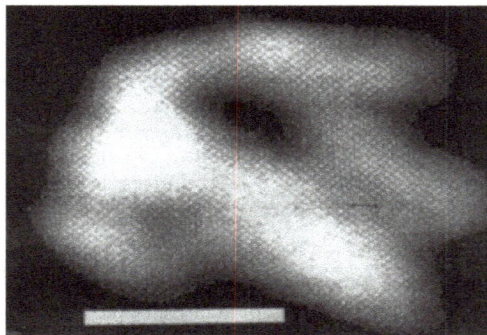

Fig. 7. Dust particles complicated internal structure, formed in the fuel-bearing materials. The length of the white line - 80 nm.

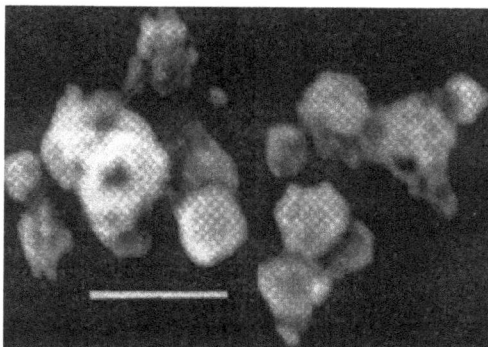

Fig. 8. Spontaneous destruction of the fuel particles to the submicron particles of "hot particles" as a result of processes of radiation defect in the fuel-bearing materials. The length of the white line – 80 nm.

The concentration of radioactivity elements occurs in the silts as the drying pools during the summer-autumn period. Thus silts in "Shelter" represent a real risk as a source of radioactive aerosols. Different kind of microorganisms are found in water accumulations at zero marks of "Shelter" – bubbler pool (Figures 9 - 12) and on the surface of the walls (Figures 13 - 16) [Rybalka V.B., Rybalko S.I. & Zimin Yu.I. 2001, Zhdanova N.N., Zaharchenko V.A. & Tugaj T.I. 2005, Rybalka V.B., Smirnova G.F. & Petelin G.I. 2005, Petelin G.I., Zimin Yu.I., Tepikin V.E. 2003, Rybalka V., Klechkovskaja E., Serbinovich V. 2001]. The experimental investigation of water samples of the Shelter are presented in Figures 9 - 13.

As follows from figures 9 and 10, in the waters of the Shelter are present microbial community in various forms and sizes. In some samples well-reviewed spores size of about 0,1 micron. There is a large number of different units and formations. The biomass of the samples has a very high level of specific activity (^{137}Cs to 3,9 x10^{10} Bq/m^3; ^{90}Sr to 7, 9 x10^9 Bq/m^3, $^{238+239+240}$Pu to 1,1 x10^5 Bq/m^3).

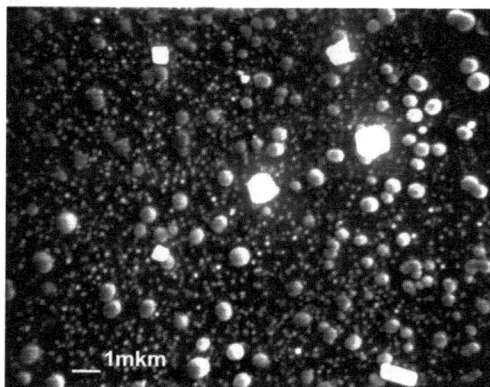

Fig. 9. The appearance of content drops of water from bubbler pool.

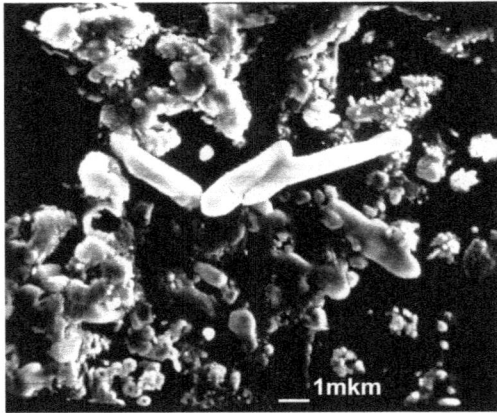

Fig. 10. The appearance of content drops of water from bubbler pool.

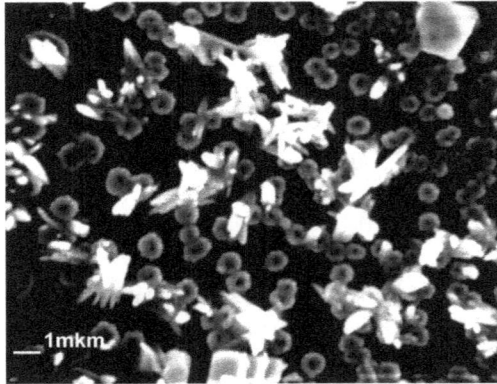

Fig. 11. The appearance of content drops from water from bubbler pool.

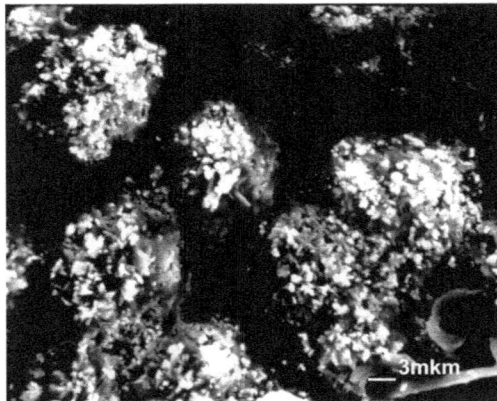

Fig. 12. The appearance of content drops from water "Shelter

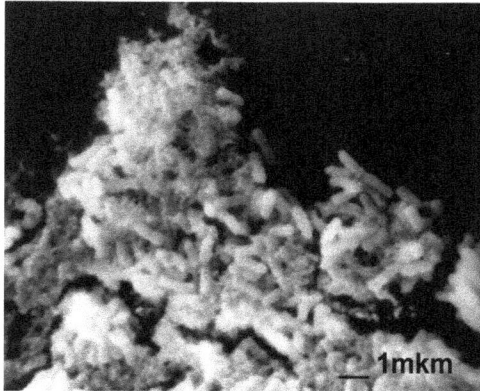

Fig. 13. The appearance of content drops of water from bubbler pool.

Fig. 14. The appearance of content drops of water from bubbler pool.

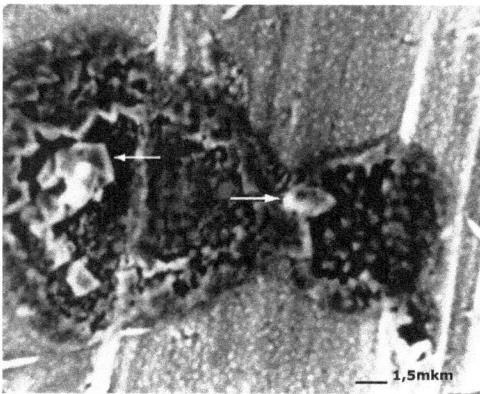

Fig. 15. The fuel particles from swabs from the walls after the burning of organic matter in a vacuum.

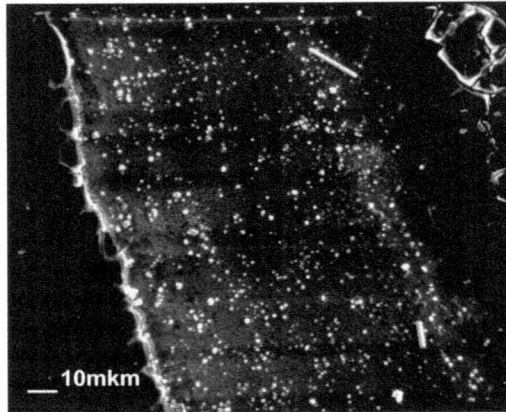

Fig. 16. The appearance of the microbes isolated from swabs of material from the walls after drying a drop of suspension on the glass

Microscopic siza fungi widely exist in the microbiota of the Shelter together with bacteria represented. It is shown that the defeat of microscopic fungi growing in areas with low levels of contamination (from one to 100 mR/hr) [Zhdanova N.N., Zakharchenko V.A. & Tugay T.I. 2005]. The life-cycle reduction, increased radioresistance, increasing the frequency of occurrence of positive radiotropic reactions and radiostimulation, high photosensitivity, that correlated with radiotropic response are typical for fungi-extermophyles isolated from such premises. In these premises increasing the risk of fungal biodegradation, increasing the probability of development of active agents of onychomycosis, skin lesions, lung infections, otitis, invasive fungal infections, as well as the selection of radio resistant organisms with unpredictable invasive characteristics. Lack in human natural immunity to these microbes, resistance of new cultures to the action of traditional medicines, high-speed distribution of microbes (biomass doubling from 15 minutes to 2 hours), the possibility of "transfer" the genetic information with hazardous properties to other bacteria, protozoa, algae, fungi and higher organisms can provide a very serious threat for people. It is known that in the light roof of sarcophagus has a large number of defects and inside the "Shelter" are circulating constant vertical wind currents. So it is quite realistic assumption that some of these microorganisms, especially in the form of spores, can potentially overcome the filters and act to the respiratory tract and lungs of people working inside this object at one side and at another side may carry away outside "Shelter".

2.2 Elucidation of radiobiological effects that can influence the formation of the structure of morbidity

The small doses of radiation can increase the likelihood of developing cancer [Brenner D.J., Doll R. & Goodhead D.T. 2003, NCR 2006] and is possibility appearing morbidity non-cancer origin [Hildebrandt G. 2010]. The main outcome of the 25-year study of morbidity in different categories of exposed persons in connection with the Chernobyl catastrophe is a significant increase in primary morbidity is not associated with tumor pathology. Not found increased risk of leukemia even among those engaged in reconstruction work. Only recently become apparent relationship of radiation exposure with increasing number of non-oncological pathology such as cardiovascular diseases [Preston D.L, Shimuzu Y. & Pierce D.A. 2003]

Statistically significant increase in the spread of non-neoplastic diseases is shown in [Bouzounov O.V., Tereshchenko V.M. 2010]. At the same time structure of morbidity leading place is occupied by diseases of digestive, circulatory and nervous system The maximum level and the largest number with the non-neoplastic diseases statistically confirmed by link determined in the dose > 0,25 Gy for subcohort study group. Long-term effects of chronic low-intensity exposure to low doses for exposed persons in this category until the end is unclear.

The rapid growth of non-tumor morbidity in exposed populations reflects certain changes in the systems that control the growth, development and aging, namely stereotypes matching sequence and intensity of reading the genetic information in different cells. It is shown that this task cannot perform any neural mechanism, or hormonal agents with their ability to alter the rate of metabolic processes. Neurotransmitters, hormones and their receptors do not possess sufficient ontogenetic variability and dispersal [Poletayev A.B. 2008].

According to modern ideas exactly a physiological autoimmunity throughout life provides a readout of genetic information in different cells of the whole organism [Churilov L.P. 2008, Maltsev V.N. 1983, Zaichik A.Sh. & Churilov L.P. 2008]. Singularity of the function of autoantibodies (AuAB) compared with other regulatory substances is considerably longer their half-lives ranged from 10 to 50 days. Therefore the system of autoimmunity has greater inertia. Autoantibodies regulate slow the physiological processes that continue some days and weeks [Ashmarin I.P., Freidlin V.P. 2005]. It is postulated that a mild autoimmune response to their own antigens is a necessary condition for the normal functioning of the immune system and is a prerequisite for the normal regulation and synchronization of cellular functions and morphogenesis [Churilov L.P. 2008]. Additional conditions for unmasking antigens of tissues and organs and represent them immune cells with subsequent increased production of specific AuAB appear as a result of exposure to ionizing radiation.

It is well known that oxygen absorbed by the mitochondria is converted into adenosine triphosphoric acid (ATP). About 5% of oxygen consumed by tissues is converted into free radicals such as superoxide, hydrogen peroxide, hydroxyl radical, singlet oxygen, peroxynitrite (reactive oxygen species, ROC) with unpaired valence electrons. Most of the ROS are produced continuously in cells as byproducts of normal cellular metabolism (mainly due to a small leakage of electrons to the mitochondrial respiratory chain, as well as other reactions in the cytoplasm), and do not cause damage to cells. An excess of ROS under intense ionizing radiation exceeds the protective capabilities cells and can cause serious cell disorders (eg, depletion of ATP). The increase of free radical molecules and their products have a place in the development of the state of homeostasis, has been called oxidative stress. Slow development of oxidative stress triggers apoptosis, and its intensive development leads to necrosis. Postradiation apoptosis is characterized by maintaining the integrity of the cell plasma membrane and the lack of exposure of intracellular contents from cells of the immune system. In the end, remnants of apoptotic cells are removed by exfoliation in intraorganic space and subsequently excreted from the body. But the shortage of ATP, in particular after irradiation, the energy dependence of the mechanisms of apoptosis are disabled, and the cell dies with loss of cell membrane integrity and release of macromolecular components (eg, ALT, AST, etc.) into the intercellular space. Necrosis caused an immune response in the form of inflammation - leukocyte infiltration of the affected tissue, interstitial fluid accumulation and subsequent induction of specific immune responses (specifically sensibilized T-lymphocytes and autoantibodies) to the unmasked and recognized by lymphocytes of intracellular components. According to data of many researchers the AuAB are primarily the attribute of the norm. They can be identified in healthy individuals. [Churilov L.P. 2008, Zaichik A. Sh. & Churilov L.P. 2008b, Poletayev A.B. 2005, Cohen I.R. 2005, Harel M. & Shoenfeld Y. 2006, Shoenfeld Y. 2008]. The AuAB involved in the process of apoptosis, cleaning the body from catabolic products, modulation of the activity of many enzymes and hormones, as well as perform the transport function [Poletayev A.B. 2008a]. It was shown that antibodies against nuclear antigens can penetrate into the cell nucleus in vivo and stimulate the synthesis of RNA and DNA in target cells [Zaichik A. Sh. & Churilov L.P. 1988].

Notkins in 2007 hypothesized that natural AuAB can be very informative not only precursors of autoimmune diseases, but also a variety of somatic diseases and syndromes [Notkins A.L. 2007]. It is important that changes in the content of organ natural AuAT in most cases, ahead of the clinical manifestation of appropriate forms of pathology. If for example the content of "cardiotropic", hepatotropic", "neurotropic" AuAB a concrete person within the borders of the norm, this suggests that the intensity of apoptosis, respectively, cells of the heart, liver or nerve tissue does not go beyond the norm. Persistent changes, for example, from "hepatotropic" AuAB should be regarded as a sign of the possible formation of a pathological process in liver tissue, even if at the time of examination are no clear clinical symptoms or specific biochemical changes [Poletayev A.B. 2008a, Churilov L.P. 2008, Notkins A.L. 2007, Zedman, A.J.W. & Vossenaar E.R. 2004].

The important role of the abolition of immune tolerance in the occurrence of non-viral hepatitis research shows serum levels of antibodies to liver-specific lipoprotein (LSP) for various categories of people. [Kovalev V.A. & Senyuk O.F., 2008].

The LSP which was first isolated by Meyer zum Burschenfelde and Miescher in 1971 is considered to be specific poly antigen for the liver. Native LSP is mixture of antigenic determinants of the substrate from the membranes of hepatocytes and contains soluble and membrane components were isolated by gel filtration (chromatography) supernatant after ultra- centrifugation of liver homogenate [Manns M, Gerken G. & Kyriatsoulis A, 1987, Ballot E., Homberg J. C. & Johanet C. 2000]. It is known that LSP is found not only in the liver, but also in some other organs [Garcia-Buey., Garcia-Monzon C. & Rodriguez S. 1995]. Therefore, the total increase in antibody levels to the PSL can be seen as a sign of abolition of immune tolerance to many organs and tissues of the human body [Kovalev V.A. & Seniuk O.F., 2008].

There are three outcomes for the cell, if the cellular radiation damage is not adequately repaired. The cell may die, or will delay it or keep playing with the viability of new qualities, or mutations as the basis for the development of remote descendants (See figure 17). The consequences of the first approach in the development of cells after irradiation described below.

The linear non-threshold concept is used as the primary standard for radiation protection and risk assessment for many years. It suggests that damage induced by low doses of radiation do not contribute significantly to increased risk of disease because a significant amount of endogenous genome damage occurs during life and they are restored in cells with high probably. In fact endogenous damages (ED) constantly appear in the cells. Some of them are due to thermodynamic processes, the hydrolysis reaction, while others arise from the effects of free radicals generated by cell during its life and still others are a necessary component of metabolism (DNA breaks accompany the process of differentiation, recombination, etc.) [Lindahl T. 1993, Bont R.D. & Van Larebeke N. 2004]. According to [Lindahl T., 1993] per one day in the DNA of one cell may have more than 50,000 endogenous damages as single-strand breaks (SSB) and 10 ones as double-strand breaks (DSB).

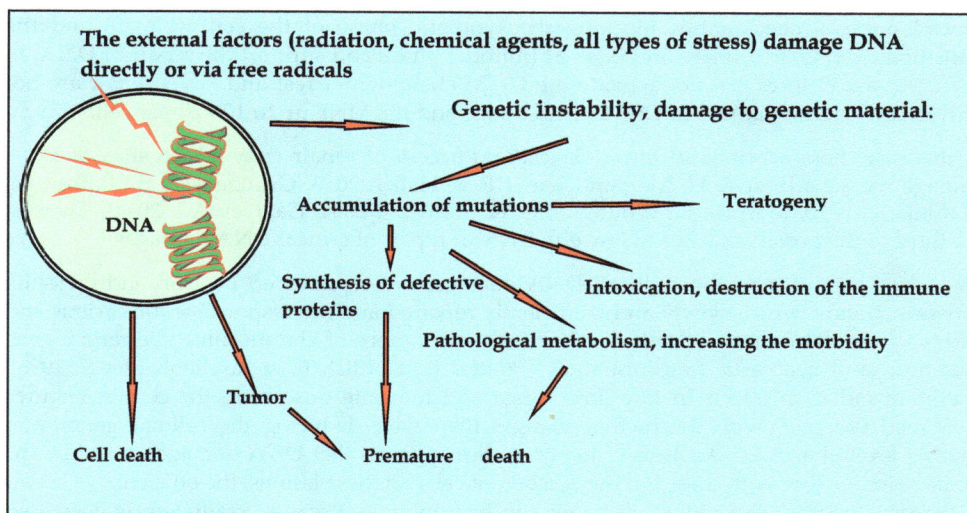

Fig. 17. Effects of radiation exposure on the genetic apparatus of cells.

Many studies suggest that most of the radiation damage (RD) occurring in the DNA of cells differ significantly in their chemical nature from the ED. The main difference between DNA damage induced by ionizing radiation, from the ED was its the complexity of their chemical nature, and clustering. The proportion of complex, critical for the fate of the cell damage is much higher when exposed to IR.

When treatment of mammalian cells to H_2O_2 ratio of DSB to SSB is 1: 10 000 [Bradley M.O. & Kohn K.W. 1979] while under the influence of IR is much higher -1:20 [Shikazono N., Noguchi M., & Fujii K. 2009]. Many of the DNA RD are not accidental, are located in close proximity and have the cluster grouping. They are formed as a result of coincidence of two or more single damages within 1-2 rotation the DNA helix [Hada M., Georgakilas A.C. 2008, Sutherland B.M., Bennet P.V. & Sidorkina O. 2000]. Especially the massive clustering of DNA damage occurs when the ionization tracks pass along chromatin fiber. In this case, they may cover DNA regions with an average size of about 2000 base pairs [Radulescu I., Elmroth K. & Stenerlow B. 2004]. At the same time, we know that the probability of occurrence of endogenous clustered DNA damage is extremely low [Bennett P.V., Cintron N.S. & Gros L. 2004]. Accumulation of ED does not occur because in the cells are constantly functioning mechanisms of reparation, specifically targeted at removing various types damages [Friedburg E.C., Walker G.C. & Siede W. 2006].

The complex nature of RD of DNA and the presence of cluster groups can be regarded as the first cause, which creates difficulties for repairing systems cells. Damages repair processes within the cluster can break down in various stages of excision repair and lead to the formation of additional or inaccurate DSB repair, important for cell survival, mutagenesis, and the risk of malignancy [Ide H., Shoulkamy M.I. & Nakano T. 2010]. The second reason for the low efficiency of repair of RD may be a relatively low amount of DNA damage. Therefore, radiation effects in low dose range and low power are certain features associated not only with destructive modifications but with deducing the cellular genome at a different level of activity. High doses of ionizing radiation via activation of cell cycle control points - checkpoints, blocking the synthetic phase of the cell cycle (S) and the transition from G_1 to S phase and G_2 to M (mitotic) phase and support the repair of DNA. In this case, small doses can not activate the G_2/M chekpoynt-arrest and DNA repair are not activated when the number of DSB DNA damage and the MNF up to 10 - 20 per cell.

In this case, heterochromatin little relaxes, and access of repair enzymes to sites of DNA damage worsens [Fernet M, Megnin-Chanet F. & Hall J. 2010, Grudzenskia S., Rathsa A., Conrada S., 2010, Marples B., Wouters B.G. & Collis S.J. 2004, Gaziyev A.I. 2011]. There is the third of the possible causes of low efficiency of repair of critical DNA damages.

The major part of radiation induced DNA damages are represented by DSB and crosslink between strands, which slowly and inefficiently repaired and are responsible for various end effects - from the radiation death of cells to the appearance of chromosome aberration, gene mutations and neoplastic transformation. [Pfeiffer P., Gottlich B. & Reichenberger S. 1996]. Therefore radiation effects in low dose range and low rate dose have the certain features associated not only with destructive changes than with deducing the cellular genome at another level of activity. Analysis of many studies suggests that DNA damage caused by the IR, increase linearly with dose, but the reaction of cells on these lesions, the efficiency of repair of the most complicated critical damages can be nonlinear. Because irradiation is decreased expression of proteins, enzymes of different systems providing the stability of DNA as a result of the accumulation of DNA damage is not recovered or recovered with errors and fix mutations. Dysfunction of many genes, regulatory systems and cellular processes that are ultimately linked to the development of various pathologies, including carcinogenesis. A significant part of replication errors - spontaneous mutations - can be harmful to an organism. There are the basis for the occurrence of hereditary diseases, carcinogenesis, etc. were revealed.

2.3 The study of the radiological effects of low and middle doses in model experiments on linear animals

It is known that some of the surviving cells after irradiation can produce functionally modified descendants, who have for many generations with a high frequency occur *de novo* chromosome aberrations and predominantly point mutations, in certain cases increase cell death by apoptosis [Little J.B. 2000, Seymour C.B., Mothersill C. & Alper T. 1986, Baverstock K. 2000]. The accumulation of such mutations suggests a high probability disorders in the bases caused by oxidative stress [Little D.B. 2007]. It is known that even in non-irradiated cells residing in close proximity to irradiated cells, induction of chromosomal instability is possible.

Placing of non-irradiated cells (NirC) in the culture medium of irradiated cells (IrC), soon leads to the appearance in the descendants of (NirC) of chromatid and chromosome aberrations, micronuclei, gene mutations that increase the content of the transformed cells

[Pfeiffer P., Gottlich N. & Reichenberger S. 1996, Seymour C.B., Mothersill C. 1997, Johansen C.O. 1999]. The phenomenon of transfer of an altered state from the modified cells under damage factors to unmodified cells was named "bystander effect" (BSE). It was first described in Chinese hamster [Nagasawa H, Little J.B. 1992] cells and was later found in different types of cells after exposure to damaging factors of different nature [Azzam E.I., Little J.B. 2003, Azzam, E.I., de Toldeo, S.M. & Little, J.B. 2003, Chakraborty A, Held K.D. & Prise K. 2009, Mothersill, C & Seymour, C. 2001]. Thus, cells that have not been laid through the tracks of IR are "bystanders" of radiation injuries caused by other irradiated cells. Reactive oxygen species (ROS) play an important role in the mechanisms of signal transmission to "bystander" cell [Kudryashov Yu.B. 2004]. Reinforced ROS production in NirC incubated in medium with serum, irradiated α-particles, or with supernatant of the suspension of IrC [Narayanan P.K., Googwin E.YH, Lehnert B.E. 1997] or after contact by IrC [Grosovsky A.J. 1999] is shown.

Irradiated cells produce several "bystander" signals - cytokines, fragments of DNA (from apoptotic cells) or other factors of protein nature. These factors cause a change in oxidative metabolism and gene expression profiles in IrC, and induce enhanced production of highly ROS [Watson G.E., Lorimore S.A., Macdonald D.A., 2000, Ermakov A.V., Kon'kova M.S., Kostyuk S.V. 2009, Snyder A.R. 2004, Lorimore S.A., Coates P.J., Scobie G.E. 2001]. In addition to chemical modification of DNA nucleotides the formation of radicals can also lead to changes in the higher levels of organization structure of the molecule to the secondary, tertiary and quaternary conditions [Nobler M.P. 1969]. Therefore the level of oxygen and antioxidants influence the quantitative yield and quality of damaged bases of DNA in IrC.

Signal transmission of such lesions through the culture medium is typical for BSE induced by IR from low linear energy transfer [Mothersill C., Seymour C.B. 1998]. In our study we used a hypothesis about the ability of γ-rays to generate in a organism of mammals part of "bystander" signals that can be distributed in the environment of the body and affect distantly unexposed tissue. At the same time we take as an axiom that some of these of signals has a free radical nature, in particular, can be represented by the ROS. Scheme of the origin and development of BSE corresponding to this hypothesis is shown in Figure 18.

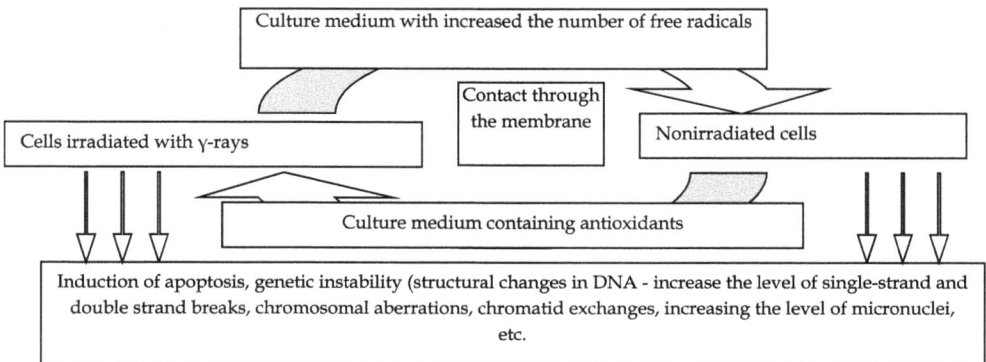

Fig. 18. Elucidation scheme of the experiments

We used a biological model, which allows you to multiply the amount of cellular material to strengthen the "bystander" signals in the intercellular space. For this purpose females of mice Balb/c and C57bl/6 were irradiated. It is known that mice C57Bl/6 is moderately sensitive to ionizing radiation, $LD_{50/30}$ = 6.70 Sv, and Balb/c line of mice is highly sensitive to IR, $LD_{50/30}$ = 5.85 Sv [Blandova Z.K., Dushkin V.A. & Malashenko A.N. 1983, Storer J.B 1966]. After exposure to γ-fields from organs and tissues of the mice (peripheral blood, spleen, liver, bone marrow and brain - astroglia) cell suspensions were obtained.

Protocol modes of mice irradiation:

1. Total one-time 16-hour exposure γ-radiation from small samples of nuclear fuel IV unit of ChNPP modified during the accident in 1986, which were evenly distributed under the cages with the exhibited animals (30 mice per line) and formed a fairly uniform horizontal γ- field with the exposure dose rate of about 8.7 • 10^{-4} Gy sec, allowed to reach a total dose of about 5.0 Sv (85% of the $LD_{50/30}$).
2. The total external γ-radiation exposure from a specially constructed flat concrete bars, containing soil from the Red Forest with a specific activity of 30 kBq/kg and creating exposure power ~ 52.2 mcSv/h. They were placed under cages with animal at 231 days. Total external dose was about 0.29 Sv. Radioactive soil we previously burned for the destruction of organic compounds.
3. Long-term (over 74 days) incorporation of ^{137}Cs with a drink (at the rate of 6.0 kBq per mouse per day) led to accumulation radioactivity from 14 to 24 kBq in the mouse body, which was identified in the γ-spectrometer CP-4900V (Nokia, Finland). The average total estimated activity was 17.0 ± 1.0 kBq per animal.

Identification model of "bystander" signals

Replacing the living environment of mice

Non-irradiated and irradiated cells obtained from organs and tissues of mice kept in a nutrient medium RPMI-1640 supplemented with 5% syngenic serum for 3 hours, after that the NirC placed in culture medium of IrC (Figure 19)

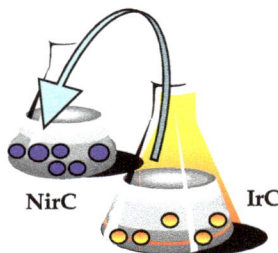

NirC   IrC

Fig. 19. Scheme of reproduction "bystander effect" in the transmission of signals through the culture medium.

We evaluated the ability of the culture medium of exposed animal cells to induce an increased number of single-strand DNA breaks (SSB DNA) in the same cells obtained from unexposed animals. Level of SSB DNA in vitro in different kind of obtained cells was determined using the method described of labeling of DNA by fluorescent dye picogreen with subsequent evaluation of the rate of its splitting [Elmendorff-Dreikorn K., Chauvin C. & Slor H. 1999]. Results were presented as the coefficient of unwinding of the DNA helix (SSF), which was calculated at 20-minute exposure the DNA double helix (dsDNA) with untwine the buffer as follows:

$$SSF = \log \ (\% \ dsDNA \ in \ sample \ / \% \ ds \ DNA \ in \ the \ control).$$

Confirmation of participation molecules free-radical nature in the transfer of "bystander" signals

Melanin-glucan complex (MGC) received from higher basidiomycetes *Fomes fomentarius*, was used as an antioxidant to confirm the involvement of molecules of free-radical nature to the transmission of signals from IrC to NirC after external influence of γ-rays (the model of the single and chronic exposure of mice).

Melanins are amorphous pigment of dark brown and black, they are widespread in nature and are found in virtually all groups of organisms. Melanins contain carboxyl, carbonyl, hydroxyl, amine and phenol functional groups. Because of this molecule melanin can simultaneously interact with both anions and cations with, ie to be donors or acceptors of free (See Figure 20) electrons and thus carry out electron transport functions. Melanin is also able to absorb photon's energy [Riley P.A. 1997]. These substances are characterized by the presence within their structure of unpaired electrons and possess the properties of stable free radicals. Melanin is not only absorb a variety of radiation, but also neutralizes and eliminates harmful for the cells of free radicals formed under action of ionizing radiation and some chemical substances on living organisms.

Study protector ability of microorganisms the presence of melanin in them almost all the basic mechanisms of reparative DNA repair revealed. Experimental results showing the increase in DNA polymerase and DNA ligase activity melanin mushrooms under UV irradiation were obtained by [Sidorik E.P., Druzhina M.O. & Burlaka A.P. 1994]. Melanin pigment has a high gene protective activity in acute exposure to ionizing radiation in a wide range of doses. It was established that melanin effectively reduces the frequency of mutations induced by ionizing radiation in both somatic and germ cells [Mosse I.B. 2002]. Ability of melanin to reduce almost to control level frequencies to genetic lesions, which are transmitted from generation to generation and accumulation in populations in the form of 'genetic load" is unique [Mosse I.B., Lyach I.P. 1994]. For the first time is shown the principle possibility of effective protection of animal populations irradiated over many generations by means of melanin (in Drosophila studied 150 generations, in mice - 5) [Mosse I.B., Dubovic B.V. & Plotnikova S.I. 1996]. Radioprotective efficiency of melanin was higher in the chronic exposure than for acute irradiation conditions [Mosse I.B., Kostrova L.N., & Dubovic B.V., 1999]. Daily oral administration of melanin in a dose of 10 mg/kg to pregnant females rats eliminated the functional deficiency of physical and emotional development, detectable in the progeny at the antenatal γ-irradiation at a dose of 1.00-1.25 Gy for the entire period of pregnancy. Conclusion of the radioprotective effect of melanin on cytogenetic and embryotoxic effects of low doses of ionizing radiation have been done on the basis of the data [Mosse I.B., Zhavoronkov L.P. & Molofey V.P. 2005]. Melanins are capable of forming complexes with metals, including radioactive elements. The ability of synthetic melanin to accumulate radioactive elements such as ^{111}In, [225]Ac and [213]Bi had shown [Howell C.R., Schweitzer A.D. & Casadevall A. 2008]. The possibility of the creation of radioprotective agents on the basis of melanin in the result of research was mentioned [Dadachova E, Ruth

A. & Bryan R.A. 2008, Schweitzer A. D., Robertha C. Howell R.C. 2008]. For our experiments used the melanin-glucan complex whith a large number of paramagnetic centers (17 • 10^{17} spin/g). This substance shows armipotent antioxidant properties in interaction with various types of free radicals [Seniuk O., Gorovoj L. & Zhidkov A. 2005].

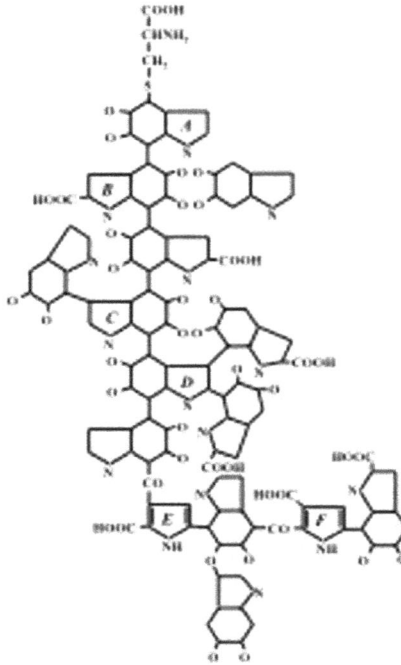

Fig. 20. The structural formula of one of fungal melanin by [Riley P.A. 1997].

Results of the study remote signaling between cells in the external effects of prolonged γ-radiation

The first attempts to search for distant signals generated by irradiation *in vivo* cells were taken from Balb/c mice that have known greater sensitivity to the action of IR. Quantitative analysis of DNA damage in a mammalian cell immediately after exposure to IR with low LET in a dose of 1 Gy has shown that it generates approximately forty DSB and MPS (DNA-DNA) 150 DNA-protein crosslinks about 2000 modification bases around 3000 AP-sites of damaged deoxyribose residues, SSB and alkali-labile sites [Von Sonntag C. 2006, Ward J.F. 1988]. It is believed that the sharp increase in single-strand DNA breaks (SSB) is directly correlated with the number of DSB. The ratio of SSB to DSB under the action of IR can correspond to the values 10 - 50 depending on exposure conditions and cell types of radiation [Oxidative Stress 1991]. Indicator of SSB DNA was used to assess the effects of ionizing radiation. In turn, the level of SSB DNA was determined by the value of the coefficient a DNA double helix unwinding in an alkaline environment.

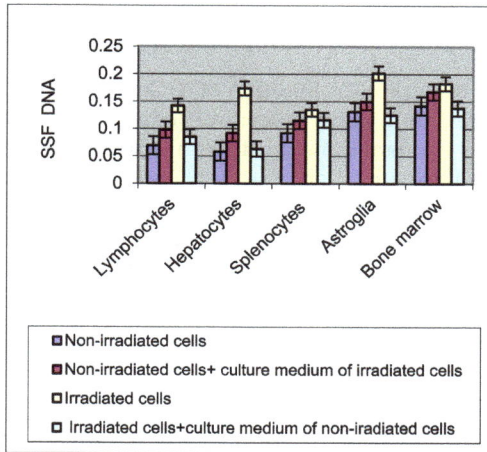

Fig. 21. Effect of living medium of exposed in vivo Balb/c mice cells to levels of SSB DNA at the first day after exposure to γ-fields with the accumulation of total external dose of about 5 Sv.

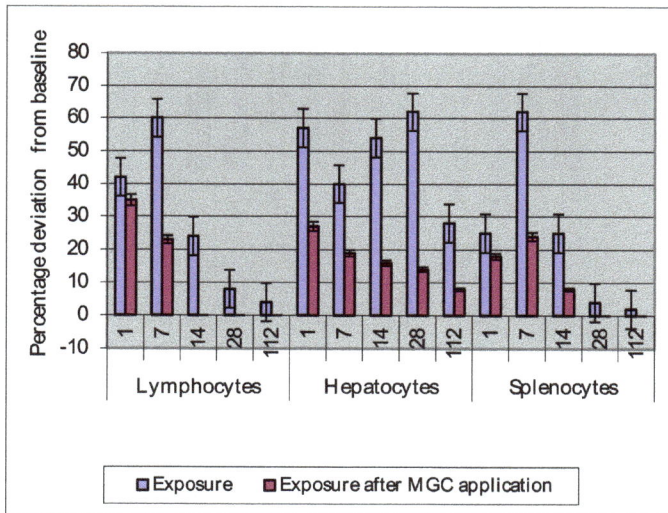

Fig. 22. Influence of living medium from cells irradiated mice and from cells irradiated mice that received intraperitoneal injection of MGC before exposure on the levels of SSB DNA at cells non-irradiated mice (1st day after exposure at dose 5 Sv

As we can be seen from Figure 21. data in all types of cells used in the first day after irradiation dose of 5 Sv significantly increases the level of SSB of DNA. At the same time it is shown that the living medium obtained after three hours stay of IrCs on the first day after exposure is able to induce an additional level of SSB DNA in different types of cells from not-irradiated mice Balb/c. In certain cases, the change of this indicator reaches significant differences, particularly in lymphocytes, hepatocytes and hematopoietic cells in bone marrow. While in splenocytes and astroglial cells revealed a clear trend towards increased levels of SSB DNA.

As shown in Figure 22 intraperitoneal injection of MGC, which has powerful antioxidant properties, into mice before irradiation procedure reduces the of BSE in all types of test cells and can serve as an argument in favor of the hypothesis about the important role of free radical molecules in the realization of this phenomenon. At the same time, it was shown that the transfer of "bystander" signals inside the irradiated Balb/c mice gradually decreases in all kinds of investigated cells during the first month after exposure and practically not detected 3,5 months after exposure (Figure 22). Receiving MGC before irradiation procedure is associated with a lower intensity "bystander" signaling in the period after exposure. The results of a comparative study of induction "bystander" signals in mice with different genetically determined radiation sensitivity under the same conditions of irradiation are shown in Figure 23.

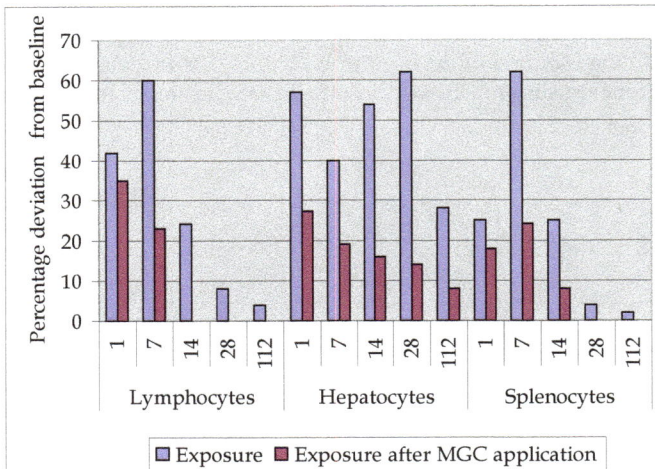

Fig. 23. Dynamics of development "bystander" effect for 112 days after MGC application before irradiation at a dose of 5.0 Sv

At least in mice with lower level of $LD_{50/30}$ (Balb/c) induction of SSB in cellular DNA after exposure in living environment of the irradiated lymphocytes in various periods after of irradiation ranged from 150 to 200% when lymphocytes of mice more resistant to the effects of IR (C57Bl/6 with higher level of $LD_{50/30}$) index of induction SSB in DNA of lymphocytes, respectively, was 2.5 and 5 times lower (Figure 24).

A correlation between higher index of $LD_{50/30}$ and low induction as SSB DNA after irradiation, and induction of SSB DNA in modeling "bystander" effect in mice of C57Bl/6, can be explained by the presence in the cells of these animals melanized structures providing the black skin and fur. Antioxidants, photo- and radioprotective properties are a direct consequence of the free-radical structure of melanins, providing the opportunity to participate in electronic exchange of redox and radical processes. An attempt to identify "bystander" effect in vivo in mice of these lines under conditions of prolonged (over 231 days) of exposure on concrete bars with radioactive soil, have accumulated a total dose of external irradiation on the level of 0.29 Sv also been made. The data are shown in Figure 25.

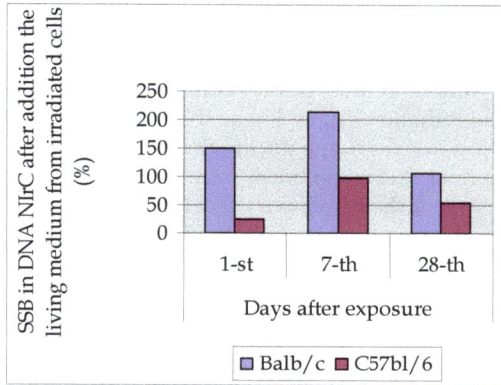

Fig. 24. Different influence of the living medium of IrC on NirC obtained from mice with differ in genetically determined level of sensitivity to IR.

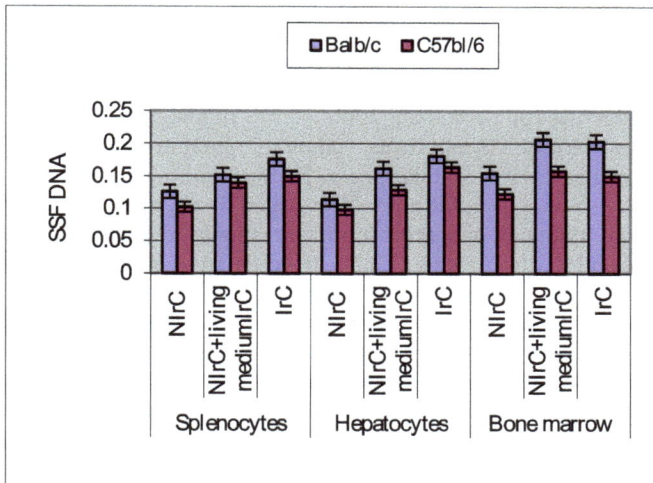

Fig. 25. Comparison levels of induction "bystander" signals in mice with different genetically determined radiation sensitivity under exposure during 231 day at dose 0.29 Sv.

Analysis of the results of this study convinces us that prolonged exposure of linear mice under the effect of γ-fields of low intensity is associated with the induction of additional level of SSB in the DNA of cells of different origin. In this case, the phenomenon of amplification of radiological effects by means of "bystander" signals occurs. In this case, as with single-dose irradiation of 5.0 Sv "bystander" effect more clearly defined in biological mediums of mice with high sensitivity to the effects of ionizing radiation - line Balb/c. So, placing the NirC into living environment of the IrC Balb/c mice induces an increase in the level of SSB DNA in different cell types from 33 to 44%, while a similar replacement of living medium in cells of mice C57Bl/6 changes the indicated index in the lower range - from 27 to 32%.

Detecting the ability of IrC to change the state of neighboring cells, through which not been laid tracks of ionizing particles, can in some extent explain the peculiarities of clinical realization of radiation effects at low doses of ionizing radiation. Identification BSE in liver tissue is an additional argument in favor of accepting the reality of the existence of diseases of hepato-biliary system of radiation origin. It was showed the development of the processes of autoimmune dysfunction in liver tissue during three-month exposure with accumulation total dose at 10.0 cSv in model studies on linear mice Balb/c [Kovalev V., Krul N., Zhezhera V. 2010]. It is known that one of the responses the cell to irradiation is the destruction of the cells with the loss of a specific morphology and functional activity.

Nowadays are known several forms of death in cells that depend on the production of ATP - apoptosis, necrosis and reproductive death. Postradiation apoptosis is characterized by maintaining the integrity of cell plasma membrane and the lack of contact, the intracellular content with cells of the immune system. In the end, remnants of apoptotic cells in tissues are removed by brushing up exfoliation in intraorganic space and subsequently excreted from the body. Mitochondria are the most sensitive to radiation cell organelles for several reasons: the practical absence of reparation and histone proteins that protect DNA, as well as the minimal activity of the enzymes to "cut out" and replace oxidized DNA regions [Anderson S, Bankier A.T. & Barrell B.G. 1981, Berehovskaya N.N., Savich A.V. 1994]. In the cell inhibits the synthesis of ATP because degradation and loss of mitochondria. But at lack of ATP, in particular energy-dependent mechanisms of apoptosis may be switched off in connection with exposure to IR. In this case cell necrotizing with the loss of integrity of the cell membrane and release of macromolecular components (alanine aminotransferase (ALT), aspartate aminotransferase, etc.) in the intercellular space. Necrosis caused an immune response in the form of inflammation - leukocyte infiltration of damaged tissue, the accumulation of interstitial fluid with subsequent induction of specific immune response to unmasked and recognized by lymphocytes intracellular components. Thus, the cells from renewable tissues which are sensitive to IR (epithelium of the gastrointestinal and urogenital tract, respiratory tract), regularly die as a result of intense radiation exposure. Their contents are subsequently released in extracellular space and blood. Taking into account the above mentioned thoughts we additionally determined the levels of intracellular enzymes ALT levels in peripheral blood of different animal groups - the control and chronically exposed mice, and mice immunized with liver-specific lipoprotein (LSP). The preparation of liver-specific lipoprotein (LSP) has been isolated from the liver of syngeneic mice by the method described by McFarlane I G. [McFarlane I.G., Wojicicka B.M., & Zucker G.M., 1977]. It is a mixture of antigenic determinants of the substrate from the membranes of hepatocytes. Because of their lability, some proteins that are part of the LSP, in particular the asialoglycoprotein receptor [Treichel U., Schreiter T. & Zumbuschenfelde K.H.M, 1995] under certain conditions, including under the influence of small doses of radiation may acquire properties of autoantigens.

The one group of mice was immunized LSP to confirm the immunogenicity of the resulting substance. The preparation of LSP has been isolated from the liver of syngeneic mice by the method described by McFarlane I.G. [McFarlane I.G., Wojicicka B.M. & Zucker G.M. 1977]. The final concentration of protein in a preparation isolated by the method of Bradford [Bradford M.M. 1977] was 2.8 mg/ml. The immunization scheme described in [Ryabenko D.V., Sidorik L.L., Sergienko O.V. 2001] was used. Increased serum level of this enzyme is considered a sign of inflammation in the liver - hepatitis, because a large amount of ALT released from the destructive cells of the body. Data obtained in this experiments are show in Figure 26.

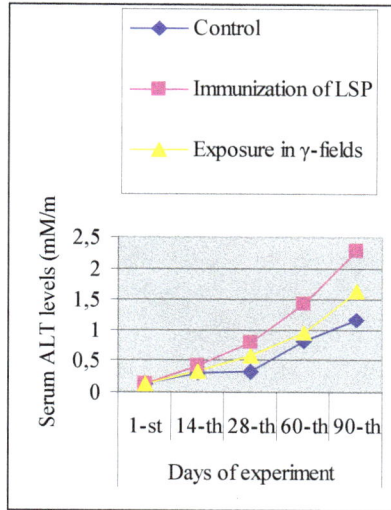

Fig. 26. Dynamics of serum levels of alanin aminotransferase in control, irradiated and immunized of LSP of mice Balb/c.

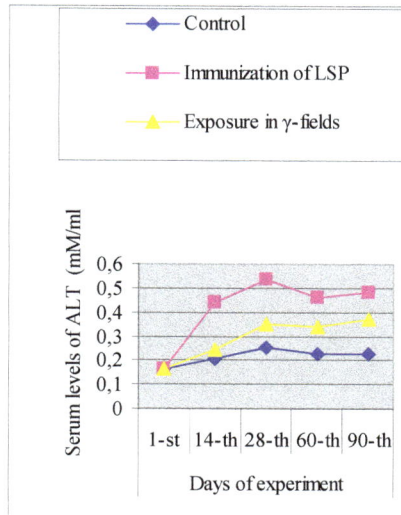

Fig. 27. Dynamics of serum levels of AuAB to LSP in control, irradiated and immunized of LSP of mice Balb/c.

As shown by data presented in this Figures 26 and 27 a gradual increase in the concentration of ALT in the bloodstream is registered in all groups of animals observed. Increased serum levels of ALT in the mice of the control group reached significant difference in the fourth month of life and reflects a legitimate age-related changes of the liver associated with aging.

A clear tendency to increase this index in the group of animals treated with an immunogenic complex of proteins from the membranes of hepatocytes (LSP), when compared with that of control animals may be explained by stimulation of the processes of destruction of hepatocytes, mediated by immune mechanisms. More significant increase in the rate of ALT in blood is detected in irradiated mice (a significant increase in ALT is determined at the end of the first month of observation), and especially in mice immunized against membrane liver antigens contained in the LSP. Serum ALT levels increase in this group of experimental mice at the end of the second week after immunization began. Noteworthy the fact that the 100-day low-intensity radiation fields of fuel "hot particles" is quite effective in serum levels of ALT. Mice immunized with only one week ahead of irradiated mice to achieve a statistically significant increase of levels of this index compared with baseline. As follows from these data that a certain baseline level of AuAB to LSP is detected in the sera of control animals, and shows a tendency to increase during aging in mice. Amount of AuAB to LSP in immunized animals progressively increased and by the end of the first month of immunization reaches a peak and then decreases slightly and stabilizes over the next two months. In irradiated animals also determined by positive changes of serum levels of AuAB to LSP in the process of dose accumulation. In this group levels AuAB LSP to gradually increase and a tendency to increase AuAB remains at least during the observation period.

Registered positive changes of serum levels of ALT and of AuAB to LSP correlated with pathomorphological changes in liver tissue. In Figure 28. shows the morphological pattern of liver healthy 2-month old animal from the control group. Globular structure of the organ with well-differentiated trabeculae, which extend radially from the portal vein, clearly visible in the picture. The boundaries of cells and nuclei are well differ over the whole area of slice. In Figure 29 shown a slice of liver tissue with beginning an inflammatory process that is still impossible to differentiate as an autoimmune process. In figures 30 and 31 shown

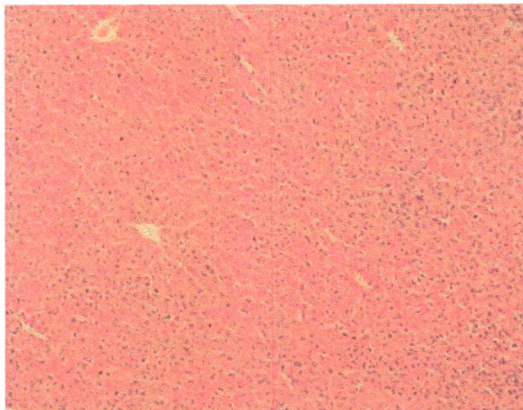

Fig. 28. Morphological picture of liver healthy 2-month old animal from the control group.

deeper degenerative processes in the parenchyma with formation of lymphoid infiltrates. In a more favorable course of the process next sections of lesions appear regeneration areas. At high intensity the process changes in the tissues may be irreversible.

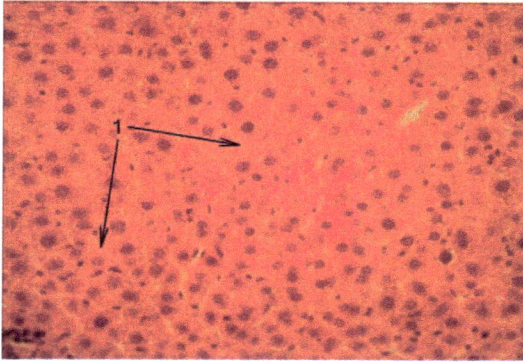

Fig. 29. Dystrophia of hepatocytes in violation of the architectonics of organ and the destruction of the nuclei. 1 - a violation of the architectonics of the lobules, blurred boundaries (irradiated mouse).

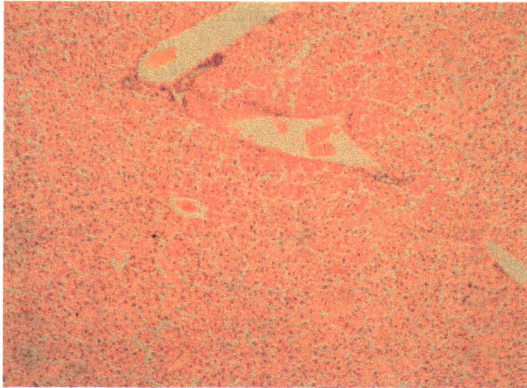

Fig. 30. Around blood vessels (or periportal) small clusters of mostly round the nuclear cell (lymphoid) elements. Stasis of blood within the vessel. More intensively painted the nucleus of cells (irradiated mouse).

Thus it was shown that exposure to IR can modify the immune status of the organism with the breakdown of immune tolerance and the subsequent emergence of autoimmune hepatitis. Its characteristic feature is the launch of autoimmune reactions against their own membrane antigens not only in the liver but also in other organs. The presence of long-term in excess of the background radiation pressure associated with the abolition of immune tolerance and the creation of conditions for the development of autoimmune reactions, in particular against antigens of liver tissue. As well known in turn the activity of autoimmunity is a favorable condition for the transfer of persistent infections in the active state and stimulation of vegetation as a saprophyte and pathogenic microorganisms was increased.

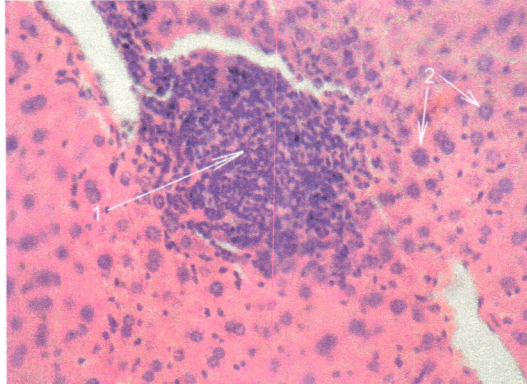

Fig. 31. Granulomas in the periportal areas of the parenchyma. 1 - Formation of granuloma on site infiltrate, 2 - Regeneration and repair of hepatocytes. (Irradiated mouse).

The impact of single external irradiation at a dose of 5.0 Sv for the transfer of intercellular signals in the bone marrow of mice Balb/c.

Nowadays it is known that hematopoietic stem cells (HSC) have higher radiation sensitivity than other cellular self-renewal systems. This conclusion was made by scientists on the basis of experimental works that showed the presence of the damage to hematopoietic progenitor cells not only due to the effects of the high doses but also upon action of the low doses of IR [Muksinova K.N., Mushkacheva G.S. 1995, Serkiz Ya.I. 1992].

It was shown that regeneration of the maturing cell pool and renewal of their quantity in the peripheral blood is determined by the completeness of the progenitor cell clone recovery. During the bone marrow regeneration period it was determined that both the proliferation of HSC is increased and the transition time of the maturing cell elements shortens [Bilko N.M. 1998, Bilko N.M., Klimenko S.V., Velichko E.A. 1999, Tavassoli M.V. 2008]. After acute period of the damage to the hematopoietic system regeneration phase follows, which is dependent on the viability of the stem cells, their migration ability in to the most affected areas of the hematopoietic tissues, time of proliferation and maturation of the committed progenitors and quantity of the functional mature cells [Shouse S.S., Warren A.S.L., Whipple G.H. 2004]. Critical for recovery of the hematopoiesis are quantity and quality of the HSC that recovered after irradiation. Possibility of hematopoiesis recovery is observed if more than 5% of the stem and progenitor cells remain intact and carry on proliferation and differentiation [Bond V.P., Fliedner T.M., Archambeau J.O. 2007]. If their level falls below this critical value, hematopoietic system can be exhausted due to lack of the stem cells capable of regeneration [Down J., Van Os. R., Ploemacher R. 1991]. Main proliferation stimulating factor of the HSC, which remain in the dormant state, is reduction of their quantity [Serkiz Ya.I., Pinchuk L.B. 1992]. Decisive role in the regulation of the recovery of the polipotent hematopoietic progenitors belongs to the microenvironment that upon interaction with HSC supports stability of their quantitative parameters in the physiological conditions and supports its recovery in case of injures [Hall E.J. 1991, Hall Mauch P., Constine L., Greenberger J. 2005]. Increase in the proliferation activity of the HSC is observed starting after irradiation exposure at the doses of 0.2 – 0.3 Gy [Grande T., Varas F., Bueren J.A. 2000].

The mechanism of the microenvironment influence on the hematopoietic system is still not fully determined. However, today it is known that its elements control the processes of hematopoiesis via production of the cytokines as well as by direct cell-to-cell contacts between HSC and microenvironment. Membrane-associated contacts serve for the transfer of the required molecules, homing and migration of the progenitor cells to the specific sites of the hematopoietic tissue and transport of the hematopoietic growth factors [Cronkite E.P., Inoue T., Hirabayashi Y. 2003]. Cultural investigations of the bone marrow (BM) indicated that despite the normalization of the quantitative parameters changes of the ability of hematopoietic elements to colony-forming had reduced character during prolonged periods of time with prevailing eosinophilic and neutrophilic colonies [Bilko N.M. 1998].

For the determination of the distant intercellular transfer of the post-radiation signals between the cells of irradiated animals a novel method of in vivo culture using diffusion capsules (DC) was described [Bilko N.M., Votyakova I.A., Vasylovska S.V., 2005]. Investigations were done of the Balb/c mice (Figure 32). The 16-hour model of exposure was used (see *Protocol modes of* mice irradiation). Animals were divided in to three groups: 1st group was irradiated without the use of radioprotector, 2nd group of animals received melanin – glucan complex prior to irradiation and the 3rd group was non-irradiated control.

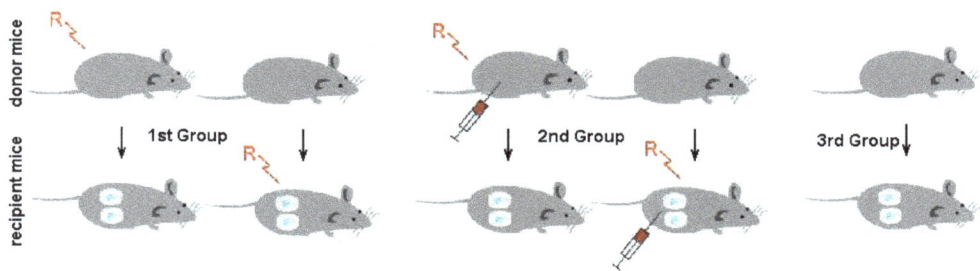

Fig. 32. Groups of experimental animals

Further each group was separated into the subgroups of the donors (3 animals) and recipients (3 animals for each donor and 2 capsules per recipient). Donor animals were sacrificed on the day 1, day 7, and day 30 after exposure and bone marrow cells were extracted from the femur. In each case, colony-forming activity (CFU) in the culture was determined by injection of the 1×10^5 cells into the inner cavity of the diffusion capsule in the semisolid (0.33%) agar Difco. Diffusion chambers (DC) permit free diffusion of the peptide factors; however, they allow avoiding any contact of the cultured material with the immune system of the recipient. Each animal was implanted with two DC into the peritoneal cavity under Sagatal narcotization. Animals were retained in the conventional vivarium conditions with 12-hour light/dark cycle illumination and free access to food and water. After 12 days implantation DC were extracted from the recipients and investigated under the inverted microscope for the CFU activity as indicated by formation of the colonies or clusters of the proliferated HSC.

The groups of cells less than 20 cells were considered as clusters, while groups of cells from 20-40 were considered as large clusters, and all cellular aggregates above 40 cells were counted as colonies (see Figure 33 and 34).

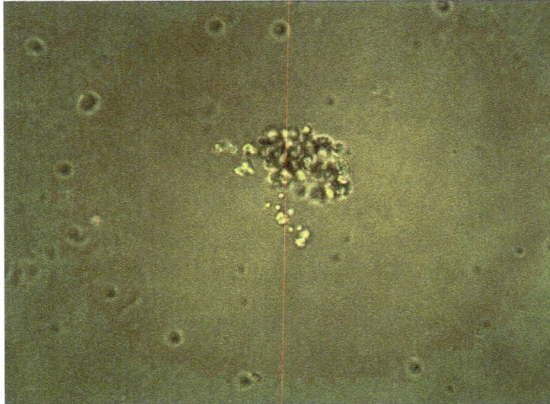

Fig. 33. Granulocyte-macrophage colony of mouse bone marrow culture of Balb/c, irradiated with a dose of 5.0 Sv. Inverted microscope, increase 200.

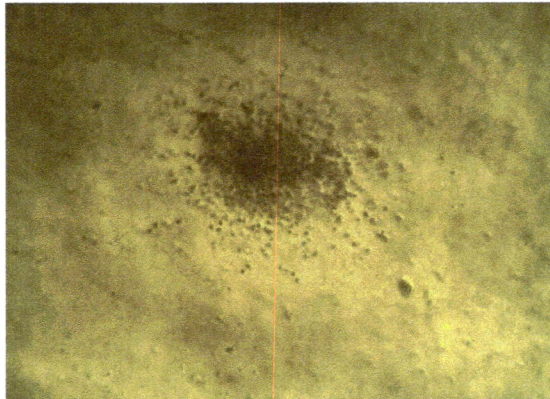

Fig. 34. Granulocyte cluster of mouse bone marrow culture of Balb/c, irradiated with a dose of 5.0 Sv. Inverted microscope, increase 400.

Cultured material was extracted from the inner cavity of the DC and individual colonies were picked up for the preparation of the cytospin slides and Pappenheim staining for identification of the cell types. Obtained data indicated that BM of the animals of the 1st group was affected by the IR. Cell aggregate numbers were on average 24 colonies and 48 clusters in the cultures of the 1st day after exposure. Almost no colony-forming activity was observed in the culture of 7th and 30th day after exposure, that was indicative of significant suppression of the bone marrow function by the IR. At the same time in the culture of bone marrow cells that were obtained from the animals, which received MGC prior to exposure, the average colony count was 32 with 86 clusters at the 1st day after exposure, 45 colonies and 112 clusters on the 7th day after exposure, and 80 colonies and 136 clusters on the 30th day after exposure.

These results may indicate that MGC is able to protect the population of the bone marrow stem cell population for the influence on the IR and stimulate recovery after irradiation at dose comparable to the $LD_{50/30}$. This conclusion can also be supported by increase of the quantity of the CFU in comparison to the 1st group of animals that have not received MGC. Implantation of the normal BM into the organism of the irradiated recipients at the 1st day after exposure after culturing resulted in formation of 114 colonies and 386 clusters on average. In the cultures of the 7th and 30th days such proliferation activity was observed, that it was not possible to determine individual colonies or clusters. There is indicative of a significant stimulation of the release of the compensatory signal substances by the radioresistant stromal cells of the BM, which highly stimulated the recovery of the radiation-damaged BM of the recipient. Normal bone marrow cells implanted into the irradiated mice treated with MGC yielded in 84 colonies and 192 clusters in the cultures of the 1st day after exposure, 52 colonies and 548 clusters on the 7th day, and 106 colonies and 302 clusters in the 30th day post exposure cultures (Figure 35).

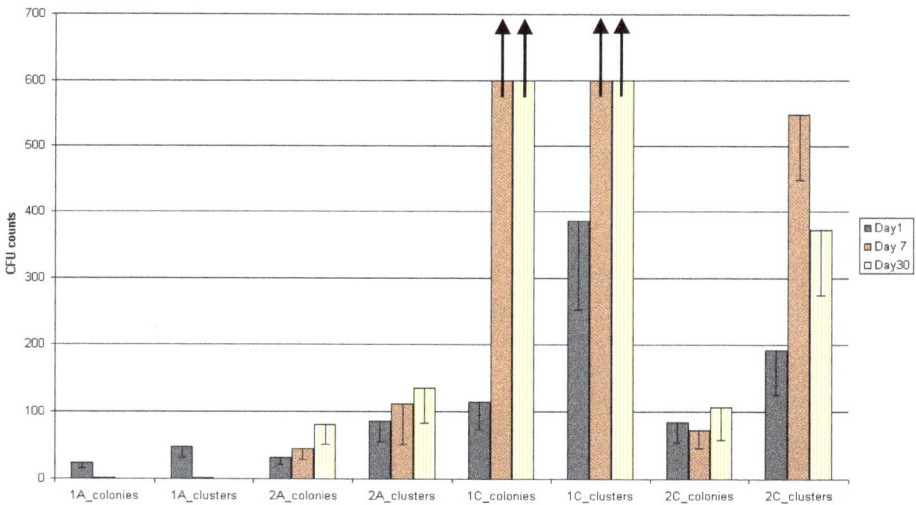

Fig. 35. Determination of the CFU activity of the bone marrow derived stem cells: Group 1A - irradiated bone marrow culture in normal recipients; Group 2A - irradiated BM protected with MGC in normal recipients; Group 1C - normal bone marrow culture in irradiated recipients; Group 2C - normal bone marrow culture in irradiated recipients treated with MGC.

Such decrease in the CFU activity is indicative of less apparent stimulation of the compensatory factors release as a result of the radiation exposure due to protective effects of the MGC on the bone marrow cells so that less factors are required for reparation and therefore less factors are available for stimulation of the CFU activity in the DC.

Results of the *in vivo* culture of the bone marrow cells indicated that colony-forming activity of the hematopoietic progenitor cells of the animals non-treated with MGC prior to radiation exposure was significantly lower if compared to the control while treatment of the animals with MGC had increased the functional activity of the bone marrow cells.

Therefore upon irradiation exposure quantity of the hematopoietic stem cells is decreased and as a result, stromal component of the BM starts to secrete large quantities of cytokines and growth factors, which are able to stimulate HSC to active proliferation [Goldberg E.D., Dygai A.M. 2001]. Clearly, in animals irradiated at doses of 5.0 Sv, production of growth factors was blocked by excessive amounts of free radicals which are formed when passing through biological tissue of ionizing particles [Timofeev-Resovskii N.V. 1963]. At the same time, free radicals can quickly neutralized by the MGC in mice receiving radioprotector. Improved products of growth factors caused the hyperproduction of bone marrow stem cells. On the 7th day after exposure in the bone marrow culture of the irradiated animals, a deep depression in the CFU activity was observed. Finally pretreatment of animals with MGC resulted in the significant increase of the CFU activity. These data indicates that in the BM of irradiated animals the pool of later mononuclear cells was rapidly exhausted that resulted in the decrease of the CFU activity [Bilko N.M., Klimenko S.V., Velichko E.A. 1999]. In the group of animals pretreated with MGC prior to exposure, CFU activity in the BM increased during the experiment. On the 30th day post exposure, CFU activity of the bone marrow stem cells remained low and in the animals that have not received MGC, and in the animals treated with this radioprotector these values were significantly higher.

Obtained data may be indicative of the gradual recovery of the BM by implementation of the so called "golden reserve" of the stem cells that during radiation exposure were positioned in the crypts of the stroma and therefore remained undamaged [Tsyb A.F., Budagov R.S., Zamulaeva A.I. 2005]. Such significant effects of the MGC on the numbers of colonies and clusters in the DC cultures indicates decrease quantity of the stimulating bystander signals and strong radioprotective properties of the MGC, however the exact mechanism of action remains to be fully elucidated.

Concluding on the obtained experimental data it is clear that MGC is a strong radiation protector that helps to avoid consequences of the $LD_{50/30}$ dose of irradiation at the level of HSC by affecting quantity of available growth stimulating factors that are commonly associated with radiation-induced damage to the BM.

2.4 Influence radionuclide fallout to plant grown around object "Shelter" in Chernobyl alienation zone

Contaminated of the wide territories in Ukraine not only with radionucludes ^{137}Cs and ^{90}Sr, and with fission products of uranium and transuranium elements is an essential consequence of the accident at the IV block of Chernobyl Nuclear Power Plant that is classified as a global ecological catastrophe. The biota behaviors and adapt in this areas captured dose from radionuclide with long half-value period decay isotopes. As dose related amount of the isotopes ^{137}Cs and ^{90}Sr during long time after accident were decreased. But only the amount of the radioactive isotope [241]Am depend of time is increasing exactly in environmental alienation zone of Chernobyl. Radionuclide [241]Am as α-emitter is a daughter product of [241]Pu isotope appeared after β-decay. The activity in environment of the isotope [241]Am is increasing with during time owing to β-decay of the [241]Pu isotope. The biota behavior in this areas captured dose from radionuclide with long half-value period decay isotopes. The peculiarity of radionuclides contamination associated with the Chernobyl accident is verified of physical and chemical forms of radioactivity elements through out into the environment [Rashydov N.M. 1999, Rashydov N.M., Konoplyova A.A., Grodzinsky D.M. 2004, Rashydov N.M., Kutsokon N.K. 2005, Rashydov N.M., Grodzinsky D., Berezhna V. 2006]. A part of the radioactivity isotopes is registered in water soluble droplets-liquid state, an other part – as "hot" particles, the interrelation between there forms being unstable and change under the influence of biotic and abiotic environmental factors. As rule in this conditions accumulation of radionuclides in plants which occurs mainly at the expense of their water-soluble and exchangeable forms, reflects rather complicated transitional processes in the soil, the rate and direction of these ones is determined by biological activity of all component of the plant rhizosphere inhabited layer of the soil. After Chernobyl accident already during 25 year a lot of "hot" particles transferred into fine dispersive conditions, which easy movements in outdoors where captured by biota which could characterize by help of transfer coefficient (TC) radionuclide ongoing. The transfer coefficient is ratio specific activities (kBq/kg) of plant to specific activity of soil (kBq/kg) where its grow that characterize go over a radionuclide from soil to vegetative plant on experimental plot. Necessary mentioned that the TC not constant and it differed on depend of parts of plant were determinate. For radionuclide ^{241}Am observed the value TC a lot of plants and mushrooms several order less than for isotopes ^{137}Cs, ^{90}Sr. Especially for matured seed the value of the TC observed less than for other vegetative parts of the plant [Rashydov N., Berezhna V., Kutsokon N. 2007, Rashydov N.M., Kutsokon N.K. 2008, Rashydov N.M., Berezhna V.V., Grodzinsky D.M. 2009, Rashydov N., Berezhna V. 2010, Rashydov N.M. 2010, Rashydov N.M. 2011]. To study of the TC peculiarity modification is reason elucidation of our field research in alienation Chernobyl zone around object "Shelter".

Contamination of plant in natural experimental fields at the alienation zone of Chernobyl significant added by flying dust with very small size radioactivity particles less than "hot particles" in environment. The results received for plants soybean (content: ^{137}Cs - 3.6 kBq/kg and ^{90}Sr – 11.84 kBq/kg) and flax (content: ^{137}Cs - 0.78 kBq/kg and ^{90}Sr – 3.55 kBq/kg) which grown in Chistogalovka (specific activity of soil is 20.65 kBq/kg and 5.18 kBq/kg for radionuclide ^{137}Cs and ^{90}Sr, accordingly) and Chernobyl confirmed this hypothesis. The value TC for above mentioned seeds specimens collected from plant which grow on Chistogalovka was approximately 22.3 (soybean) and 6.63 (flax) times (for isotope ^{137}Cs) and 13.97 and 4.71 (for isotope ^{90}Sr) times higher by comparison with control variants which grown in Chernobyl where specific activity was 1.41 kBq/kg for radionuclide ^{137}Cs and 0.55 kBq/kg for isotope ^{90}Sr, correspondingly.

The peculiarity distribution in controlled laboratory conditions the radionuclide ^{241}Am in *Arabidopsis thaliana* plant on high level first layer leaves, in petiole and in carry out fascicles of the leaves significantly that go into this isotope from root system to top of plant very slow and membrane of cells played as discrimination barrier in this processes as mentioned in our previously investigations [Rashydov N.M., Berezhna V.V., Grodzinsky D.M. 2009].

In laboratory conditions for autoradiography investigation purpose the seedlings *Arabidopsis thaliana* were aseptically grown in hard agar cultured medium containing ^{241}AmCl$_3$ in concentration with specific activity 50 kBq/kg. After 25 days some leaves and top of stems of plants witch had not direct contact with medium were carefully cut off so that to avoid contact with medium. Selected parts of plants *Arabidopsis thaliana* settled down on the microscopic glass slides and dried a few days. During this process they were gluing to the slides themselves. The slides with parts of plants were coated with photo emulsion LM-1 in gel (Amersham – Biosciences UK) and exposure during time 20 days at temperature +4^0 C. After development the samples of slides were observed of the track of α-particles from radionuclide ^{241}Am with light microscope. A lot of datum confirmed that the coefficient uptake very small for radionuclide 241Am and this element maldistribution by organs and tissues. We observed that accumulation the radionuclide of 241Am depended of carry out fascicles system of the leaves and localization of the layer leaves not far from length root collar of plant which grow in laboratory conditions. The first layer leaves were taken up high-level amount radionuclide 241Am. As result the capture dose also may tissues of plant distribute no uniform. It is known that mineral nutrients are transported apoplastically, i.e. in the wall system outside the plasma membrane, or symplastically, i.e. in the cytoplasm from cell to cell deal with through plasmodesmata. The nutrient elements that penetrate into the cytoplast can also be shuttled into the vacuole via various mechanisms depending of biological function in cell life behaviors for mentioned isotope.

For field experiments we use plant white blow (*Erophila verna (L.) Bess.*) for autoradiography investigation from Chistogalovka and Yaniv contaminated soil sites the distribution radionuclide essential differs in spite of above-mentioned experiment. On the top shoot apex leaves and flower observed a lot of tracks of the particles α- and β- decays [Rashydov N.M., Berezhna V.V. 2010] (Figure 36).

Our experimental data confirms that radioactivity fallout in environment essentially differed important amendment of the TC. Thus extra-root nutrition that included micro- or nano- size "hot" particles had essential role of plant behavior in environment. But for plants that harvested from contaminated sites distribution of the radionuclide 241Am by tissue and organs essentially differed from plants which grown in laboratory conditions.

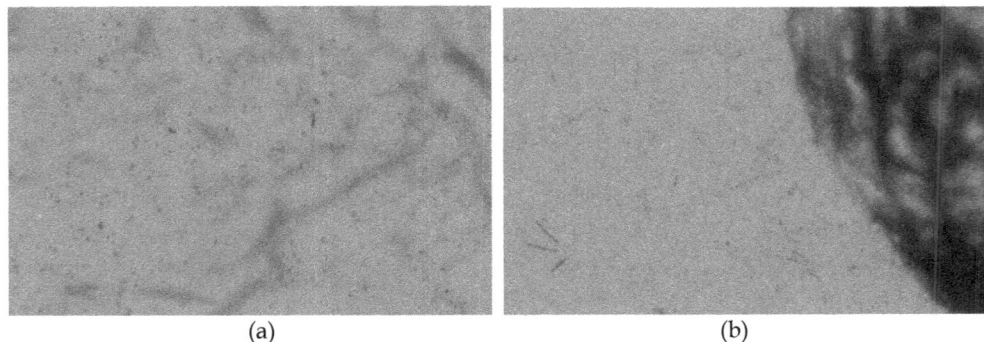

(a)   (b)

Fig. 36. Appear inside radioactivity of petal of bud (a) and also tracks of α-particles on surface of sepal (b) flower of the plants white blow (*Erophila verna (L.) Bess.*).

Contamination with radionuclide in natural experimental fields significant added tracks elementary particles from flying in air very small dust such as nano- and micro-size with radioactivity similarly "hot" particles in environment by help foliar pathway uptake into top leaves and aboveground apical apex of plants, especially around the object "Shelter".

2.5 Flowering flax plant under chronic irradiation

Flowering plant and seed development have long held the interest of general biologists because they represent a critical sensitivity changes phases in the pattern of shoot development and have significant consequences in creating yield seed [Hopkins W.G.,

Huner N.P.A., 2009]. The generative phase growth of plant include appear the floral organ with flower and filling seed event is involve important question: Why do plant this sensitivity processes controlled in shot during flowering spring and/or summer under stress factors? The flax plant flowering under chronic irradiation is a complex event that involves genome destabilizations as well as posttranslational regulation, signal transduction and epigenetic regulation metabolism inside living cell [Kutsokon N.K., Rashydov N.M., Grodzinsky D.M. 2007, Kutsokon N., Rashydov N.M. & Grodzinsky D. 2003, Kutsokon N., Rashydov N.M. & Berezhna V. 2004, Kutsokon N., Lazarenko L.M. & Bezrukov V.F. 2004]. For shed light this problem we carry out especially experiment by help of pretreatment by melanin-glucan complex flax seeds which growth under chronic irradiation in Chernobyl zone. As well known the melanin-glucan complex has high gene protective response against of chronic IR in a wide range of doses due to capturing free radical regulation and influencing on epigenetic changes in cells. The curve bloom rate of flax plant depend of during of term observe and treatment by the melanin-glucan complex under influenced chronic irradiation shown in figure 37.

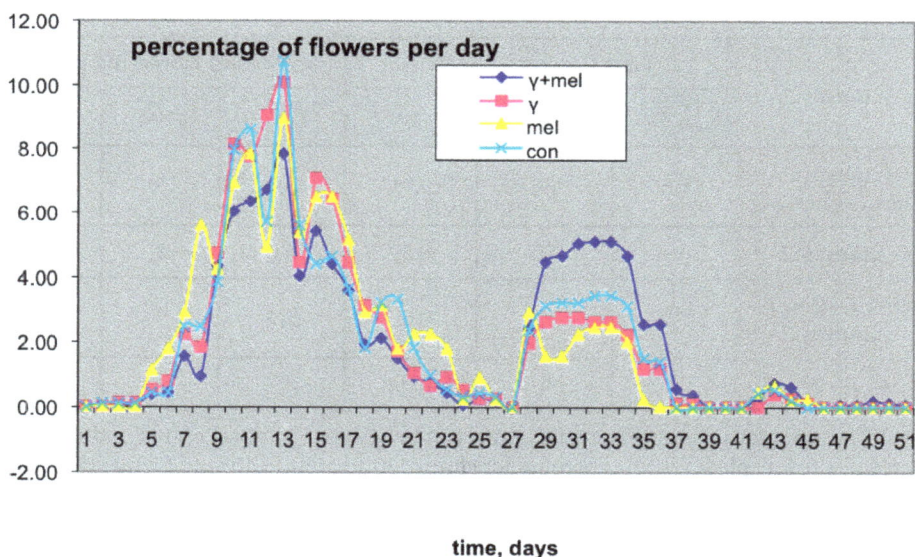

time, days

Fig. 37. Flowers per day depend of observed time and approach of treatment flax plants: with melanin-glucan complex plus chronic irradiation (γ+mel), irradiated by only chronic radiation (γ), with melanin-glucan complex (mel) and control variant (con) without any treatments.

On base above given curves we were calculated for all variants for the experiment curve parameters that characterized the altitude rate bloom and the time of flower appear necessary for realization of the half of peak flowering. The height of the altitude curve of bloom H (%) and half height of the peak flowering term L (days) of first and second bloom peak of flax plants which grown under different specific chronic irradiation with rate 2.6±0.3 mcSv/h and 25.4±0.4 mcSv/h to versus of control variants depend of melanin-glucan complex treatments shown at tables 1 and 2.

Variants	First flowering peak			Second flowering peak		
	H, %	L, days	R^2	H, %	L, days	R^2
γ + melanin-glucan complex	7.2	14	0.91	3.3	15	0.96
γ - irradiated	9.4	12	0.95	2.1	11	0.96
melanin-glucan complex	9.2	13	0.89	1.5	12	0.94
Control	9.2	12	0.92	1.5	12	0.96

Table 1. The altitude curve of bloom H (%) and half height time L (days) of first and second flowering peak flax plants which grown under chronic irradiation with specific rate of radiation 2.6±0.3 mcSv/h and control variants depend of melanin-glucan complex treatments

Variants	First flowering peak			Second flowering peak		
	H, %	L, days	R^2	H, %	L, days	R^2
γ + melanin-glucan complex	6.6	9	0.91	5.5	8	0.96
γ - irradiated	8.4	10	0.89	3.0	8	0.96
melanin-glucan complex	7.2	12	0.87	2.4	7	0.74
Control	7.6	10	0.92	3.7	8	0.94

Table 2. The altitude curve of bloom H (%) and half height time L (days) of first and second flowering peak flax plant which grown under chronic irradiation with specific rate of radiation 25.4±0.4 mcSv/h and control variants depend of melanin-glucan complex treatments

As shown from figure 37 after 10-15 days since started flowering period the percent of flax flower per day increased until maximal magnitude and the curve passed first peak. During 10-12 day bloom probable decrease until zero and keeping in spontaneous level a few days. But after that during 10-12 days probable of bloom increased and on the curve appear second late flowering peak. For treatment melanin-glucan complex reveal more quantity flower release during second bloom period depend of chronic irradiation: yield proximally 3.3% under chronic irradiation with specific dose 2.6 mcSv/h, in case specific dose 25.4 mcSv/h it increased until 5.5%, accordingly. Increasing the second peak inflorescence appear for both curves variant treatment with melanin-glucan complex at dose 2.6 mcSv/h as well as under chronic irradiation specific dose 25.4 mcSv/h deal with involve this subtract in epigenetic changes in genome regulations flowering process during ontogenesis of flax plants. Necessary mentioned that the bloom of second peak for several plant usually excited appear sterility flowers but for flax was appeared yield in second peak bloom a lot of fertile seeds in generally with germination rate proximally 97%. However the average numbers of seeds per seed box or per infructescence of plant flax grown in several conditions do not differed from control variant significantly.

As known epigenetic events regulate the activities of gene which take part in vegetative and generative phases of growth plant by help of processes methylation of DNA, acetylation, phosphorylation of the histone tails in chromatin fiber. The heritability of DNA methylation, which often occurs in the early and late stages of vegetative development of the ontogenesis plant, allows cells to keep irrelevant genes silenced in successive generations of embryo cell. However some genes – such as plant genes that govern cell dormancy and spring time flowering – require silenced genes to be reactivated. Evidence is beginning to emerge those different classes by help of micro ribonucleic acid (miRNA) and on noncoding RNAs regulate these protein synthesis [Danchenko, M., Škultéty, L. & Berezhna, V.V. 2008, Danchenko M, Skultety L. & Rashydov N.M. 2009, Danchenko M, Klubicova K. & Skultety L. 2011]. Thus the seedlings since germination until flowering, filling seed and matured seed during all ontogenesis term under chronic irradiation perceived some signal transduction events which accumulate permanently in plants as epigenetic changes transient heritability.

3. Conclusion

Quality and qualitative changes in the structure of the radiation factor of object "Shelter" occurred over twenty-five years of its existence. Inside the "Shelter" the spontaneous destruction of the fuel-containing materials in IV Unit of the ChNPP which arose out of nuclear fuel in the acute phase of the disaster under the influence of intense ionizing radiation and high temperatures was developing. As a result of a new type of radioactive aerosol particles of nanometer size were appeared as called fuel hot particles. Accumulation of radionuclides with long half-life, in particular, radioactivity isotopes of strontium and plutonium and their decay products in the tissues of plant and animal organisms is especially dangerous. The fuel-containing materials are dissolved under the action of microorganisms and a new compound of radioactivity isotopes with organic matter was appeared. They are potentially more mobile and more ecologically dangerous around and inside "Shelter" environments. The radiation fields with different intensity inside object "Shelter" have contributed to increasing the mutation rate and breeding of radiation-resistant microorganisms with new aggressive properties. A lot of number particles of micrometer and nanometer size which consisting the cells of microorganisms and spores in the investigated samples of aerosols originated from "Shelter" premises. It is known that the light roof of sarcophagus has a large number of defects and some of these microorganisms, especially in the form of spores, may potentially carry away outside of the object "Shelter".

The hot particles of fuel contain gamma-, beta- and alpha- irradiators, such as radioisotopes of strontium and plutonium and their decay products continue irradiating organism late times even after termination of introduce to them. When IR from similarly radioactivity isotopes interaction with living matter they pass it relatively high energy. It is being spent on the excitation of atoms or ions and changes the chemical properties of matter and additional radicals may be appeared in living cells.

As rule the linear non-threshold concept suggests that DNA damage induced by low doses of radiation do not contribute significantly to increased risk of disease because a significant amount of endogenous genome damage occurs and is restored in cells constantly. The main difference between DNA damage induced by IR, from the endogenous damaged - it's clustering and the complexity of their chemical nature. The proportion of complex, critical for the fate of the cell damage is much higher when exposed to chronic IR. Especially the massive clustering of DNA damage occurs when the ionization tracks pass along chromatin fiber. At the same time it is known that the probability of occurrence of endogenous clustered DNA damage is extremely low. Accumulation a lot of endogenous damages does not occur because in the cells are constantly functioning mechanisms of reparation, specifically targeted at removing various types damages. Low doses of IR can not activate the G_2/M chekpoynt-arrest and DNA repair are not activated when the number of DSB DNA damage and crosslinks between the fibers under 10 - 20 per cell.

There are four outcomes for the cell fate, if the cellular radiation damage is not adequately was repaired. The cell may die or will delay it or keep playing with the viability of new qualities or mutations as the basis for the development of remote descendants.

The energy dependence of the mechanisms of apoptosis is disabled after irradiation, and the cell dies with loss of cell membrane integrity and its release of macromolecular components, such as intracellular enzymes etc., into the intercellular space. These substances cause an immune response in the form of inflammation as leukocyte infiltration of the affected tissue, interstitial fluid accumulation and subsequent induction of specific immune responses (specifically sensible T-lymphocyte and autoantibody) to the unmasked and recognized by lymphocytes of intracellular components.

We detected a significant increase in the rate of ALT levels in chronically irradiated by low doses mice Balb/c compared with non-irradiated. Increased serum levels of this enzyme are considered a sign of inflammation in the liver because a large amount of ALT is released from the damaged cells of this body. Positive changes of serum levels of AuAB to liver-specific protein (LSP) that is a mixture of antigenic determinants of the substrate from the membranes of hepatocytes are also determined in the process of dose accumulation at irradiated animals. Because of their labiality, some proteins that are part of the LSP in particular the asialoglycoprotein receptor under certain conditions including under the influence of small doses of radiation may acquire properties of auto antigens.

Thus chronic exposure to low doses of ionizing radiation of low intensity may lead to the abolition of immune tolerance, not only in the liver but also in other organs. As is known in turn increased the activity of autoimmunity is a favorable condition for the transfer of persistent infections in the active state and for stimulation of vegetation as saprophyte and pathogenic microorganisms.

It is known that the IR increases the level of signal transduction in the infected cell and can activate the promoter of certain genes of the virus and subsequent synthesis of the corresponding proteins. As a result of these events, the cell does not die, as is the case with an active infection, and changes its functional properties. In this changing antigenic repertoire of cell membranes was described. Antigens become available for immune recognition. As a result by an autoimmune reaction and changes the pathogenesis of a wide range of infectious diseases to cancer stimulated.

The main outcome of the 25-year study of morbidity in different categories of exposed persons in connection with the Chernobyl catastrophe is a significant increase in primary morbidity is not associated with tumor pathology. An additional mechanism of radiobiological effects in mammals, which can have a direct influence on the biomedical effects of radiation, was described. This is a phenomenon of production by irradiated cells of signals that cause lesions similar to radiation damage in no irradiated neighbor cells as "bystander effect" (BSE) was named. We used index of SSB DNA as indicator to assess the effects of ionizing radiation. It was shown that the living medium obtained after three hours stay of IrCs on the first day after exposure at dose 5.0 Sv for 16 hour is able to induce an additional level of SSB DNA in different types of cells from not-irradiated mice Balb/c - in lymphocytes, hepatocytes, bone marrow hematopoietic cells, splenocytes and astroglial cells. Identification BSE in liver tissue is an additional argument in favor of accepting the reality of the existence of diseases of hepato-biliary system of radiation origin.

At the same time it was shown that the transmission "bystander" signals within the irradiated BALB/c mice were observed for at least one month after exposure. We also found a correlation between high rate of $LD_{50/30}$ and low induction SSB DNA in non-irradiated cells after placing them in the living environment of cells from irradiated animals. In mice with lower level of $LD_{50/30}$ (Balb/c) induction of SSB in cellular DNA non-irradiated lymphocytes after exposure in living environment of the irradiated cells in various periods after of irradiation was higher in comparison with similar index of C57Bl/6 mice with higher level of $LD_{50/30}$. We observed BSF in mice of two lines with different modes and doses: the external influence of gamma-ray dose, 5.0 Sv (for 16 hours), and external chronic exposure for 231 days with a cumulative dose of 0.29 Sv, with internal radiation for 74 days with the accumulation of activity in the mouse body about 18 kBq.

The reactive oxygen species (ROS) play an important role in the mechanisms of signal transmission to "bystander" cells. Reinforced ROS production in non-irradiated cells incubated in medium from serum, irradiated gamma-particles, or from supernatant of the suspension of irradiated cells was shown by many authors. Irradiated cells produce a spectrum of "bystander" signals - cytokines, fragments of DNA (from apoptotic cells) or other factors of protein nature. These factors cause a change in oxidative metabolism and gene expression profiles in irradiated cells, and induce enhanced production of highly reactive oxygen species. In addition to chemical modification of DNA nucleotides, the formation of radicals can also lead to changes in the higher levels of organization structure of the molecule - the secondary, tertiary and quaternary.

Therefore the melanin-glucan complex (MGC) from higher mushroom basidiomycetes - *Fomes fomentarius*, was used as an antioxidant to confirm the involvement of molecules of free-radical nature to the transmission of signals from irradiated cells to non-irradiated cells after external influence of gamma-rays (the model of the single and chronic exposure of mice). Antioxidants, photo-and radio-protective properties are a direct consequence of the free-radical structure of melanin providing the opportunity to participate in electronic exchange of redox in cell and radical recombination processes around DNA damages. We have shown that intra peritoneal injection of MGC, which has powerful antioxidant properties, to mice prior to irradiation at a dose procedure 5.0 Sv decreases from BSE in all kinds of testing cells and increase the number of colony forming units in bone marrow of irradiated mice. These results may serve as an argument in favor of the hypothesis about the important role of free radicals, molecules in the implementation of the phenomenon of BSE. Concluding on the obtained experimental data it is clear that melanin-glucan complex is a strong radiation protector that helps to avoid consequences of the $LD_{50/30}$ dose of irradiation at the level of HSC by affecting quantity of available growth stimulating factors that are commonly associated with radiation-induced damage to the bone marrow.

In the experiments on vegetative plant test-system evidence that under influence chronic irradiation were formed radiobiological reactions living organisms which include several level functions and structure organization processes since metabolic pathways and cellular systems until genetically and population changes. We observed that accumulation the radionuclide of ^{241}Am depended of carry out fascicles system of the leaves and localization of the layer leaves not far from length root collar of plant which grow in laboratory conditions. The first layer leaves were taken up high-level amount radionuclide ^{241}Am. The peculiarity distribution radionuclide ^{241}Am in Arabidopsis thaliana plant on high level layer leaves, in petiole and in carry out fascicles of the leaves significant that go into this radioactivity isotope from root system to top of plant very slow and membrane of cells played as discrimination barrier in this processes as mentioned in our previously investigations. It is known that mineral nutrients are transported apoplastically, i.e. in the wall system outside the plasma membrane, or symplastically, i.e. in the cell wall system outside the plasma membrane, or symplastically, i.e. in the cytoplasm from cell to cell deal with through plasmodesmata. Elements that penetrate into the cytoplast can also be shuttled into the vacuole via various mechanisms depending of element. But for plants that harvested from contaminated sites distribution of the radionuclide ^{241}Am by tissue and organs essentially differed from plants which grown in laboratory conditions. When we observed plant white blow (*Erophila verna (L.) Bess.*) for autoradiography investigation from Chistogalovka and Yaniv contaminated soil sites the distribution tracks of radionuclide essential differ in compare of above mentioned experiment. Contamination with radionuclide in natural experimental fields significant added tracks elementary particles from flying in air very small dust such as nano- and micro-size with radioactivity less than "hot" particles in environment by help foliar pathway into top leaves and apex of plant. Existence of nano- and micro- size dust as radioactive aerosols with specific properties around object "Shelter " circulate an environment represents potential risk radiobiological consequences of plant that grow in alienation Chernobyl 30-km zone.

The chronic irradiation around object "Shelter" was including instability genome, which carried out genetic and somatic changers in late heritable on perennials as well as in several generation annual plants. Proteomics analysis observes; bloom, filling and maturing seeds evidence that there is induce epigenetic changes of genome that excited late genetic changers on progenies and somatic changes of flax plant. Treatment by the melanin-glucan complex from fungus *F. fomentarius* before sowing seed procedure increase second peak on flowering curve depend of ontogenesis flax under influence chronic irradiation.

It is known that under chronic irradiation the epigenetic events regulate the activities or inactivate of genes by help of several mechanisms which may be include miPNA inter action deal with process that take part in vegetative and generative phases (bloom, filling and maturing seed) of growth plant by help of processes methylation of DNA, acetylation, phosphorylation of the histone tails in chromatin fiber of chromosome. However some genes – such as plant genes that govern cell dormancy and spring time flowering – require silenced genes to be reactivated. The heritability of DNA methylation, which often occurs in the early and late stages of vegetative development of the ontogenesis plant, allows cells to keep irrelevant genes silenced in successive generations of embryo cell. Evidence is beginning to emerge those different classes by help of micro ribonucleic acid (miRNA) and on noncoding RNAs regulate these protein synthesis.

Our experiments carry out with chronic irradiation revealed that in mammalian and plants cells excided not only direct damage DNA or appear mutation event as well as were observed complex responsible processes including genome instability, signal transduction, bystander effect between sells and epigenetic changes several profiles protein synthesis depend of activity or silent some genes which adjusted by help of treatment melanin-glucan complex.

4. Acknowledgment

The authors thanks to deputy director of the Nuclear Research Institute of National Academy of Sciences of Ukraine Dr. Tryshyn V.V. help us with maintenance radioactivity measure of several contaminated samples by radionuclides from Chernobyl zone.

5. Keywords

Chernobyl, object "Shelter", radioactive isotopes, "hot" particles with micro- and nano-sizes, radioactivity aerosol, embryogenesis, signal system, autoradiography, single strand break (SSB) DNA, chronic irradiation, bystander effect, autoimmunity, alanine aminotransferase (ALT), liver-specific lipoprotein (LSP), radiation hepatitis, bone marrow, hematopoietic stem cells, medical consequence, melanin-glucan complex, flowering of flax, epigenetic.

6. References

Anderson S, Bankier A.T., Barrell B.G., de Bruijn M.H., Coulson A.R., Drouin J, Eperon I.C., Nierlich D.P., Roe B.A., Sanger F, Schreier P.H., Smith A.J., Staden R, Young I.G. Sequence and organization of the human mitochondrial genome// Nature. 1981 Apr 9; 290(5806): p.457-65.

Ashmarin I.P., Freidlin V.P. History and the practical aspects of a new understanding of the role of autoimmunity. Abstracts 1st Moscow International Conference "Natural autoimmunity in health and disease. " Moscow. 15 - 17 September, 2005, 245p.

Azzam E.I., Little J.B. The Radiation-induced Bystander effect: Evidence and Significance. In: Bystander effects and the dose response. Belle News-letter/ Biological Effects of Low level Exposure. 2003. V.11.N 3 P .2 – 6.

Azzam, E.I., de Toldeo, S.M., Little, J.B. Stress signaling from irradiated to non-irradiated cells// Current Cancer Drug Targets. 2004. Vol.4 N 1. P. 53-64.

Eallot E., Homberg J. C., Johanet C. Antibodies to soluble liver antigen: an additional marker in type 1 auto-immune hepatitis // J. Hepatology, – 2000. – 33, – P. 208 – 215.

Baryakhtar V.G., Gonchar V.V., Zhidkov A.V., Klutchnikov A.A. Dust productivity of damaged irradiated fuel and lava-like fuel-containing materials of "Shelter" object. Preprint of ISTC "Shelter" of Natl. Acad. Sci. of Ukraine, 1997, No.97-10, p.197 (in Russian).

Baryakhtar V., V.Gonchar, A.Zhidkov, V.Zhydkov Radiation damages and self-sputtering of high-radioactive dielectrics: spontaneous emission of submicronic dust particles // Condensed Matter Physics, 2002, Vol.5, No.3(31), p.449-471.

Baverstock K. Radiation-induced genomic instability: a paradigm-breaking phenomenon and its relevance to environmentally induced cancer // Mutat. Res. 2000. V.454. No 1-2. P.89-109.

Bennett P.V., Cintron N.S., Gros L., Laval L. And Sutherland B.M. Are endogenous clustered DNA damage induced in human cells? // Free Radic. Biol. Med. 2004. V. 37. P.488-499.

Berehovskaya N.N., Savich A.V. Radiation damage of mitochondrial genome and its role in long-term consequences of irradiation // Radiat. Biol. Radioekol. 1994. V 34. No 3 P.349-351.

Bilko N.M., Klimenko S.V., Velichko E.A., Zhazhal E..I Peculiarities of hemopoiesis in reconvalescents of acute radiation sickness in a period of long-term effects / Proceedings of scientific papers KIAPE. - Kiev, 1999. - Issue8. - P.40-43.

Bilko N.M. Hemopoietic progenitor cells in radiation exposure (an experimental and clinic study): Author. dis. Dr. med. Sciences: 03.00.01 / NAS, Inst Exp. pathologies. oncology. i radiobiol. - K., 1998. - 31 p.

Bilko N.M., Votyakova I.A., Vasylovska S.V., Bilko DI.Characterization of the interactions between stromal and haematopoietic progenitor cells in expansion cell culture models. //Cell Biol Int. 2005 Jan; 29(1): p. 83-86.

Blandova Z.K., Dushkin V.A., Malashenko A.N. Lines of laboratory animals for biomedical research / M. "Science. " 1983. 180 p.

Bond V.P., Fliedner T.M., Archambeau J.O. Mammalian Radiation Lethality. // New York: Academic Press. – 2007. – 159 p.

Bondarenko O.A., Aryasov P.B., Melnichuk D.V., Medvedev S.Y. Analysis of aerosol distribution inside the object "Shelter" at the Chornobyl reactor site // Health Physics, 2001, vol.81, No.2, p.114-123.

Bont R.D., Van Larebeke N. Endogenous DNA damage in humans: a review of quantitative data. Mutagenesis. 2004. V19. P. 169-185.

Bouzounov O.V., Tereshchenko V.M. Mortality from non-neoplastic diseases of participants aftermath of the Chornobyl nuclear power plant in a post-accident period, doze dependent effects. In the book. Problems of radiation medicine and radiobiology. Proceeding of scientific papers. Ed Bebeshko V.G. Issue 15., Kiev, 2010. p. 416.

Bradford M.M. A rapid and sensitive method for quantitation and microgramm quantities of protein utilizing the principle of protein binding / / Anal.Biochem. - 1977 - 1986. - P. 193-200.

Bradley M.O., Kohn K.W. Fluid mechanisms of DNA double-strand filter elution / / Nucleic Acid Res. 1979. Oct 10, 7 (3). P.793-804.

Brenner D.J., Doll R., Goodhead D.T., Hall EJ, Land CE, Little JB, Lubin JH. Cancer risk attributable to low dose of ionizing radiation: assessing what we really know // Proc. Natl. Acad. Sci. USA. 2003. - V. 100.-P. 13761-13766.

C. Garcia-Buey L., Garcia-Monzon C., Rodriguez S., Bourque M.J., Garcia-Sancherz A., Iglesias R., De Castro M., Mateos F.G., Vicario J.L, Baslos A. Latent autoimmune hepatitis triggered during interferon therapy in patients with chronic hepatitis C // Gastroenterology –1995. – №.108. (6) –P. 1770-1771.

Chakraborty A, Held K.D., Prise K.M., Liber H.L., Redmond R.W. Bystander effects induced by diffusing mediators after photodynamic stress// Radiation Research, 2009. V.172. N.1. P. 74-81.

Churilov L.P. Autoimmune regulation of cellular functions, human antigenome and autoimmunomic: a paradigm shift // Medicine twenty-first century. 2008, 4 (13). p. 10-20.

Cronkite E.P., Inoue T., Hirabayashi Y., Bullis J. Are stem cells exposed to ionizing radiation in vivo as effective as nonirradiated transfused stem cells in restoring hematopoiesis? // Exp. Hematol. – 2003. - Vol. 21. - P. 823-842.

Dadachova E, Ruth A. Bryan R.A, Howell R.C., Schweitzer A.D., Aisen P., Nosanchuk J.D., Arturo Casadevall A.C. The radioprotective properties of fungal melanin are a function of its chemical composition, stable radical presence and spatial arrangement. Pigment Cell & Melanoma Research. Vol. 21, Issue 2, April 2008. P. 192–199.

Danchenko M, Skultety L, Rashydov N.M., Berezhna V.V, Mátel L, Salaj T, Pret'ová A, Hajduch M. Proteomic analysis of mature soybean seeds from the Chernobyl area suggests plant adaptation to the contaminated environment// J. Proteome Res. 2009, 8: 2915-2922.

Danchenko, M., Škultéty, L., Berezhna, V.V., Rashydov, N.M. , Pret'ová, A. , Hajduch, M. Twenty one years since Chernobyl disaster; What seed protein can tell us? Cordova, Spain, 2008, p.134

Danchenko M, Klubicova K, Skultety L, Berezhna V, Rashydov N, Hajduch M. Concept of crop adaptation to the Chernobyl environment based on proteomic data//Climate Change: Challenges and Opportunities in Agriculture, Budapest, Hungary, 21-23/03/ 2011, p. 151-154.

Down J., Van Os. R., Ploemacher R. Radiation sensitivity and repair of long-term repopulating bone marrow stem cells. // Exp. Hematol. - 1991. – Vol. 19. - P. 474 – 486.

Elmendorff-Dreikorn K., Chauvin C, Slor H., Joachim Kutzner, Werner Muller E.G. Assessment of DNA damage and repair in human peripheral blood mononuclear cells using a novel DNA unwinding technique // Cellular and Molecular Biology. 1999. Vol.45 (2), P. 211-218.

Ermakov A.V., Kon'kova M.S., Kostyuk S.V., Ershov E.S., Egolina N.A., Veiko N.N. Fragments of DNA from the extracellular medium of incubation of human lymphocytes irradiated at low doses, trigger the development of oxidative stress and adaptive response in unexposed bystander lymphocytes / / Radiats. Biol. Radioekol. 2009. T. 48, № 5. P. 553-564.

Fernet M, Megnin-Chanet F, Hall J, Favaudon V. Control of the G2/M checkpoints after exposure to low doses of ionising radiation: implications for hyper-radiosensitivity// DNA Repair (Amst). 2010, 9(1) p.48-57.

Friedburg E.C., Walker G.C., Siede W, Wool R.D, Schult R.A., Ellenverger T. DNA repair and mutagenesis. Washington, DC.: ASM Press. 2006. 1118 p.

Gaziyev A.I. Low efficiency of repair of critical DNA damage caused by low doses of radiation. 2011 (in press).

Goldberg E.D., Dygai A.M. Dynamic theory of regulation of hematopoiesis and the role of cytokines in the regulation of gemopoeza / / Medical Immunology. - 2001. - Volume 3, № 4. P. 487-497.

Grande T., Varas F., Bueren J.A. Residual damage of lymphohematopoietic repopulating cells after irradiation of mice at different stages of development. // Exp. Hematol . – 2000. – Vol. 28. – P. 87.

Grosovsky A.J. Radiation-induced mutations in unirradiated DNA // Proc. Natl. Acad. Sci. 1999. V.96. N.10. P.46-53.

Grudzenskia S., Rathsa A., Conrada S., C.E. Rubeb, M. Lobrich, Inducible response required for repair of low-dose radiation damage in human fibroblasts. PNAS, 2010, 107, 32,14205–14210.

Gusev M.V., Mineeva L.A. Microbiology. 3rd ed. M: publ. MGU, 1992.-448 p.

Hada M., Georgakilas A.C. Formation of clustered DNA damage after high-LET irradiation: a rewiev// J. Radiat. Res. 2008. V.49. P.203-210.

Hall E.J. Radiology for the Radiologist. Philadelphia: LippincottCo. – 1991. - 334 p.;

Hall Mauch P., Constine L., Greenberger J., Knospe W., Sullivan J., Liesveld J.L., Deeg H.J. Hematopoietic stem cell compartment: acute and late effects of radiation therapy and chemotherapy. // Int J Radiat Oncol Biol Phys. – 2005. –Vol. 3. - P.1319-39.

Harel M., Shoenfeld Y. Predicting and preventing Autoimmunity, myth or reality// Ann. N.Y. Acad. Sci. 2006. 1069: 322-346.

Hildebrandt G. Non-cancer diseases and non-targeted effects. //Mutat. Res. 1996, 687; 73-77.

Hopkins W.G., Huner N.P.A. Introduction to Plant Physiology. 2009, Wiley, USA, 503 p.

Howell C.R., Schweitzer A.D., Casadevall A., Dadachova E. A. Chemosorption of radiometals of interest to nuclear medicine by synthetic melanins. Nucl Med Biol. 2008 April; 35(3): 353–357.

Ide H., Shoulkamy M.I., Nakano T., Miyamoto-Matsubara M, Salem AM. Repair and biochemical effects of DNA-protein crosslink. // Mutat. Res. 2010, V704, P.172-183.

Johansen C.O. Cellular telephones, magnetic field exposure, risk of brain tumors and cancer at other sites in cohort study //Radiat. Brot. DOSIM. 1999. V.83. N.1-2. P. 155-157.

Kovalev V., Krul N., Zhezhera V., Seniuk O. Autoimmune hepatitis as a result of chronic exposure to low doses of radiation. /Uzhgorod University Scientific Bulletin Series of Biology Issue 27. 2010: P.245-249.

Kovalev V.A., Senyuk O.F. The state of tolerance under the influence of ionizing radiation "Chernobyl of the spectrum" /Ecological Bulletin, Minsk. 2008. № 2 (5). p. 36-42.

Kudryashov Yu.B. Radiation Biophysics (Ionizing radiation). Ed. By Mazurik V.K., Lomanova M.F. 2004. 448. www.medliter.com. (in Russian).

Kutsokon N., Lazarenko L.M., Bezrukov V.F., Rashydov N.M., Grodzinsky D. The number of aberrations per cell is an index of chromosome instability. 2. Comparative analysis of effects produced by factors of various natures. Cytology and Genetics 2004, Jan-Feb; 38(1), p. 49-57.

Kutsokon N., Rashydov N.M., Berezhna V., Grodzinsky D. Biotesting of radiation pollutions genotoxicity with the plants bioassays. Radiation safety problems in the Caspian region. Kluwer Academic Publishers. 2004. P. 51 – 56

Kutsokon N., Rashydov N.M., Grodzinsky D., Bezrukov V.F., Lazarenko L.M. Number of aberrations in an abnormal cells as a parameter of chromosomal instability. 1. Characteristics of dosage dependences. Cytology and Genetics 2003, Jul-Aug; 37(4), p. 17-24.

Kutsokon N.K., Rashydov N.M., Grodzinsky D.M. Cytogenetic effects of ^{241}Am in Allium-test. In Book "Int. Conf. "Genetic Consequences of Emergency Radiation Situations", Moscow, 2002. - P.146 – 148.

Lindahl T. Instability and decay of the primary structure of DNA. // Nature. 1993. V362. P. 709 – 715.

Little D.B. Not target effects of ionizing radiation: findings in relation to low-dose effects. //Radiation Biology. Radioecology. 2007, V.47, N 3, p. 262-272.

Little J.B. Radiation carcinogenesis. // Carcinomogenesis, 2000. V.21. N3. P.397-400.; Little J.B. Genomic instability and bystander effects: A historical perspective. // Oncogene. 2003. V.22. P.6978-6987.

Lorimore S.A., Coates P.J., Scobie G.E., Milne G., Wright E.G. 1nflammatory-type responses after exposure to ionizing radiation in vivo: a mechanism for radiation induced bystander effects? Oncogene. 2001. N.20. P.708-595.

Maltsev V.N. Quantitative regularities of radiation immunology. -M.: 1983. ???? p.

Manns M, Gerken G, Kyriatsoulis A, Staritz M, Meyer zum Büschenfelde K-H. Characterisation of a new subgroup of autuimmune chronic active hepatitis by autoantibodies against a soluble liver antigen // Lancet. – 1987. – P. 292 – 294.

Marples B., Wouters B.G., Collis S.J., Chalmers AJ, Joiner MC. Low-dose hyper-radiosensitivity: a consequence of ineffective cell cycle arrest of radiation-damaged G2-phase cells. Radiat Res 161:247–255; 2004.

McFarlane I.G., Wojicicka B.M., Zucker G.M., Eddleston AL, Williams R. Purification and Characterization of human liver-specific membrane lipoprotein (LSP) / / Clin. Exp. Immunol. - 1977. - 27, № 3. - P. 381-390.

Mosse I.B. Modern Problems biodosimetry// Radiation Biology. Radioecology. 2002., 42 . № 6. pp. 661-664 (in Russian).

Mosse I.B., Dubovic B.V., Plotnikova S.I., Kostrova L.N., Subbot S.T. Melanin decreases remote consequences of long-term irradiation. // Proc.f 9th Int. Congr. on Radiation Protection. Vienna, Austria, 14–19 April 1996. – Vienna, 1996. - V. 2. - P. 127-129.

Mosse I.B., Lyach I.P. Influence of melanin on mutation load in Drosophila population after long-term irradiation. // Radiation Research. - 1994. - V.139. - №3. - P.356-358.

Mosse I.B., Zhavoronkov L.P., Molofey V.P., Izmestieva O.S., Posad V.M., Izmest'ev V.I. Development on the basis melanin means of prevention of genetic and ontogenetic effects of irradiation. Vestnik VOGiS, 2005, Vol. 9, № 4, p. 527-533.

Mothersill C., Seymour C.B. Cell–cell contact during gamma irradiation is not required to induce a bystander effect in normal human keratinocytes: evidence for release during irradiation of a signal controlling survival into the medium// Radiat. Res. 1998. V.149. P. 256-262.

Mothersill, C. Seymour, C. Radiation-induced bystander effects: past history and future direction. Radiat. Res. 2001. V. 155, P. 759–767.

Muksinova K.N., Mushkacheva G.S. Cellular and molecular basis of the restructuring of hematopoiesis in long-term radiation exposure. Moscow: Energoatomizdat, 1995. – 161 p.

Nagasawa H, Little J.B. Induction of sister chromatid exchanges by extremely low doses of a-particles. *Cancer Res* 1992;V.52. P.6394–6396.

Narayanan P.K., Googwin E.Y, Lehnert B.E. Alpha particles initiate biological production of superoxide anions and hydrogen peroxide in human cells // Cancer Res. 1997. Vol. 57. N 18. P.3963-3971.

NCR (National Research Council of the National Academies). // Health Risk from Exposure to Low Level of Ionizing Radiation (BEIR VII, Phase II). National Academies Press, Washington, DC, 2006. P. 386.

Nobler M.P. The abscopal effect in malignant lymphoma and its relationship to lymphocyte circulation. Radiobiology. 1969. V.93. P. 410-412.

Notkins A.L. New predictors of disease // Scientific American 2007, V. 296. No 3. P.72-80.

Oxidative Stress: Oxidants and Antioxidants /Ed. Sies H. N.Y.: Academic, 1991. 546 p.

Pazuhin E.M., Krasnov V.A., Lagunenko A.S., B.I. Ogorodnikov, N.I.Pavljuchenko, V.E. Khan, A.A.Odintsov, V.B.Rybalka, G.I.Petelin, Yu.I.Zimin, I.N. Kantseva. Studying of physical and chemical properties of nuclear danger divising materials influencing a degree of nuclear, radiating and ecological safety of object "Shelter". // Problems of Chornobyl. Issue 14. 2004. p. 129 - 136.

Petelin G.I., Zimin Yu.I., Tepikin V.E., V.B. Rybalka, E.M. Pazuhin. "Hot" particles of nuclear fuel of Chernobyl emission in a retrospective estimation of emergency processes on 4-th block ChNPP. Radiochemistry, 2003, V.45, No 3, PP 278-281;

Pfeiffer P., Gottlich B., Reichenberger S., Feldmann, E.; Daza, P.; Ward, J. F.; Milligan, J. R.; Mullenders, L. H. F.; Natarajan, A. T., DNA Lesions and Repair. Mut. Res. Rev. Gen. Tox. 1996, 366 (2) 69-80.

Poletayev A.B. About the "difficult issues" of autoimmunity, or as the concept of immunkuluma can become the foundation of preventive medicine. // Medicine XX I. 2008. № 2 (11). P.84-91. Cohen I.R. The Immunological gomunculus spears in microarray. Natural autoimmunity in physiology and pathology abstracts. 2005. P.13.

Poletayev A.B. Immunophysiology and immunopathology. - Moscow: MIA, 2008. 208 p.

Preston D.L, Shimuzu Y, Pierce DA, Suyama A and Mabuchhi K. Studies of mortality of Atomic Bomb survivors. (2003) Report 13: Solid Cancer and Non-cancer Disease Mortality: 1950-1997 Radiation Research Vol. 160: P. 381-407

Radulescu I., Elmroth K., Stenerlow B. Chromatin organization contributes to non-randomy distributed // Radiat. Res. 2004. V.161. P.1 -8.

Rashydov N., Berezhna V. Distribution [241]Am by organs and tissues of plants. In: Nuclear Track Detectors: Design, Methods and Applications ISBN: 978-1-60876-826-4 (Editor: Maksim Sidorov and Oleg Ivanov), Nova Science Publishers, Inc., USA, chapter 7, 2010, p. 213-225.

Rashydov N., Berezhna V., Kutsokon N. Evaluation of water and soil genotoxicity with bioassays in sites contaminated with radionuclides. Proceedings "Sustainable Water Management" MEDA Water International Conference, Tunis, 2007 p. 307-313.

Rashydov N.M. Characterization risk of the contamination water pathways and determination of toxicity. Abstract book NATO Advanced research workshop on "Climate change and its effect on water resources – Issues of national and global security" Editors: Alper Baba. Goknem Tayfur, Irem Shahin (Izmir, Turkey 01-04 September 2010) p. 26.

Rashydov N.M. Risk characterization of the contamination water pathways and determination of toxicity. In book issues NATO Advanced research workshop on "Climate change and its effect on water resources – Issues of national and global security". Chapter 18, Editors: Alper B. Goknem T. & Orhan G. Springer, 2011, p. 157-165.

Rashydov N.M. Study radiosensitivity seeds Oenothera biennis L. grown in conditions chronic irradiation// In book: Study ontogenesis of plants natural and cultural flora in botanical organizations and in arboretum of Europe - Asia, Bila Tserkva, 1999, p.237-240.

Rashydov N.M., Berezhna V.V., Grodzinsky D.M. Radiobiological effects of [241]Am incorporated in cells of organism and methods of prevention of the menace of combined toxicity of the transuranic elements // In book: Aycik, Gul Asiye. New techniques for the detection of nuclear and radioactive agents. Springer, Verlag, 2009. – 347, p.313-323.

Rashydov N.M., Grodzinsky D., Berezhna V., Sakada V. The study of adaptation responses under γ-irradiation of transgenic lines Arabidopsis thaliana L. Proceedings of the 35th Annual Meeting of the European Radiation Research Society. Current problems of radiation research, Kyiv, Ukraine (Edited by D. Grodzinsky & A. Dmitriev) 2006, p. 227-235.

Rashydov N.M., Konoplyova A.A., Grodzinsky D.M. Research of possible mechanisms of morphogenetic processes suppression in tissue culture, donor plants of which were exposed under action of chronic irradiation. Scientific papers of the Institute for Nuclear Research, Kiev, Ukraine. 2004, N 2 (13), p. 122-132.

Rashydov N.M., Kutsokon N.K. Target and non-target radiobiological reactions – its threshold and non-threshold effects. Problems of nuclear power plants safety and of Chernobyl. V3, N2, 2005. p. 42-49.

Rashydov N.M., Kutsokon N.K. Effects of chronic irradiation of plants deal with radionuclides man-caused origin. D. M. Grodzinsky (Editor), Radiobiological effects chronic irradiation of plants in zone influence Chernobyl catastrophe. P.70-134, 373p. (2008) (in Ukrainian).

Riley P.A. Melanin // Int. J. Biochem. Cell. Biol. 1997. Vol. 29. No. 11. P.1235-1239.

Ryabenko D.V., Sidorik L.L., Sergienko O.V., Trunina I.V., Bobyk V.I., Fedorkova O,M., Kovenia T.V., Matsuka G.Kh. Myosin-induced myocardial damage: experimental study// Ukrainian Rheumatology Journal – 2001. – 3, №1. – P. 58-61.

Rybalka V., Klechkovskaja E., Serbinovich V., Petelin G., Kantzeva I. The microbiological factor and possibilities of radionuclides outwash control, changing of its uptake by plants on an example of the contaminated soils of Chernobyl zone. International Conference, Prague, 11-13 September 2001 Crop Science on the verge of the 21-th Century - Opportunities and Challenges/Proceedings, P.164-167.

Rybalka V.B., G.F. Smirnova, G.I.Petelin, I.N.Kantseva, V.V. Serbinovich, V.A. Krasnov, V.E. Khan. The microbic factor, fuel containing materials and formation of submicronic particles in object "Shelter". Problems of safety of nuclear power Plants and of Chornobyl. Issue.3. P.1. 2005. P. 87-97.

Rybalka V.B., Rybalko S.I., Zimin Yu.I., Petelin G.I., Rybakova E.A., Tepikin V.E. "Hot" particles of the Chernobyl exclusion zone in the esearch by electron microscopy. Atlas. ISBN 966-7654-41-9. Chornobyl Center radioecological, Chernobyl, 2001, 52 p.

Schweitzer A. D., Robertha C. Howell R.C., Jiang Z., Ruth A. Bryan R.A., Gerfen G., Cher. Ch.Ch., Mah D., Cahill S., Casadevall A, Dadachova E. Physico-Chemical Evaluation of Rationally Designed Melanins as Novel Nature-Inspired Radioprotectors. PLoS One. 2009; 4(9): p. 7229.

Senyuk O., Gorovoj L., Zhidkov A., V. Kovalev, L. Palamar, V. Kurchenko, N. Kurchenko, H-Ch. Schroeder. Genome protection properties of the chitin-containing preparation Mycoton In: Advances in chitin science Vol. VIII. Print. by ESUS, Poznan. 2005. P.430-439.

Serkiz Ya.I., Pinchuk L.B. Radiobiological effects of the accident at Chernobyl nuclear power plant. - K.: Naukova Dumka. -1992. – 172 p.

Seymour C.B., Mothersill C. Delayed expression of lethal mutations and genomic instability in the progeny of human epithelial cells that survived in a bystander-killing environment //Radiat. Oncol. Investig. 1997. V. 5. No3. P.106-110.

Seymour C.B., Mothersill C., Alper T. High yields of lethal mutations in somatic mammalian cells that survive ionizing radiation. // Int. J. Radiat. Biol. Relat. Stud. Phys.Chem. Med. 1986. V.50. N1. P. 167-179.

Shikazono N., Noguchi M., Fujii K., Urushibara A., Yokoya A. The Yield, processing and biological consequences of clustered DNA damage induced by ionizing radiation // J.Radiat.Res. 2009. V.50. p.27-36.

Shoenfeld Y. The Mosaic of Autoimmunity Predicting and treatment in autoimmune disease. IMAJ: 2008. Vol.10. P.12-19.

Shouse S.S., Warren A.S.L., Whipple G.H. Apalasia of marrow and fatal intoxication in dogs produced by roentgen irradiation of all bones. // Exp. Med. - 2004. - Vol. 53. – P. 42 – 98.

Sidorik E.P., Druzhina M.O., Burlaka A..P, Danko M.Y., Pukhov F . G. Zhdanova, N.N, Shkolniy O.T. (1994) Dokl. Ukrainian Academy of Sciences. № 9, 174-178 (in Ukrainian).

Sinitsky V.N., Kovtun T.V., Kharchenko, N.K. et al. Pathological mechanisms of maladaptation of the central nervous system in humans exposed to radiation // Fiziol. Journ. - 1995. - T. 41, № 3 - 4. - p. 55 – 66.

Snyder A.R. Review of radiation-induced bystander effects. Human & Experimental Toxicology. 2004. N. 23, P. 87 – 89, www.hetjournal.com.

Storer J.B. Acute responses to ionizing radiation//in: Biology of the laboratory mouse., Ed. E.L.Green., N.Y. 1966. P.427-446.

Sutherland B.M., Bennet P.V., Sidorkina O., Laval L. Clustered DNA induced by isolated DNA and in Human Cells by Low Doses of ionizing Radiation // Proc. Natl. Acad. Sci. USA. 2000. V.97, P.103-108.

Tabachnikov S.I., Shtengelov V.V., Danilov V.M., O.F.Seniuk, V.N. Oskina. Characteristics of brain bioelectric activity, psychological and emotional sphere, the cytokine network and biogenic amines in conditions of prolonged exposure of the complex factors the "Shelter" / Problems of Chornobyl. Issuence 10, Part.II., Chornobyl 2002 .- PP. 252 – 261.

Tavassoli M.V. Bone marrow: the seedbed of blood. In: Wintrobe MM, ed. Blood, pure and eloquent: a story of discovery, of people, and of ideas. // New York: McGraw-Hill. - 2008. - P. 57-79.

Timofeev-Resovskii N.V. The primary mechanisms of biological action of ionizing radiation. - M.: AScUSSR - 1963. P. 162-164.

Tsyb A.F., Budagov R.S., Zamulaeva A.I. Radiation and pathology. - M.: Vysshaya. shkola. 2005. - 314 p. (in Russian).

Von Sonntag C. / Free-Radical-induced DNA Damage and its Repair. Heidelberg. Springer. Verlag. 2006. P465.

Walker C. Radiation enhanced diffusion and gas release from recrystallized UO_2 grains in high burn-up water reactor fuel. //Annual report of Institute for Transuranium elements, Karlsruhe, JRC, EC, 2000, Report EUR 19812 BN, sect. 3.1.4, p.86-87.

Ward J.F. DNA damage produced by ionizing radiation in mammalian cells: identities, mechanisms of formation and reparability.// Prog. Nucleic Acid Res. Mol. Biol. 1988. V.35 P.95-125.

Watson G.E., Lorimore S.A., Macdonald D.A., Wright EG. Chromosomal instability in unirradiated cells induced in vivo by a bystander effect of ionizing radiation. Cancer Res 2000. Oct.15.V.60. N, 20.P. 5608 - 5611.

Zaichik A. Sh, Churilov LP Fundamentals of general embryology. St. Petersburg: ELBI-St. 1988. 440 p.

Zaichik A. Sh., Churilov L.P. Total pathophysiology with the basics of immunopathology. Ed. 4-e. St. Petersburg: 2008b. ELBI-St. p.490-491.

Zaichik A.Sh., Churilov L.P. Natural autoimmunity as a synchronization system of genetically determined processes. Immunophysiology. Natural autoimmunity in health and disease. Moscow. 2008a. P.73-91.

Zhdanova N.N., Zakharchenko V.A., Tugay T.I., Yu.V. Karpenko, L.T. Nakonechnaya, A.K. Pavlichenko, V.A. Zheltonozhsky, A.V. Zhydkov, O.F. Seniuk. Fungal damage of "Shelter" // Problems of safety of nuclear power Plants and of Chornobyl. Issue.3. P.1. 2005. P. 78-86.

Zedman, A.J.W., Vossenaar E.R. Autoantibodies to circulated (poly) peptides: a key diagnostic and prognostic marker for rheumathoid arthritis. Autoimmunity. 2004. 37. P. 295-299.

10

Analysis of Primary/Containment Coupling Phenomena Characterizing the MASLWR Design During a SBLOCA Scenario

Fulvio Mascari[1], Giuseppe Vella[1], Brian G. Woods[2],
Kent Welter[3] and Francesco D'Auria[4]
[1]Department of Energy, The University of Palermo
[2]Department of Nuclear Engineering and Radiation Health Physics
Oregon State University
[3]NuScale Power Inc.
[4]San Piero a Grado - Nuclear Research Group (SPGNRG), University of Pisa
[1,4]Italy
[2,3]USA

1. Introduction

Today considering the world energy demand increase, the use of advanced nuclear power plants, have an important role in the environment and economic sustainability of country energy strategy mix considering the capacity of nuclear reactors of producing energy in safe and stable way contributing in cutting the CO_2 emission (Bertel & Morrison, 2001; World Energy Outlook-Executive Summary, 2009; Wolde-Rufael & Menyah, 2010; Mascari et al., 2011d). According to the information's provided by the "Power Reactor Information System" of the International Atomic Energy Agency (IAEA), today 433 nuclear power reactors are in operation in the world providing a total power installed capacity of 366.610 GWe, 5 nuclear reactors are in long term shutdown and 65 units are under construction (IAEA PRIS, 2011).

In the last 20 years, the international community, taking into account the operational experience of the nuclear reactors, starts the development of new advanced reactor designs, to satisfy the demands of the people to improve the safety of nuclear power plants and the demands of the utilities to improve the economic efficiency and reduce the capital costs (D'Auria et al., 1993; Mascari et al., 2011c). Design simplifications and increased design margins are included in the advanced Light Water Reactors (LWR) (Aksan, 2005). In this framework, the project of some advanced reactors considers the use of emergency systems based entirely on natural circulation for the removal of the decay power in transient condition and in some reactors for the removal of core power during normal operating conditions (IAEA-TECDOC-1624, 2009; Mascari et al., 2010a; Mascari et al., 2011d). For example, if the normal heat sink is not available, the decay heat can be removed by using a passive connection between the primary system and heat exchangers (Aksan, 2005; Mascari et al., 2010a, Mascari, 2010b). The AP600/1000 (Advanced Plant 600/1000 MWe) design, for example, includes a Passive Residual Heat Removal (PRHR) system consisting of a C-Tube type heat exchanger immersed in the In-containment Refueling Water Storage Tank (IRWST) and connected to one of the Hot Legs (HL) (IAEA-TECDOC-1391, 2004; Reyes, 2005c; Gou et al., 2009; Mascari et al., 2010a). A PRHR from the core via Steam Generators (SG) to the atmosphere, considered in the WWER-1000/V-392 (Water Moderated, Water Cooled Energy Reactor) design, consists of heat exchangers cooled by atmospheric air, while the PRHR via SGs, considered in the WWER-640/V-407 design, consists of heat exchangers immersed in emergency heat removal tanks installed outside the containment (Kurakov et al., 2002; IAEA-TECDOC-1391, 2004; Gou et al., 2009; Mascari et al., 2010a). In the AC-600 (Advanced Chinese PWR) the PRHR heat exchangers are cooled by atmospheric air (IAEA-TECDOC 1281, 2002; Zejun et al., 2003; IAEA-TECDOC-1391, 2004; Gou et al., 2009; Mascari et al., 2010a) and in the System Integrated Modular Advanced Reactor (SMART) the PRHR heat exchangers are submerged in an in-containment refuelling water tank (IAEA-TECDOC-1391, 2004; Lee & Kim, 2008; Gou et al., 2009; Mascari et al., 2010a). The International Reactor Innovative and Secure (IRIS) design includes a passive Emergency Heat Removal System (EHRS) consisting of an heat exchanger immersed in the Refueling Water Storage Tank (RWST). The EHRS is connected to a separate SG feed and steam line and the RWST is installed outside the containment structure (Carelli et al., 2004; Carelli et al., 2009; Mascari, 2010b; Chiovaro et al., 2011). In the advanced BWR designs the core water evaporates, removing the core decay heat, and condenses in a heat exchanger placed in a pool. Then the condensate comes back to the core (Hicken & Jaegers, 2002; Mascari et al., 2010a). For example, the SWR-1000 (Siede Wasser Reaktor, 1000 MWe) design has emergency condensers immersed in a core flooding pool and connected to the core, while the ESBWR (Economic Simplified Boiling Water Reactor) design uses isolation condensers connected to the Reactor Pressure Vessel (RPV) and immersed in external pools (IAEA-TECDOC-1391, 2004; Aksan, 2005; Mascari et al., 2010a).

The designs of some advanced reactors rely on natural circulation for the removing of the core power during normal operation. Examples of these reactors are the MASLWR (Multi-Application Small Light Water Reactor), the ESBWR, the SMART and the Natural Circulation based PWR being developed in Argentina (CAREM)(IAEA-TECDOC-1391, 2004, IAEA -TECDOC-1474, 2005; Mascari et al., 2010a). In particular the MASLWR (Modro et al., 2003), figure 1, is a small modular integral Pressurized Water Reactor (PWR) relying on natural circulation during both steady-state and transient operation.

In the development process of these advanced nuclear reactors, the analysis of single and two-phase fluid natural circulation in complex systems (Zuber, 1991; Levy, 1999; Reyes & King, 2003; IAEA-TECDOC-1474, 2005; Mascari et al., 2011e), under steady state and transient conditions, is crucial for the understanding of the physical and operational phenomena typical of these advanced designs. The use of experimental facilities is fundamental in order to characterize the thermal hydraulics of these phenomena and to develop an experimental database useful for the validation of the computational tools necessary for the operation, design and safety analysis of nuclear reactors. In general it is expensive to design a test facility to develop experimental data useful for the analyses of complex system, therefore reduced scaled test facilities are, in general, used to characterize them. Since the experimental data produced have to be applicable to the full-scale prototype, the geometrical characteristics of the facility and the initial and boundary

conditions of the selected tests have to be correctly scaled. Since possible scaling distortions
are present in the experimental facility design, the similitude of the main thermal hydraulic
phenomena of interest has to be assured permitting their accurate experimental simulation
(Zuber, 1991; Reyes, 2005b; Reyes et al., 2007; Mascari et al., 2011e).

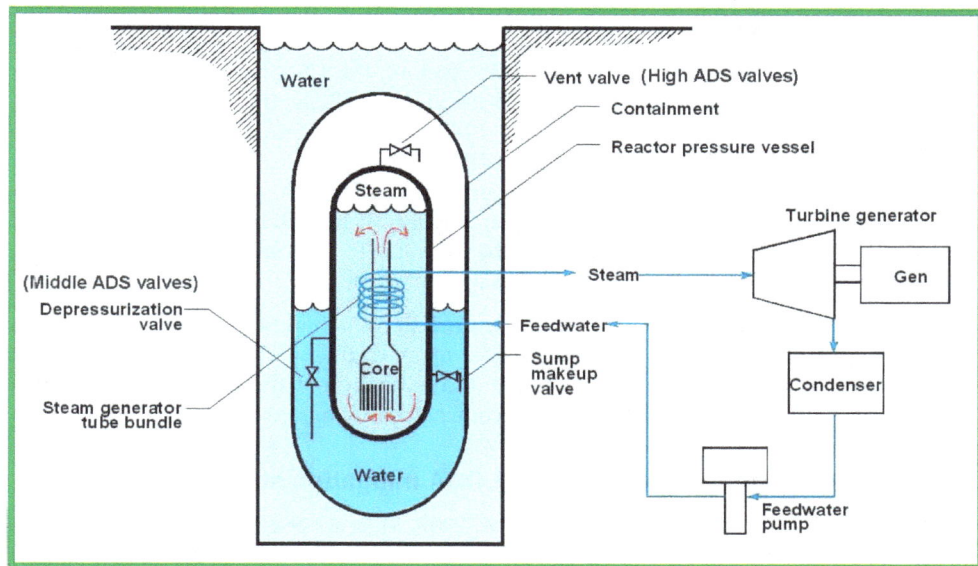

Fig. 1. MASLWR conceptual design layout (Modro et al, 2003; Reyes et al., 2007; Mascari et
al., 2011a).

Different computer codes have been developed to characterize two-phase flow systems,
from a system and a local point of view. Accurate simulation of transient system behavior of
a nuclear power plant or of an experimental test facility is the goal of the best estimate
thermal hydraulic system code. The evaluation of a thermal hydraulic system code's
calculation accuracy is accomplished by assessment and validation against appropriate
system thermal hydraulic data, developed either from a running system prototype or from a
scaled model test facility, and characterizing the thermal hydraulic phenomena during both
steady state and transient conditions. The identification and characterization of the relevant
thermal hydraulic phenomena, and the assessment and validation of thermal hydraulic
systems codes, has been the objective of multiple international research programs (Mascari
et al., 2011a; Mascari et al., 2011c).

In this international framework, Oregon State University (OSU) has constructed, under a
U.S. Department of Energy grant, a system level test facility to examine natural circulation
phenomena of importance to the MASLWR design. The scaling analysis of the OSU-
MASLWR experimental facility was performed in order to have an adequately simulation of
the single and two-phase natural circulation, reactor system depressurization during a
blowdown and the containment pressure response typical of the MASLWR prototype
(Zuber, 1991; Reyes & King, 2003; Reyes, 2005b). A previous testing program has been

conducted in order to assess the operation of the prototypical MASLWR under normal full pressure and full temperature conditions and to assess the passive safety systems under transient conditions (Modro et al. 2003; Reyes & King, 2003; Reyes, 2005b; Reyes et al., 2007; Mascari et al., 2011e). The experimental data developed are useful also for the assessment and validation of the computational tools necessary for the operation, design and safety analysis of nuclear reactors.

For many years, in order to analyze the LWR reactors, the USNRC has maintained four thermal-hydraulic codes of similar, but not identical, capabilities, the RAMONA, RELAP5, TRAC-B and TRAC-P. In the last years, the USNRC is developing an advanced best estimate thermal hydraulic system code called TRAC/RELAP Advanced Computational Engine or TRACE, by merging the capabilities of these previous codes, into a single code (Boyac & Ward, 2000; TRACE V5.0, 2010; Reyes, 2005a; Mascari et al., 2011a). The validation and assessment of the TRACE code against the MASLWR natural circulation database, developed in the OSU-MASLWR test facility, is a novel effort.

This chapter illustrates an analysis of the primary/containment coupling phenomena characterizing the MASLWR design mitigation strategy during a SBLOCA scenario and, in the framework of the performance assessment and validation of thermal hydraulic system codes, a qualitative analysis of the TRACE V5 code capability in reproducing it.

2. MASLWR conceptual design and SBLOCA mitigation strategy

The MASLWR (Modro et al., 2003; Reyes et al. 2007, Mascari et al., 2011a; Mascari et al., 2011e), figure 1, is a small modular integral PWR of 35 MWe developed by Idaho National Engineering and Environmental Laboratory, OSU and Nexant–Bechtel. During steady state condition the primary fluid, in single phase natural circulation, removes the core power and transfers it to the secondary fluid through helical coil SG. In transient condition the core decay heat is removed through a passive primary/containment coupling mitigation strategy based on natural circulation. The use of natural circulation reduces the number of active components simplifying the configuration of nuclear steam supply system. The reactor core and a helical coil SG are both located within the RPV. The integrated SG consists of banks of vertical helical tubes located in the upper region of the vessel outside of the HL chimney.

Its small size considered the prototypical MASLWR relatively portable and thus well suited for employment in smaller electricity grids but take into account its design simplicity, its simplified parallel construction, the consequent reduction of the capital costs, reduction of construction time, reduction of finance and operation cost, recognizes it to be able to reach larger electricity market in developing and developed regions (Modro et al., 2003; Reyes & Lorenzini, 2010; Mascari et al., 2011d).

As it is shown in figure 1, the primary coolant flows outside the SG tubes, and the Feed Water (FW) is fully vaporized resulting in superheated steam at exit of the SG. The safety systems are designed to operate passively. The RPV is surrounded by a cylindrical containment, partially filled with water. This containment provides pressure suppression and liquid makeup capabilities and is submerged in a pool of water that acts as the ultimate heat sink. The MASLWR steady-state operating conditions are reported in the table 1.

Reactor Thermal Power	150 MW
Primary Pressure	7.60 MPa
Primary Mass Flow Rate	597 kg/s
Reactor Inlet Temperature	491.80 K
Reactor Outlet Temperature	544.30 K
Primary Side Saturation Temperature	565 K
Secondary Side Steam Pressure	1.50 MPa
Secondary Side Steam Outlet Quality	1
Secondary Side Steam Temperature	481.40 K
Secondary Side Saturation Temperature	471.60 K
Feedwater Temperature	310 K
Feedwater Flowrate	56.10 kg/s

Table 1. MASLWR steady-state operating conditions (Modro et al., 2003; Reyes & King, 2003; Reyes, 2005b).

The RPV, figure 1, can be depressurized using the Automatic Depressurization System (ADS), consisting of six valves discharging into various locations within the containment. In particular two independent vent valves (high ADS valves), two independent depressurization valves (middle ADS valves) and two independent sump recirculation valves are considered in the MASLWR design.

The integral arrangement of the plant allows avoiding pressurized primary components outside the RPV eliminating the possibility of large break Loss of Coolant Accident (LOCA) and reducing the Small Break LOCA (SBLOCA) initiating event. Of particular interest is the SBLOCA mitigation strategy typical of the MASLWR design. Following, for example, an inadvertent opening of an ADS valve, a primary side blowdown into the pressure suppression containment takes place. The RPV blowdown causes a primary pressure decrease and a consequent containment pressure increase causing a safety injection signal. It automatically opens, figure 1, the high ADS valves, the middle ADS valves and the sump recirculation valves. As the primary and the containment pressures become equalized, the blowdown is terminated, and a natural circulation flow path is established. Infact, when the sump recirculation valves are opened the vapor produced in the core goes in RPV upper part and through the high ADS valve goes to the containment where it is condensed. At this point through the sump recirculation lines and the down comer the fluid goes to the core again. The pressure suppression containment is submerged in a pool that acts as the ultimate heat sink. This mechanism, based on natural circulation, permits the cooling of the core (Modro et al., 2003; Reyes & King, 2003; Reyes, 2005b; Reyes et al., 2007; Mascari et al., 2011).

The MASLWR concept design and its passive safety features were tested in a previous test campaign developed at the OSU-MASLWR experimental facility (Modro et al., 2003; Reyes & King, 2003; Reyes et al., 2007; Mascari et al., 2011a; Mascari et al., 2011e), figure 2. The planned work related to the OSU-MASLWR test facility will be not only to specifically investigate the MASLWR concept design further but also advance the broad understanding of integral natural circulation reactor plants and accompanying passive safety features as well. Furthermore an IAEA International Collaborative Standard Problem (ICSP) on the "Integral PWR Design Natural Circulation Flow Stability and Thermo-Hydraulic Coupling of Containment and Primary System During Accidents" is being hosted at OSU and the experimental data will be collected at the OSU-MASLWR facility. The purpose of this IAEA ICSP is to provide experimental data on single/two-phase flow instability phenomena under natural circulation conditions and coupled containment/reactor vessel behavior in integral-type reactors (Woods & Mascari, 2009; Woods et al.; 2011).

3. OSU-MASLWR test facility

The OSU-MASLWR test facility (Modro et al., 2003; Reyes & King, 2003; Reyes et al., 2007; Galvin, 2007; Mascari et al., 2011a, 2011b, 2011c, 2011d, 2011e), figure 2, is scaled at 1:3 length scale, 1:254.7 volume scale and 1:1 time scale, is constructed entirely of stainless steel and it is designed for full pressure and full temperature prototype operation.

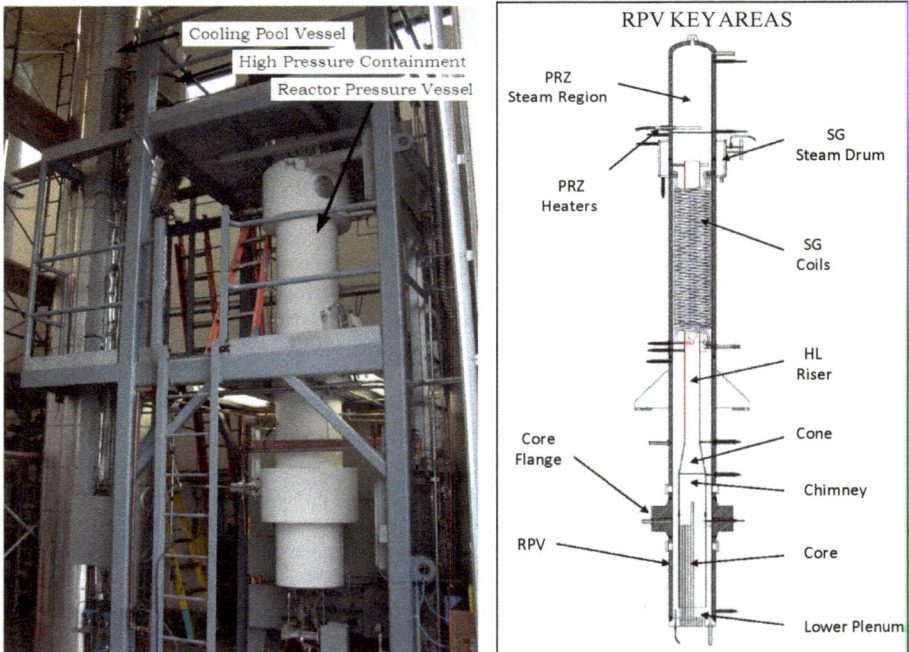

Fig. 2. OSU-MASLWR test facility layout (Reyes et al., 2007; Mascari et al., 2011a; Mascari et al., 2011e).

The facility includes the primary and secondary circuit and the containment structures. Two vessels, a High Pressure Containment (HPC) vessel and a Cooling Pool Vessel (CPV) with an heat transfer surface between them to establish the proper heat transfer area, are used to model the containment structures, in which the RPV sits, as well as the cavity within which the containment structure is located. The two middle ADS lines, two high ADS lines and the two ADS sump recirculation lines are modelled separately. In addition to the physical structures that comprise the test facility, there are data acquisition, instrumentation and control systems.

The facility is instrumented for capturing its thermal hydraulic behaviour during steady and transient conditions; in particular thermocouples are used to measure fluid, wall and heater temperature; pressure transducers are used to measure pressure; differential pressure cells are used to measure water level, pressure loss and flow rate; flow Coriolis meters are used to measure mass flow rate; vortex flow meters are used to measure steam mass flow rate, pressure and temperature; power meters are used to measure heater power.

In the previous testing program four tests have been conducted: the OSU-MASLWR-001 -inadvertent actuation of 1 submerged ADS valve-; the OSU-MASLWR-002 -natural circulation at core power up to 210 kW-; the OSU-MASLWR-003A- natural circulation at core power of 210 kW (Continuation of test 002)-; the OSU-MASLWR-003B -inadvertent actuation of 1 high containment ADS valve-.

Since the target of the OSU-MASLWR-001 test was to determine the pressure behavior of the RPV and containment following an inadvertent actuation of one middle ADS valve, it gives a wide number of informations about the primary/containment coupling phenomena characterizing the MASLWR design. Therefore it is the test chosen for this analysis.

3.1 OSU-MASLWR RPV description

The internal components of the RPV, figure 2, are the core, the HL riser, the Upper Plenum (UP), the Pressurizer (PRZ), the SG primary side, the Cold Leg (CL) downcomer and the Lower Plenum (LP). The RPV shell is covered by Thermo-12 hydrous calcium silicate insulation.

The core is modelled with 56 cylindrical heater rods distributed in a 1.86 cm pitch square array with a 1.33 pitch to diameter ratio. The nominal power of each heater rod is 7.1 kW resulting in a maximum core power of 398 kW. The diameter of the core rod is 1.59 cm.

A lower core flow plate, figure 3, contains 76 auxiliary flow holes of 0.635 cm of diameter, arranged at 1.86 cm pitch square array, and 57 core rod flow holes. In order to create a flow annulus between the flow plate and the core rod, the holes of the rodded lower core flow plate are oversized at 1.72 cm.

The core is shrouded, figure 4, to ensure all flow enters the core via the bottom and travels the entire heated length. The shroud is shaped to partially block the primary coolant flow through the outermost auxiliary flow holes in order to ensure that each heated rod receives approximately equal axial coolant flow. The amount of blockage is dependent on the number and location of heated rods adjacent to each auxiliary flow hole. At mid elevation a core grid wires is considered in order to maintain the radial alignment of the core rods. Four thermocouples for measuring the core inlet temperatures are located at the bottom CL entering the core. The core heater rod temperatures are measured. Six thermocouples vertically spaced every 15.24 cm measuring water temperatures, are located in the center of core thermocouple rod. The pressure loss in the core is measured; the power to the core heater rod bundles is measured.

The HL riser, figure 2, consists of a lower region, an upper region and a transition region. The lower region consists of a pipe with an outside diameter of 20.32 cm, an inside diameter of 19.71 cm and a wall thickness of 0.305 cm. The upper region consists of a pipe with an outside diameter of 11.43 cm, an inside diameter of 10.23 cm, and a wall thickness of 0.602 cm. The transition region consists of a cone with a 0.305 cm thickness and an half angle of 20.61°. The pressure loss between core top and riser cone, the pressure loss in the riser cone and the pressure loss in the chimney, from the exit of the transition cone to the UP, are measured. Along the riser a thermocouple measures the water temperature inside chimney below SG coil and another one measures the water temperature at top of chimney. The flow rate within the HL chimney is measured with a differential pressure cell used to measure flow. The primary containment water level is measured as well.

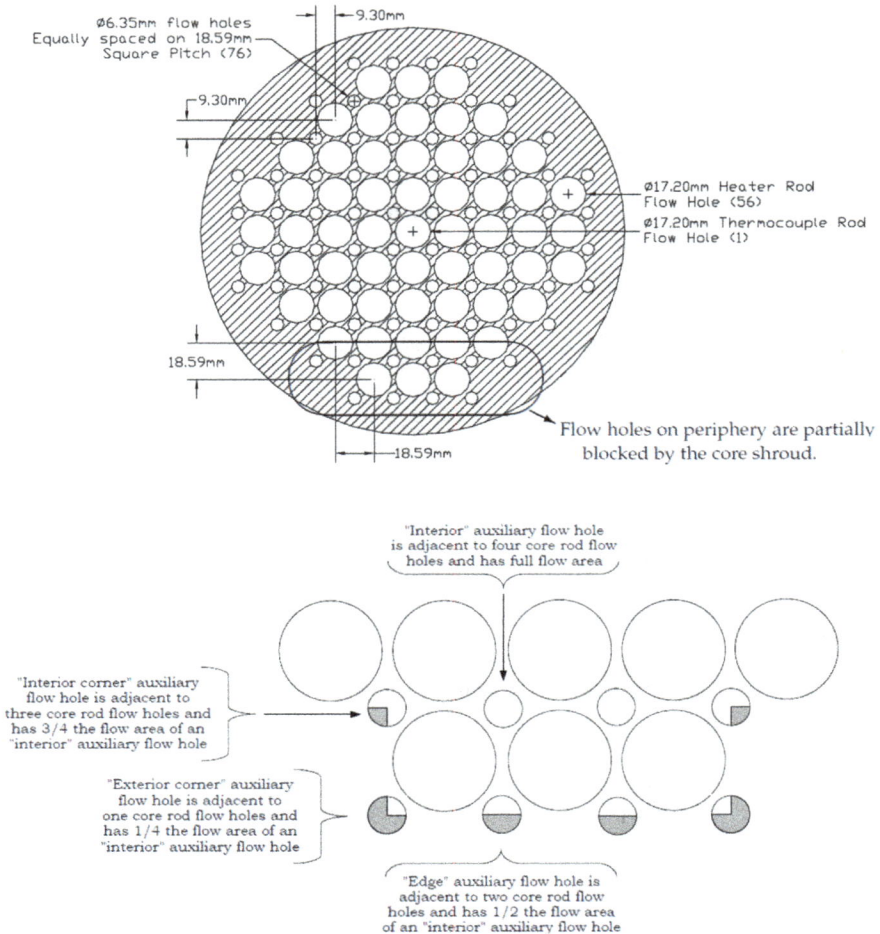

Fig. 3. Lower core flow plate layout and auxiliary flow hole blockage by core shroud (Galvin, 2007; Mascari et al., 2011e).

The UP is separated from the heated upper PRZ section by a thick baffle plate having eight 2.54 cm holes, spaced uniformly around the baffle plate periphery, which allow free communication of the PRZ to the remainder of the RPV during normal operation and for volume surges into and/or out of the PRZ due to transients.

Fig. 4. Core shroud photo (Galvin, 2007; Mascari et al., 2011e).

The PRZ is integrated in the RPV and is located in its upper part. In the PRZ are located three heater elements, each 4 kW, that are modulated by the test facility control system to maintain nominal primary system pressure at the desired value. The PRZ steam temperature and pressure are measured. The PRZ level is measured as well.

The CL downcomer region, figure 2, is an annular region bounded by the RPV wall on the outside and the HL riser on the inside, and the flow area reduces at the HL riser cone. In the SG primary side section is inserted the SG helical coil bundle. Thermocouples are located in the CL downcomer region to measure water down flow temperatures after SG coils. The primary side pressure loss across SG and the pressure loss in the annulus below SG are measured.

The SG of the facility is a once through heat exchanger and is located within the RPV in the annular space between the HL riser and the inside surface of the RPV. The tube bundle, is a helical coil consisting of fourteen tubes with a total heated length of 86 m. There are three separate parallel coils of stainless steel tubes. The outer and middle coils consist of five tubes each while the inner coil consists of four tubes. The value of the degree of the steam superheat is changed in order to control the facility. A thermocouple is located at the exit of each helical coil. The main steam pressure and temperature are measured. The main steam mass flow rate is measured with a vortex flow meter. The FW supply in the SG outer, middle and inner coil mass flow rate are measured with flow Coriolis meters. The related pressures are measured. The FW temperature is measured as well.

3.2 OSU-MASLWR containment structures description

The HPC, figure 5, consists of a lower cylindrical section, an eccentric cone section, an upper cylindrical section and an hemispherical upper end head. For scaling reasons, in order to have an adiabatic boundary condition in all the wall of the HPC except through the heat transfer plate wall where the condensation has to take place, four groups of containment heaters have been installed permitting the heat transfer takes place only between the CPV and HPC containment. These heaters are located in the exterior surface of the HPC, under the insulation, and above the containment water level. The temperatures of heaters located on the walls of the HPC and the temperatures within the walls of the HPC, between the heaters and the water, are measured. The entire HPC is covered by Thermo-12 hydrous calcium silicate insulation. The HPC level and pressure are measured.

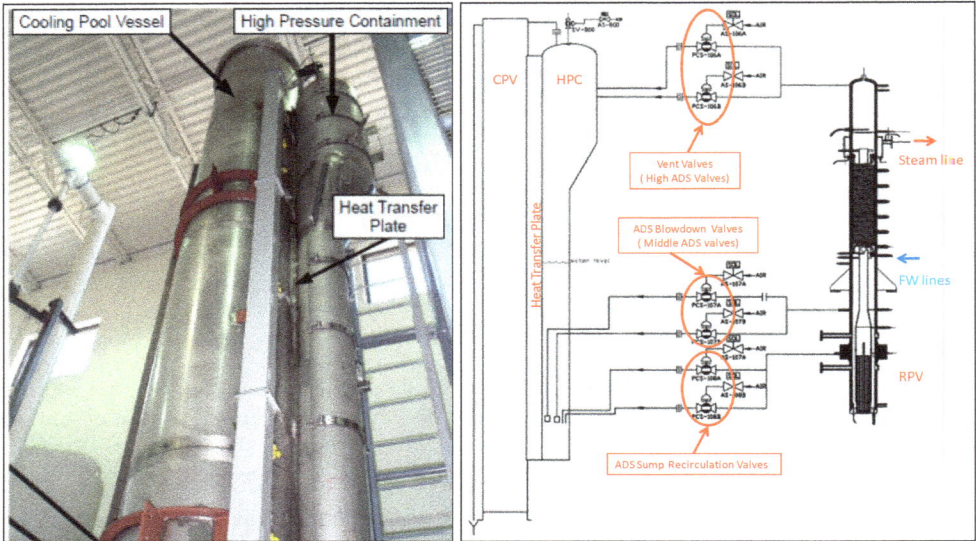

Fig. 5. OSU-MASLWR containment structures (Reyes et al., 2007; Mascari et al., 2011d; Mascari et al., 2011e).

The CPV consists of a tall right cylindrical tank covered by Thermo-12 hydrous calcium silicate insulation. The CPV level and water temperature are measured. One disk rupture is connected between the HPC and the CPV.

The heat transfer plate, having the same height of the HPC without the hemispherical head, provides the heat conduction between the HPC and CPV. The heat transfer plate is scaled in order to model the heat transfer area between the MASLWR design high pressure containment vessel and the cooling pool in which it sits.

Five thermocouples are located at six different elevations to measure the temperature distribution from the HPC to the CPV. In particular one group of thermocouples measures the water temperatures located inside the HPC near the heat transfer plate, one measures the water temperatures located inside the CPV near the heat transfer plate, one measures the wall temperatures at the midpoint of the heat transfer plate between the CPV and the HPC, one measures wall temperatures within the heat transfer plate between the CPV and the HPC nearest to the HPC and one measures wall temperatures within the heat transfer plate between the CPV and the HPC nearest to the CPV.

3.3 OSU-MASLWR ADS lines description

The high ADS lines, figure 6, are horizontally oriented and connect the PRZ steam space with the HPC. A pneumatic motor operated globe valve is located in each line. Downstream from each isolation valve is a transition piece with an internal square-edge orifice. The two high ADS lines enter the HPC above the waterline, penetrate it and then terminate with a sparger.

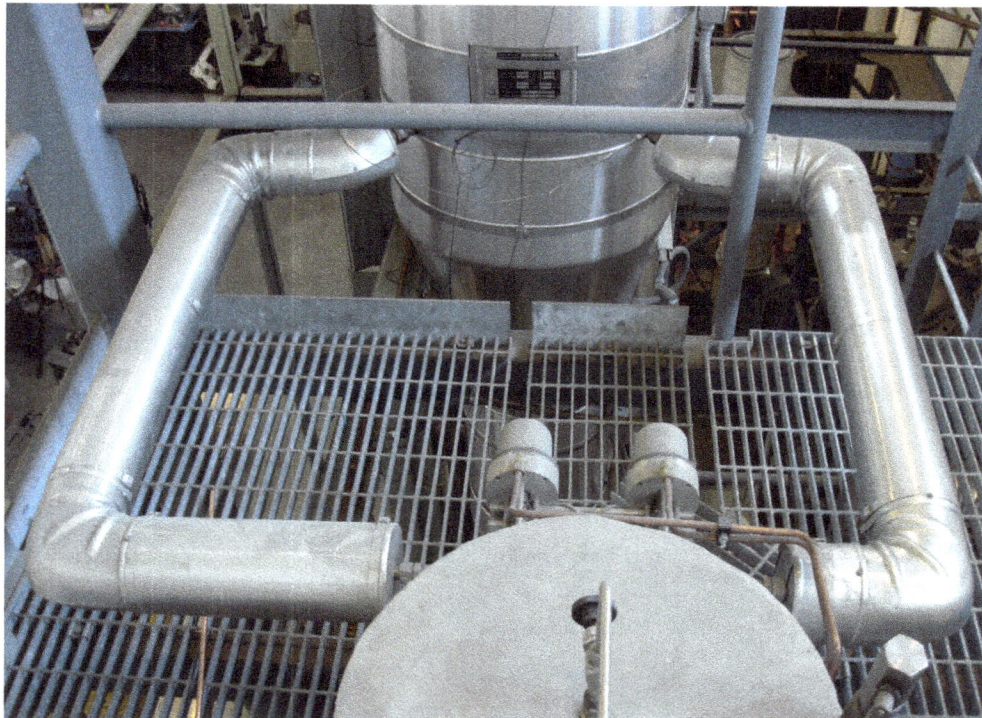

Fig. 6. High ADS lines photo.

The middle ADS lines are horizontally oriented and connect the RPV CL to the HPC. A pneumatic motor operated globe valve is located in each line. Downstream from each isolation valve is a transition piece with an internal square-edge orifice. These two lines enter the HPC, penetrate it and then turn downward before terminating below the HPC waterline. A sparger is considered at the end of these lines.

The ADS sump recirculation lines are horizontally oriented and connect the RPV lower CL to the HPC. A pneumatic motor operated globe valve is located in each line. Downstream from each isolation valve is a transition piece with an internal square-edge orifice. These two lines enter the HPC, penetrate it and then turn downward before terminating below the HPC waterline. No sparger is considered for these lines.

The water temperatures inside the ADS lines outside of the HPC are measured.

4. OSU-MASLWR-001 test description

The purpose of the OSU-MASLWR-001 test (Modro et al. 2003; Reyes et al., 2007; Pottorf et al., 2009; Mascari et al., 2011e), a design basis accident for MASLWR concept design, was to determine the pressure behavior of the RPV and containment following an inadvertent actuation of one middle ADS valve. The test successfully demonstrated the blowdown behavior of the MASLWR test facility during one of its design basis accident.

Following the inadvertent middle ADS actuation the blowdown of the primary system takes place. A subcooled blowdown, characterized by a fast RPV depressurization, takes place after the Start Of the Transient (SOT). A two-phase blowdown occurs when the differential pressure, at the break location, results in fluid flashing. A choked two-phase flow condition prevails and a decrease in depressurization rate of the primary system is experimentally observed. When the PRZ pressure reaches saturation, single phase blowdown occurs and the depressurization rate increases. The RPV and HPC pressure and the primary saturation temperature are shown in figure 7. The P_{sat}, saturation pressure, is based on the temperature at the core outlet.

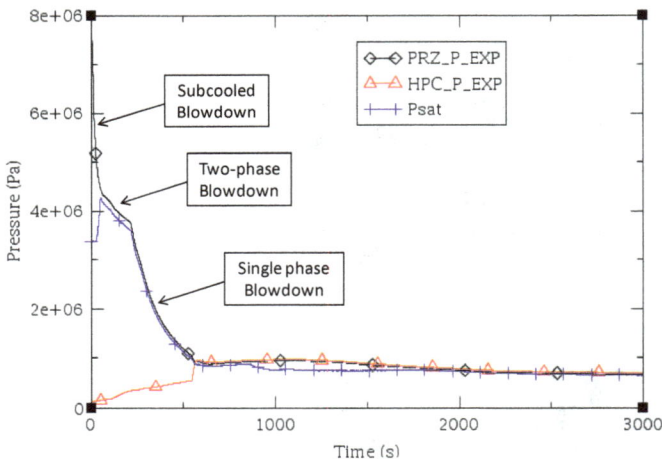

Fig. 7. RPV and HPC pressure behaviour during the OSU-MASLWR-001 test (Modro et al., 2003; Reyes et al., 2007; Mascari et al., 2011e).

At 539 s after the SOT the pressure difference between the RPV and the HPC reaches a value less than 0.517 MPa, one of the high ADS valve is opened and, with approximately 10 s of delay, the other high ADS valve is opened equalizing the pressure of the primary and HPC system.

At 561 s after the SOT the pressure difference between the RPV and the HPC reaches a value less than 0.034 MPa, one of the sump recirculation valve is opened and, with approximately 10 s of delay, the other sump recirculation valve is opened terminating the blowdown period and starting the refill period. The refill period takes place for the higher relative coolant height in the HPC compared to the RPV. Figure 8 shows the RPV level evolution experimentally detected during the test. The RPV water level never fell below the top of the core during the execution of the test 1.

During the saturated blowdown period, the inlet and the outlet temperature of the core are equal each other assuming the saturation temperature value. A core reverse flow and a core coolant boiling off at saturation is present in the facility during this period. When the refill takes place the core normal flow direction is restarted and a delta T core is observed depending on the refill rate and core power, figure 9.

When the refill of the reactor takes place the level of the coolant reaches the location of the flow rate HL measurement point, therefore an increase of the RPV flow rate is detected for this phenomenon, figure 10.

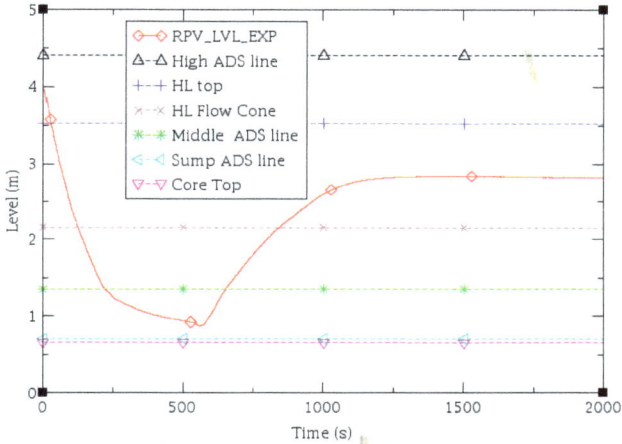

Fig. 8. RPV water level inventory behaviour during the OSU-MASLWR-001 test (Modro et al., 2003; Reyes et al., 2007; Mascari et al., 2011e).

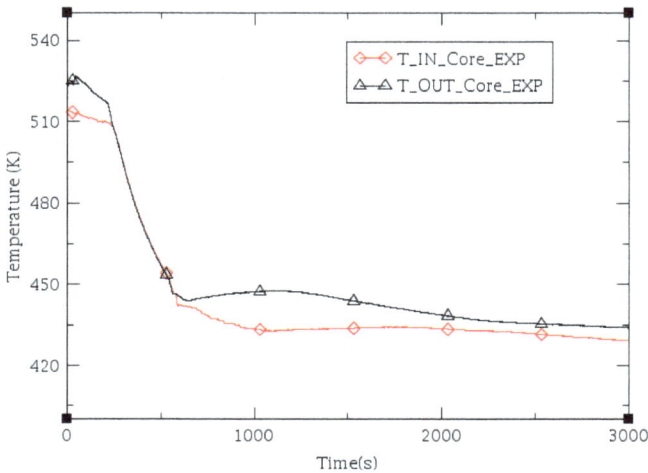

Fig. 9. Inlet/outlet core temperature behavior during the OSU-MASLWR-001 test (Modro et al., 2003; Mascari et al., 2011e).

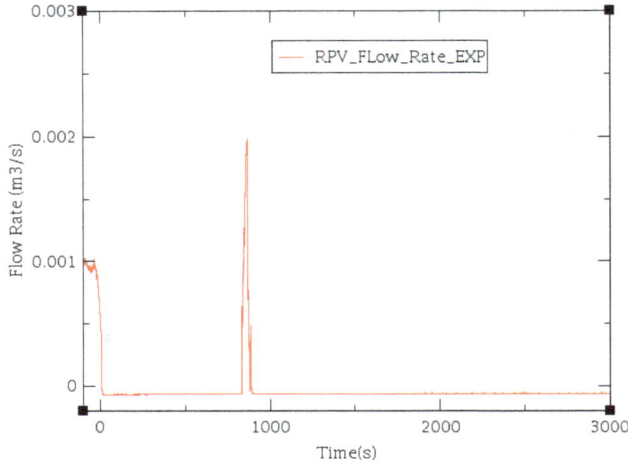

Fig. 10. RPV flow rate during the OSU-MASLWR-001 test (Modro et al., 2003; Mascari et al., 2011e).

5. Code application

5.1 TRACE code

TRACE (TRACE V5.0, 2010) is a component-oriented code designed to perform best estimate analyses for LWR. In particular this code is developed to simulate operational transients, LOCA, other transients typical of the LWR and to model the thermal hydraulic phenomena taking place in the experimental facilities used to study the steady state and transient behavior of reactor systems (Mascari et al., 2011a).

TRACE is a finite volume, two fluid, code with 3D capability. The code is based on two fluid, two-phase field equations. This set of equations consists of the conservation laws of mass, momentum and energy for liquid and gas fields (Reyes, 2005a):

- Mixture mass conservation equation:

$$\frac{\partial}{\partial t}[\rho_v \alpha + (1-\alpha)\rho_l] + \nabla \cdot [\rho_v \vec{v}_v \alpha + \rho_l \vec{v}_l (1-\alpha)] = 0$$

- Vapor mass conservation equation:

$$\frac{\partial}{\partial t}(\rho_v \alpha) + \nabla \cdot (\rho_v \vec{v}_v \alpha) = \Gamma_v$$

- Liquid momentum conservation equation:

$$\frac{\partial \vec{v}_l}{\partial t} + \vec{v}_l \cdot \nabla \vec{v}_l = -\frac{1}{\rho_l}\nabla p + \frac{c_i}{(1-\alpha)\rho_l}(\vec{v}_v - \vec{v}_l)\,|\,\vec{v}_v - \vec{v}_l\,| - \frac{\Gamma_{cond}}{(1-\alpha)\rho_l}(\vec{v}_v - \vec{v}_l) + \frac{c_w}{(1-\alpha)\rho_l}\vec{v}_l\,|\,\vec{v}_l\,| + \vec{g}$$

- Gas momentum conservation equation:

$$\frac{\partial \vec{v}_v}{\partial t} + \vec{v}_v \cdot \nabla \vec{v}_v = -\frac{1}{\rho_v}\nabla p + \frac{c_i}{\alpha \rho_v}(\vec{v}_v - \vec{v}_l)\,|\,\vec{v}_v - \vec{v}_l\,| - \frac{\Gamma_{Boiling}}{\alpha \rho_v}(\vec{v}_v - \vec{v}_l) + \frac{c_{wv}}{\alpha \rho_v}\vec{v}_v\,|\,\vec{v}_v\,| + \vec{g}$$

- Mixture energy conservation equation:

$$\frac{\partial}{\partial t}[(\rho_v \alpha e_v + \rho_l(1-\alpha)e_l] + \nabla \cdot [\rho_v \alpha e_v \vec{v}_v + \rho_l(1-\alpha)e_l \vec{v}_l] = -p\nabla \cdot [\alpha \vec{v}_v + (1-\alpha)\vec{v}_l] + q_{wl} + q_{dlv}$$

- Vapor energy conservation equation:

$$\frac{\partial}{\partial t}(\rho_v \alpha e_v) + \nabla \cdot (\rho_v \alpha e_v \vec{v}_v) = -p\frac{\partial \alpha}{\partial t} - p\nabla \cdot (\alpha \vec{v}_v) + q_{wv} + q_{dv} + q_{iv} + \Gamma_v h_v$$

The resulting equation set is coupled to additional equations for non-condensable gas, dissolved boron, control systems and reactor power. Relations for wall drag, interfacial drag, wall heat transfer, interfacial heat transfer, equation of state and static flow regime maps are used for the closure of the field equations. The interaction between the steam-liquid phases and the heat flow from solid structures is also considered. These interactions are in general dependent on flow topology and for this purpose a special flow regime dependent constitutive-equation package has been incorporated into the code.

TRACE uses a pre-CHF flow regime, a stratified flow regime and a post-CHF flow regime. In order to study the thermal history of the structures the heat conduction equation is applied to different geometries. A 2D (r and z) treatment of conduction heat transfer is taken into account as well.

A finite volume numerical method is used to solve the partial differential equations governing the two-phase flow and heat transfer. By default, a multi-step time-differencing procedure that allows the material Courant-limit condition to be exceeded is used to solve the fluid-dynamics equations.

TRACE can be used together with a user-friendly front end, Symbolic Nuclear Analysis Package (SNAP), able to support the user in the development and visualization of the nodalization, to show a direct visualization of selected calculated data, and accepts existing RELAP5 and TRAC-P input. The TRACE/SNAP architecture is shown in figure 11.

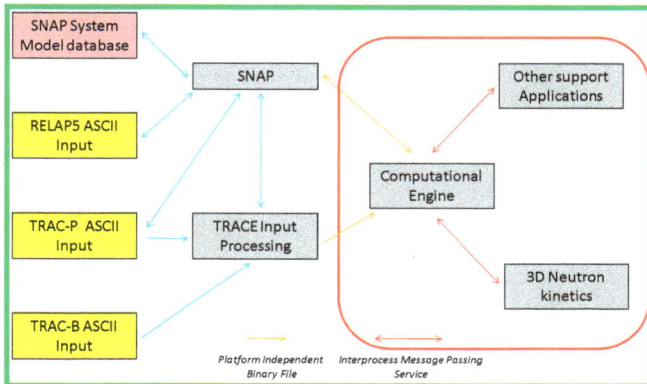

Fig. 11. TRACE/SNAP environment architecture (Staudenmeier, 2004; Mascari et al., 2011a).

SNAP (SNAP users manual, 2007) is a suite of integrated applications including a "Model Editor", "Job Status", the "Configuration Tool" client applications and a "Calculation Server". In particular, the "Model Editor" is used for the nodalization development and visualization and for the visualization of the selected calculated data by using its graphical and animation model capability. The codes currently supported in SNAP are CONTAIN, COBRA, FRAPCON-3, MELCOR, PARCS, RELAP5 and TRACE.

5.2 OSU-MASLWR TRACE model

An OSU-MASLWR TRACE model (Mascari et al., 2008, 2009a, 2009b, 2010b, 2011a, 2011b, 2011c, 2011d) is developed in order to evaluate the TRACE code capabilities in predicting the thermal hydraulic phenomena typical of the MASLWR design as natural circulation, heat exchange from primary to secondary side by helical SG in superheated condition and primary/containment coupling during transient scenario.

The TRACE nodalization, developed by using SNAP, models the primary and the secondary circuit. The containment structures consisting of the HPC, CPV and heat transfer plate are modeled as well, figure 12.

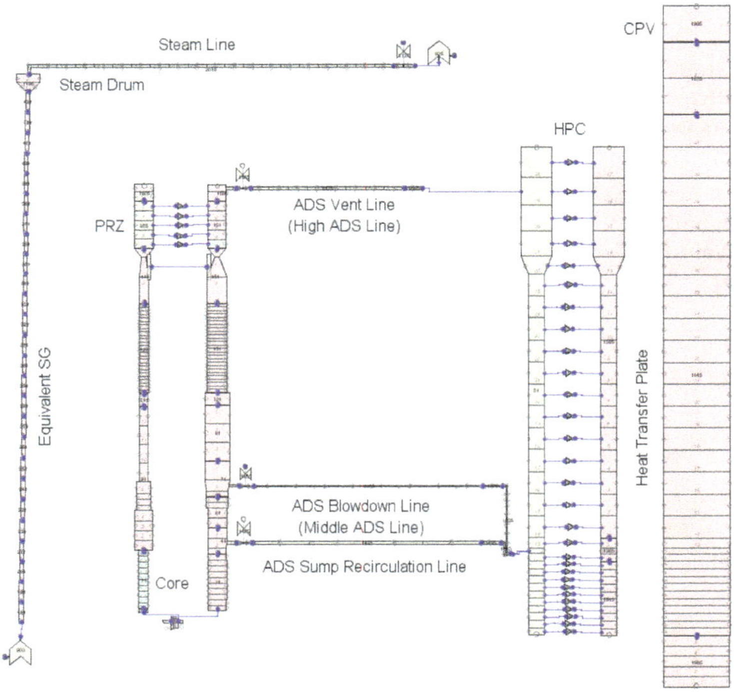

Fig. 12. OSU-MASLWR TRACE model.

The "slice nodalization" technique is adopted in order to improve the capability of the code to reproduce natural circulation phenomena. This technique consists in realizing the mesh cells of different nodalization zones, at the same elevation, with the same cell length (Mascari et al.,

2011a). In this way it is avoided the error due to the position/elevation of the cell nodalization center that can influences the results of the calculated data when natural circulation regime is present. If the "slice nodalization" technique is not used, this error has to be taken into account and its effect increases if larger nodalization cells are used. In this case it can be reduced by using a "fine nodalization". In general, its effect on the results is less important when forced circulation regimes are simulated. However the "slice nodalization" technique could require nodes of small length increasing the numerical error and the computational time. The "code user" has to take into account these disadvantages during the nodalization development.

The primary circuit of the TRACE model, comprises the core, the HL riser, the UP, the PRZ, the SG primary side, the CL down comer and the LP. After leaving the top of the HL riser, the flow enters the UP divided in two thermal hydraulic regions connected to the PRZ. After living the UP, the flow continues downward through the SG primary section and into the CL down comer region. The core is modeled with one thermal hydraulic region thermally coupled with one equivalent active heat structure simulating the 56 electric heaters. The PRZ is modeled with two hydraulic regions, connected by different single junctions, in order to allow potential natural circulation/convection phenomena. The three different PRZ heater elements are modeled with one equivalent active heat structure. The thick baffle plate is modeled as well. The direct heat exchange by the internal shell between the hotter fluid, in the ascending riser, and the colder fluid, in the descending annular down comer, is modeled by heat structures thermally coupled with these two different hydraulic regions. SG coils are modeled with one "equivalent" group of pipes, in order to simulate the three separate parallel helical coils. The steam line of the facility is modeled as well.

In order to simulate the OSU-MASLWR-001 test the HPC is divided in two thermal hydraulic regions, connected by single junctions, in order to allow the simulation of possible natural circulation phenomena. The ADS lines are modeled.

The RPV, HPC and CPV shell and the connected insulation are modeled.

5.3 TRACE model qualification process

A nodalization, representing an actual system (integral test facility or nuclear power plant), can be considered qualified when it has a geometrical fidelity with the involved system, it reproduces the measured nominal steady state conditions of the system, and it shows a satisfactory behavior in time-dependent conditions.

The OSU-MASLWR nodalization qualification process (Bonuccelli et al., 1993) is still in progress, because the facility experimental characterization will be conducted in the framework of the current IAEA ICSP. In particular several important facility operational characteristics, like pressure drop along the primary loop, at different primary side mass flow rates, and heat losses, at different primary side temperatures, determined to be of importance during the planned ICSP experiments, will be evaluated and distributed to the ICSP participants. Besides, some nodalization models, here presented, are still preliminary because some geometrical data and the complete instrument characterization and location will be delivered in the ICSP framework as well. Therefore the current results are preliminary and should not be used for the code assessment, but are able to show the TRACE capability to reproduce the primary/containment coupling phenomena typical of the MASLWR prototypical design (Mascari et al., 2009b; Mascari et al., 2011c).

5.4 Analysis of the OSU-MASLWR-001 TRACE calculated data

Starting from the calculated data developed in previous analyses (Pottorf et al., 2009; Mascari et al., 2009b; Mascari et al., 2011c) the target of this section is to give an expanded revised analyses, after a first review of the TRACE nodalization, of the TRACE V5 patch 2 code capability in predicting the primary/containment coupling phenomena typical of the MASLWR prototypical design.

The analysis of the OSU-MASLWR-001 calculated data shows that the TRACE code is able to qualitatively predict the primary/containment coupling phenomena characterizing the test. The blowdown phenomena, the refill of the core and the long term cooling, permitting of removing the decay power, are predicted by the code.

In particular, following the inadvertent middle ADS actuation, the blowdown of the primary system takes place. A subcooled blowdown, characterized by a fast RPV depressurization, is predicted by the code after the SOT.

When the differential pressure in the facility at the break location results in flashing, a two-phase blowdown, qualitatively predicted by the code, occurs. A decrease in depressurization rate of the primary system is then observed, in agreement with the experimental data.

When the PRZ pressure reaches saturation, single phase blowdown occurs and the depressurization rate increases again, in agreement with the experimental data. The RPV and HPC pressure versus code calculations are shown in figure 13.

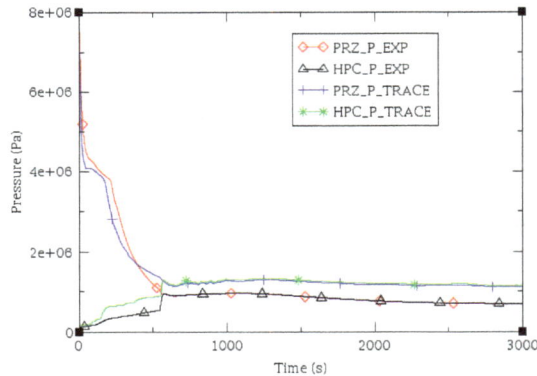

Fig. 13. Experimental data versus code calculation for PRZ and HPC pressure.

In agreement with the experimental data, when the pressure difference between the RPV and the containment reaches a value less than 0.517 MPa, the high ADS valves are opened which equalizes their pressure.

When the pressure difference reaches a value less than 0.034 MPa, the sump recirculation valves are opened and the refill period begins. The refill phenomenon is predicted by the code. As in the experimental data, the refill period takes place for the higher relative coolant height in the HPC compared to the RPV.

Figure 14 shows the RPV level evolution experimentally detected during the test versus the calculated data. In agreement with the experimental data, the RPV water level never fell

below the top of the core. Figure 15 shows the HPC level versus code calculation during the
test. The qualitative behavior is well predicted by the TRACE code.

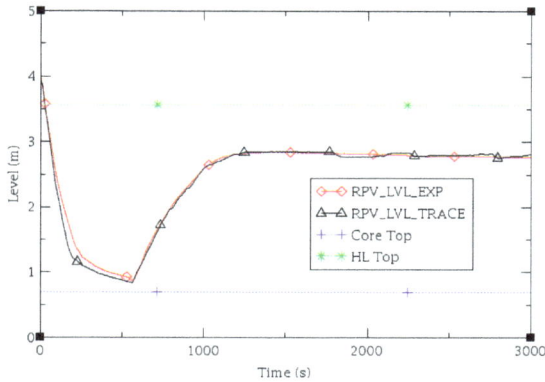

Fig. 14. Experimental data versus code calculation for RPV level.

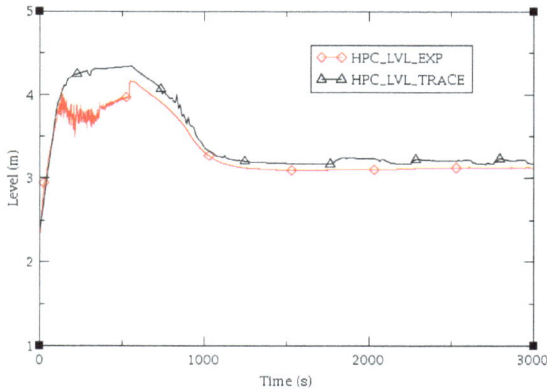

Fig. 15. Experimental data versus code calculation for HPC level.

In agreement with the experimental data, during the saturated blowdown period the inlet and
the outlet temperature of the core are equal each other assuming the saturation temperature
value. A core reverse flow and a core coolant boiling off at saturation is predicted by the code.
When the refill takes place, the core normal flow direction is restarted. Figure 16 shows the
experimental data versus code calculation for outlet core temperature.

In agreement with the experimental data when the sump recirculation valves are opened the
vapor produced in the core goes in the upper part of the facility and through the high ADS
valve goes to the HPC where it is condensed. At this point through the sump recirculation
line and the down comer the fluid goes to the core again (Mascari et al., 2011c, 2011d).
Figure 17 shows the long term cooling flow path typical of the MASLWR design.

Figure 18 shows, by using the SNAP animation model capabilities, the fluid condition of
facility, 976 s after the SOT, predicted by the TRACE code.

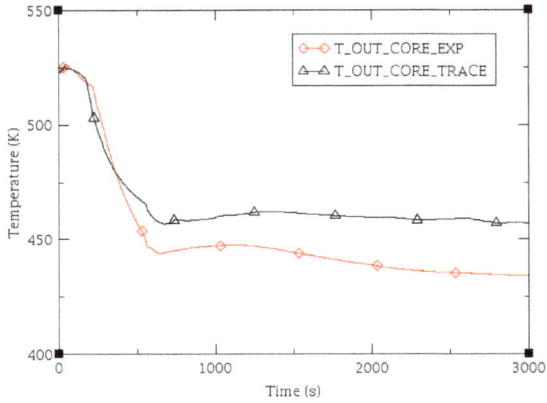

Fig. 16. Experimental data versus code calculation for outlet core temperature.

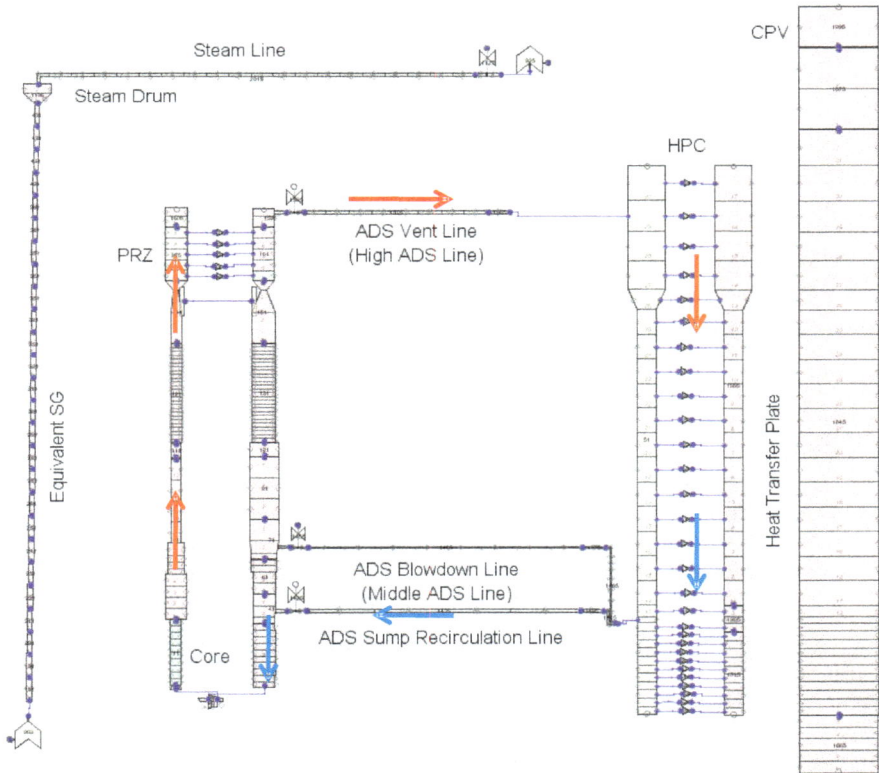

Fig. 17. Long term cooling flow path typical of the MASLWR design.

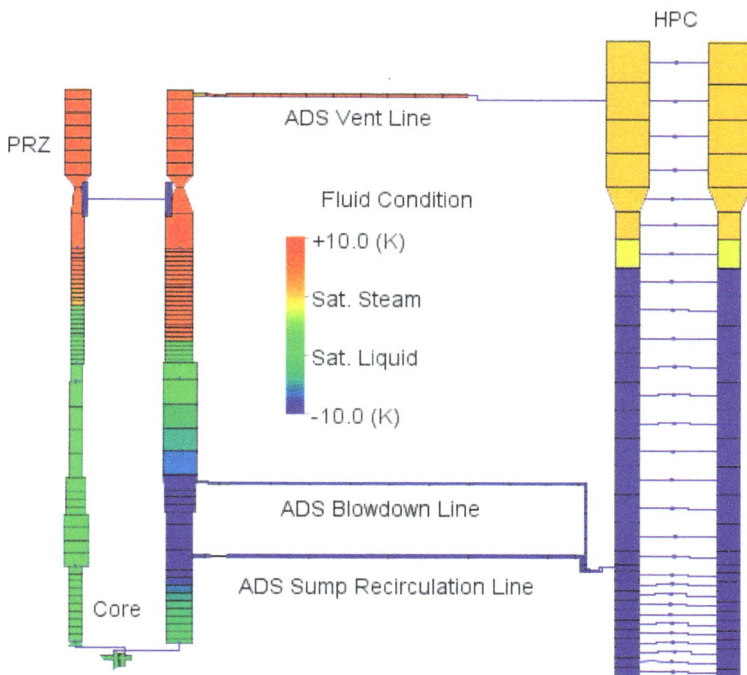

Fig. 18. SNAP animation model used to analyze the OSU-MASLWR-001 test (976s after the SOT).

From a quantitative point of view the results of the calculated data show a general over prediction compared with the experimental data. It is thought that this could be due to a combination of selection of vent valve discharge coefficients and condensation models applied to the inside surface of the containment (Pottorf et al., 2009; Mascari et al., 2011c).

However, in order to quantitatively evaluate the capacity of the TRACE code to simulate OSU-MASLWR-001 primary/containment coupling phenomena, a qualification of the TRACE nodalization against an heat losses experimental characterization at different primary side temperature is necessary. Figures 19 and 20 show the behavior of PRZ pressure and outlet core temperature respectively by increasing the heat losses of the TRACE model (TRACE_HL). A general quantitative improvement of the calculated data has been showed by TRACE_HL calculation. Therefore in order to quantitatively evaluate the capability of the TRACE code to simulate the OSU-MASLWR phenomena, and therefore use the calculated data for the TRACE code assessment, is necessary a TRACE nodalization qualification against several facility operational characteristics like heat losses at different primary side temperatures and pressure drop at different primary mass flow rate. Currently the TRACE model qualification process is in progress considering the facility characterization that will be disclosed during the current IAEA ICSP.

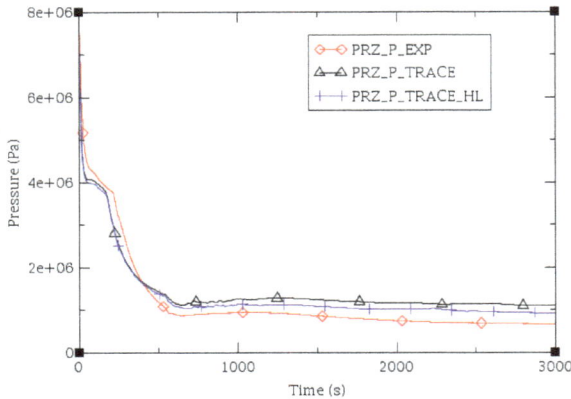

Fig. 19. Experimental data versus code calculations for PRZ pressure.

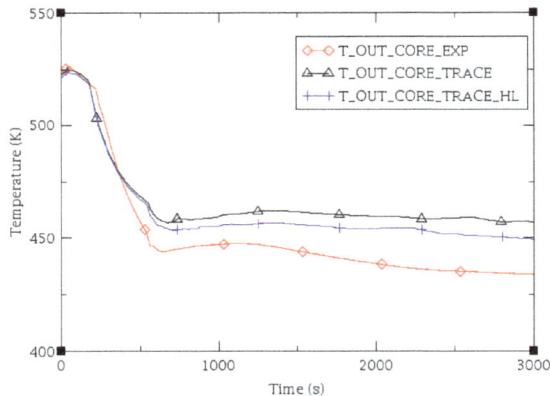

Fig. 20. Experimental data versus code calculations for outlet core temperature.

Considering the importance of the containment/reactor vessel thermal hydraulic coupled behavior for the advanced integral-type LWR, in the IAEA ICSP framework a further test simulating a loss of FW transient with subsequent ADS actuation and long term cooling, will be executed in the OSU-MASLWR facility (Woods & Mascari, 2009; Woods et al.; 2011).

6. Conclusions

The MASLWR is a small modular integral PWR relying on natural circulation during both steady-state and transient operation. During steady state condition the primary fluid, in single-phase natural circulation, removes the core power and transfers it to the secondary fluid through helical coil SG. In transient condition the core decay heat is removed through a passive primary/containment coupling mitigation strategy based on natural circulation.

The experimental analysis of the primary/containment coupling phenomena, characterizing the MASLWR design, has been developed in a first testing program at OSU-MASLWR experimental integral test facility. In particular the OSU-MASLWR-001 test determined the pressure behavior of the RPV and containment following an inadvertent actuation of one middle ADS valve. The test successfully thermal hydraulically demonstrated the passive primary/containment coupling typical of the MASLWR design SBLOCA mitigation strategy.

In the last years the USNRC has developed the advanced best estimate thermal hydraulic system code TRACE in order to simulate operational transients, LOCA, other transients typical of the LWR and to model the thermal hydraulic phenomena taking place in the experimental facilities used to study the steady state and transient behavior of reactor systems. The validation and assessment of the TRACE code against the MASLWR passive primary/containment coupling mitigation strategy is a novel effort and it is the topic of this chapter. Since the qualification process of the OSU-MASLWR TRACE nodalization is still in progress, considering the facility characterization conducted in an IAEA ICSP on "Integral PWR Design Natural Circulation Flow Stability and Thermo-Hydraulic Coupling of Containment and Primary System during Accidents", the current results are preliminary and should not be used for the code assessment, but are able to show the TRACE capability to reproduce the thermal hydraulic phenomena typical of the MASLWR primary/containment coupling SBLOCA mitigation strategy.

The analysis of the OSU-MASLWR-001 calculated data shows that the TRACE code is able to qualitatively predict the primary/containment coupling phenomena characterizing the MASLWR design. The sub-cooled blowdown, two-phase blowdown and single phase blowdown, following the inadvertent middle ADS actuation, are qualitatively predicted by the code. The refill phenomenon is qualitatively predicted by the code as well. In general the results of the calculated data show an over prediction compared with the experimental data. It is thought that this could be due to a combination of selection of vent valve discharge coefficients and condensation models applied to the inside surface of the containment. In agreement with the experimental data, the RPV water level never fell below the top of the core. However, in order to quantitatively evaluate the capability of the TRACE code to simulate the OSU-MASLWR phenomena, and therefore use the calculated data for the TRACE code assessment, is necessary a TRACE nodalization qualification against several facility operational characteristics like pressure drop at different primary mass flow rates and heat losses at different primary side temperatures. Currently the TRACE model qualification process is in progress considering the facility characterization conducted during the IAEA ICSP.

7. Abbreviations

ADS	Automatic Depressurization System;
AP600/1000	Advanced Plant 600/1000 MWe;
CAREM	Natural Circulation based PWR being developed in Argentina;
CHF	Critical Heat Flux;
CL	Cold Leg;
CPV	Cooling Pool Vessel;
EHRS	Emergency Heat Removal System;
ESBWR	Economic Simplified Boiling Water Reactor;
FW	Feed Water;

HL   Hot Leg;
HPC   High Pressure Containment;
IRWST   In-containment Refueling Water Storage Tank;
IAEA   International Atomic Energy Agency;
ICSP   International Collaborative Standard Problem;
IRIS   International Reactor Innovative and Secure;
LOCA   Loss of Coolant Accident;
LP   Lower Plenum;
LWR   Light- Water Reactor;
MASLWR   Multi-Application Small Light Water Reactor;
OSU   Oregon State University;
PRHR   Passive Residual Heat Removal;
PRZ   Pressurizer;
PWR   Pressurized Water Reactor;
RPV   Reactor Pressure Vessel;
SMART   System Integrated Modular Advanced Reactor ;
SBLOCA   Small Break Loss of Coolant Accident;
SG   Steam Generator;
SOT   Start of the Transient;
SNAP   Symbolic Nuclear Analysis Package;
SWR   Siede Wasser Reaktor;
TRAC   Transient Reactor Analysis Code;
TRACE   TRAC/RELAP Advanced Computational Engine;
UP   Upper Plenum;
USNRC   U.S. Nuclear Regulatory Commission;
WWER   Water Moderated, Water Cooled Energy Reactor.

8. Nomenclature

e Total energy;
g Gravitational force;
p Pressure;
T Temperature;
v_k Phase velocity.

9. Greek symbols

α Vapor void fraction;
Γ_k Mass generation rate per unit volume;
ρ Density.

10. References

Aksan, N. (2005). Application of Natural Circulation Systems: Advantage and Challenges II, In: *IAEA TECDOD-1474, Natural Circulation in Water Cooled Nuclear Power Plants*, IAEA, pp.101-114, ISBN 92-0-110605-X, Vienna, Austria, 2005

Bertel, E.; Morrison, R. (2001). Nuclear Energy Economics in a Sustainable Development Prospective, *NEA News 2001 - No. 19.1*

Bonuccelli, M.; D'Auria, F.; Debrecin, N.; Galassi, G. M. (1993). A Methodology For the Qualification of Thermalhydraulic Code Nodalizations, *Proc. of NURETH-6 Conference*, Grenoble, France, October 5–8, 1993

Boyack, B. E.; Ward, L.W. (2000). Validation Test Matrix for the Consolidated TRAC (TRAC-M) Code. Los Alamos National Laboratory, LA-UR-00-778, February 2000

Carelli, M. D.; Conway, L. E.; Oriani, L.; Petrovi´c, B.; Lombardi, C. V.; Ricotti, M. E.; Barroso, A. C. O.; Collado, J. M.; Cinotti, L.; Todreas, N. E.; Grgi´c, D.; Moraes, M. M.; Boroughs, R.D.; Ninokata, H.; Ingersoll, D.T.; Oriolo F. (2004). The Design And Safety Features of the IRIS Reactor. *Nuclear Engineering and Design* 230 (2004) 151–167

Carelli, M.; Conway, L.; Dzodzo, M.; Maioli, A.; Oriani, L.; Storrick, G.; Petrovic, B.; Achilli, A.; Cattadori, G.; Congiu, C.; Ferri, R.; Ricotti, M.; Papini, D.; Bianchi, F.; Meloni, P.; Monti, S.; Berra, F.; Grgic, D.; Yoder, G.; Alemberti A. (2009). The SPES3 Experimental Facility Design for the IRIS Reactor Simulation. Hindawi Publishing Corporation, *Science and Technology of Nuclear Installations*, Volume 2009, Article ID 579430, 12 pages doi:10.1155/2009/579430

Chiovaro, P.; Di Maio, P. A.; Mascari, F.; Vella, G. (2011). Analysys of The SPES-3 Direct Vessel Injection Line Break By Using TRACE Code, *Proceeding of the XXIX Congresso UIT sulla Trasmissione del calore*, ISBN: 978- 88467- 3072- 5, Torino, June 20-22, 2011.

D'Auria, F.; Modro, M.; Oriolo, F.; Tasaka, K. (1993). Relevant Thermal Hydraulic Aspects of New Generation LWRs. *Nuclear Engineering and Design* 145, 241-259 (1993)

Galvin, M. R. (2007). OSU-MASLWR Test Facility Modification Description Report. IAEA Contract Number USA-13386, Oregon State University (November 19, 2007)

Gou, J.; Qiu, S.; Su, G.; Jia D. (2009). Thermal Hydraulic Analysis of a Passive Residual Heat Removal System for an Integral Pressurized Water Reactor. Hindawi Publishing Corporation, *Science and Technology of Nuclear Installations*, Volume 2009, Article ID 473795, 12 pages, doi: 10.1155/2009/473795

Hicken, E. F.; Jaegers, H. (2002). Passive Decay Heat Removal From the Core Region, In: *IAEA-TECDOC-1281, Natural Circulation Data and Methods For Advanced Water Cooled Nuclear Power Plant Designs*, IAEA, pp. 227-237, ISSN 1011–4289, Vienna, Austria, April 2002

IAEA-TECDOC-1281. (2002). *Natural Circulation Data and Methods for Advanced Water Cooled Nuclear Power Plant Designs*. IAEA, ISSN 1011–4289, Vienna, Austria, April 2002

IAEA-TECDOC-1391. (2004). *Status of Advanced Light Water Reactor Designs 2004*. IAEA, ISBN 92–0–104804–1, Vienna, Austria, May 2004

IAEA-TECDOD-1474. (2005). *Natural Circulation in Water Cooled Nuclear Power Plants*. IAEA, ISBN 92–0–110605–X, Vienna, Austria, November 2005

IAEA-TECDOC-1624. (2009). *Passive Safety Systems And Natural Circulation In Water Cooled Nuclear Power Plants*. IAEA, ISBN 978–92–0–111309–2, Vienna, Austria, November 2009

IAEA Power Reactor Information System (PRIS). http://www.iaea.org/programmes/a2/ 2 December 2011

IAEA – TECDOC DRAFT. *Natural Circulation Phenomena and Modelling for Advanced Water Cooled Reactors*. In preparation

Kurakov, Y.A.; Dragunov, Y. G.; Podshibiakin, A. K.; Fil, N. S.; Logvinov, S. A.; Sitnik, Y. K.; Berkovich, V.M.; Taranov, G. S. (2002). Development and Validation of Natural Circulation Based Systems for new WWER Designs, In: *IAEA-TECDOC-1281*, IAEA, pp.83-96, ISSN 1011-4289, Vienna, Austria, April 2002

Lee, K. Y.; Kim M. H. (2008). Experimental and Empirical study of Steam Condensation Heat Transfer with a Noncondensable Gas in a Small-Diameter Vertical Tube. *Nuclear Engineering and Design* 238 (2008) 207-216

Levy, S. (1999). Two-Phase flow in complex systems. *A Wiley-Interscience Publication*, John Wiley & Son, Inc. (1999)

Mascari, F.; Woods, B. G.; Adorni, M. (2008). Analysis, by TRACE Code, of Natural Circulation Phenomena in the MASLWR-OSU-002 Test, *Proceedings of the International Conference Nuclear Energy for New Europe 2008*, ISBN 978-961-6207-29-4, Portoroz, Slovenia, September 8-11, 2008

Mascari, F.; Vella, G.; Woods, B.G.; Welter, K.; Pottorf, J.; Young, E.; Adorni, M.; D'Auria, F. (2009). Sensitivity Analysis Of The MASLWR Helical Coil Steam Generator Using TRACE, *Proceedings of International Conference Nuclear Energy for New Europe 2009*, ISBN 978-961-6207-30-0, Bled, Slovenia, September 14-17

Mascari, F.; Vella, G.; Woods, B.G.; Adorni, M.; D'Auria, F. (2009). Analysis of the OSU-MASLWR Natural Circulation Phenomena Using TRACE Code, IAEA, organized in cooperation with the OECD Nuclear Energy Agency and the European Commission, Technical meeting on *"Application of Deterministic Best Estimate Safety Analysis"*, University of Pisa, Pisa, Italy, September 21-25, 2009

Mascari, F.; Vella, G; Buffa, P.; Compagno, A.; Tomarchio, E. (2010). Passive Safety Systems in view of Sustainable Development. *Final Report on the Round Tables*, Erasmus Intensive Programme Project (IP) ICARO Intensive Course on Accelerator and Reactor Operation 28/02/10 – 12/03/10– Sicilia, Italia –

Mascari, F. (2010). PhD Thesis on "Natural Circulation and Phenomenology of Boron Dilution in the Pressurzied Water Reactors (Circolazione Naturale e Fenomenologie di Boron Dilution in Reattori ad Acqua in Pressione). University of Palermo, 2010

Mascari, F.; Vella, G.; Woods, B.G.; Welter, K.; Pottorf, J.; Young, E.; Adorni, M.; D'Auria, F. (2011). Sensitivity Analysis Of The MASLWR Helical Coil Steam Generator Using TRACE. *Nuclear Engineering and Design* 241 (2011) 1137-1144

Mascari, F.; Vella, G. (2011). IAEA International Collaborative Standard Problem on Integral PWR Design Natural Circulation Flow Stability and Thermo-hydraulic Coupling of Containment and Primary System during Accidents" Double Blind Calculation Results. Dipartimento dell'Energia, Università degli Studi di Palermo. Prepared for IAEA, Vienna, Austria , February 2011

Mascari, F.; Vella, G.; Woods, B. G.; D'Auria, F. (2011). Analysis of the Multi-Application Small Light-Water Reactor (MASLWR) Design Natural Circulation Phenomena, *Proceedings of ICAAP 2011*, Nice, France, 2-5 May, 2011

Mascari, F.; Vella, G.; Woods, B.G. (2011). TRACE Code Analyses For The IAEA ICSP On Integral PWR Design Natural Circulation Flow Stability And Thermo-Hydraulic Coupling Of Containment And Primary System During Accidents, *Proceedings of* the *ASME 2011 Small Modular Reactors Symposium*, Washington, DC, USA, September 28-30, 2011

Mascari F.; Vella, G.; Woods, B.G.; D'Auria F. (2011) Analyses of the OSU-MASLWR Experimental Test Facility. *Science and Technology of Nuclear Installations*, Hindawi, In Progress

Modro, S.M.; Fisher, J. E.; Weaver, K. D.; Reyes, J. N., Jr.; Groome J. T.; Babka, P.; Carlson, T. M. (2003). Multi-Application Small Light Water Reactor Final Report, DOE Nuclear Energy Research Initiative Final Report, Idaho National Engineering and Environmental Laboratory, December 2003

Pottorf, J.; Mascari, F.; Woods, B. G (2009). TRACE, RELAP5 Mod 3.3 and RELAP5-3D Code Comparison of OSU-MASLWR-001 Test. *Transactions of the ANS*, Volume 101, 2009, ISSN 0003-018X

Reyes, J. N., Jr.; King, J. (2003). Scaling Analysis for the OSU Integral System Test Facility. Department of Nuclear Engineering Oregon State University, NERI Project 99-0129, Prepared For U.S. Department of Energy

Reyes, J. N., JR. (2005). Governing Equations in Two-Phase Fluid Natural Circulation Flows, In: *IAEA TECDOD-1474, Natural Circulation in Water Cooled Nuclear Power Plants*, IAEA, pp.155-172, ISBN 92-0-110605-X, Vienna, Austria, November 2005

Reyes, J. N., Jr. (2005). Integral System Experiment Scaling Methodology, In: *IAEA TECDOD-1474, Natural Circulation in Water Cooled Nuclear Power Plants*, IAEA, pp.321-355, ISBN 92-0-110605-X, Vienna, Austria, November 2005

Reyes, J.N. Jr. (2005). AP 600 and AP 1000 Passive Safety System Design and Testing in APEX, In: *IAEA TECDOD-1474, Natural Circulation in Water Cooled Nuclear Power Plants*, IAEA, pp.357-381, ISBN 92-0-110605-X, Vienna, Austria, November 2005

Reyes, J.N., JR.; Groome, J.; Woods, B. G.; Young, E.; Abel, K.; Yao, Y.; Yoo, J. Y. (2007). Testing of the Multi Application Small Light Water Reactor (MASLWR) Passive Safety Systems. *Nuclear Engineering and Design 237*, 1999-2005 (2007)

Reyes, J.N., Jr.; Lorenzini, P. (2010). NusScale Power: A Modular, Scalable Approach to Commercial Nuclear Power. *Nuclear news*, ANS (June 2010)

Staudenmeier, J. (2004). TRACE Reactor System Analysis Code, MIT Presentation, Safety Margins and Systems Analysis Branch, Office of Nuclear Regulatory Research, U.S. Nuclear Regulatory Commission (2004)

Symbolic Nuclear Analysis Package (SNAP), 2007. Users Manual. Applied Programming Technology, Inc., Bloomsburg, PA

TRACE V5.0. *Theory and User's manuals*. 2010. Division of System Analysis, Office of Nuclear Regulatory Research, U.S. Nuclear Regulatory Commission, Washington, DC 20555-0001

Wolde-Rufael, Y.; Menyah, K. (2010). Nuclear Energy Consumption and Economic Growth in Nine Developed Countries. *Energy Economics 32*, 550-556, (2010)

Woods, B. G.; Mascari, F. (2009). Plan for an IAEA International Collaborative Standard Problem on Integral PWR Design Natural Circulation Flow Stability and Thermo-hydraulic Coupling of Containment and Primary System during Accidents. Department of Nuclear Engineering and Radiation Health Physics, Oregon State

University. Prepared for IAEA, Vienna, Austria. Manuscript completed: March 2, 2009. Data Published: April 8, 2009

Woods, B. G.; Galvin, M. R.; Jordan, B. C. Problem specification for the IAEA International Collaborative Standard Problem on Integral PWR Design natural circulation flow stability and thermo-hydraulic coupling of containment and primary system during accident, Department of Nuclear Engineering and Radiation Health Physics, Oregon State University. Prepared for IAEA, Vienna, Austria. DRAFT

World Energy Outlook, Executive Summary. IEA - International Energy Agency-, 2009

Zejun, X.; Wenbin, Z.; Hua, Z.; Bingde, C.; Guifang, Z.; Dounan, J. (2003). Experimental Research Progress on Passive Safety Systems of Chinese Advanced PWR. *Nuclear Engineering and Design* 225 (2003) 305–313

Zuber, N. (1991). Appendix D: Hierarchical, Two-Tiered Scaling Analysis, An Integrated Structure and Scaling Methodology for Severe Accident Technical Issue Resolution, U.S Nuclear Regulatory Commission, Washington, D.C. 20555, *NUREG/CR-5809, November 1991*

AREVA Fatigue Concept – A Three Stage Approach to the Fatigue Assessment of Power Plant Components

Jürgen Rudolph*, Steffen Bergholz, Benedikt Heinz and Benoit Jouan
AREVA NP GmbH
Erlangen
Germany

1. Introduction

1.1 General remarks

Within the continuously accompanying licensing process for NPPs until the end of their operational lifetime, the ageing and lifetime management plays a key role. Here, one of the main tasks is to assure structural integrity of the systems and components. With the help of the AREVA Fatigue Concept (AFC), a powerful method is available. The AFC provides different code-conforming fatigue analyses (e.g. according to the wide spread ASME code [1]) based on realistic loads. In light of the tightening fatigue codes and standards, the urge is clearly present that, in order to still be able to comply with these new boundaries, margins which are still embedded within most of the fatigue analyses in use, have to be reduced. Moreover, thermal conditions and chemical composition of the fluid inside the piping system influences the allowable fatigue levels, which have come under extensive review due to the consideration of environmentally assisted fatigue (EAF) as proposed in the report [2]. Therefore, for highly loaded components, some new and improved stress and fatigue evaluation methods, not overly conservative, are needed to meet the increasingly stringent allowable fatigue levels. In this context, the fatigue monitoring system FAMOS, central module of AFC, is able to monitor and record the real local operating loads. The different modules of the AFC are schematically represented in Figure 1.

1.2 Safety concept context of NPPs

NPPs are subject of particular safety requirements due to the increased risk potential. The utmost aim from the point of view of safety in the design as well as the plant operation is the prevention of unforeseen events or accidents.

Ageing effects may also induce unexpected events during operation. A comprehensive ageing management is required in order to avoid these a priori. Particularly regarding the aspects of new lifetime periods of nuclear power generation works (60 years of operation for

---

* Corresponding Author

new NPPs such as AREVA's EPR™) or due to lifetime extension projects (e.g. in the USA, Sweden or Switzerland) there is an increasing need of knowing the current state of the plant exactly in order to enable a qualified respective assessment. Of course, the knowledge of the occurred and expected loading of the power plant components is an essential prerequisite for such an assessment. The operation of a new NPP will extend to three generations of operational staff. Conceptual long term solutions are required in this context. I.e., all relevant load data should be recorded in a way that future staff still have access to these data and can consider them in the evaluation of the plants.

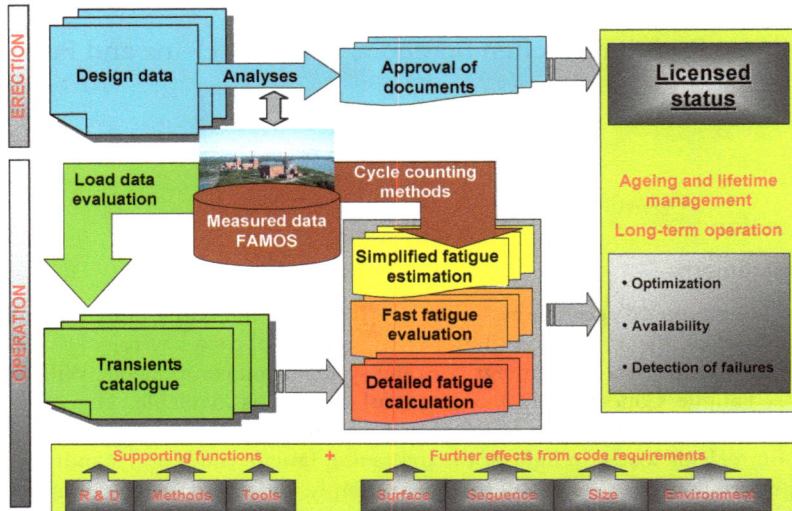

Fig. 1. Modules of the AFC

1.3 Fatigue

The fatigue check takes a central position within the ageing management. The successful fatigue check shows the design-conforming state of cyclic operational loads. During operation of NPPs, particularly the thermal cyclic loadings are fatigue relevant. They are due to transient states of operation. E.g., respective cold or hot feed conditions occur during start-up and shut-down as well as testing conditions of the safety equipment. Furthermore, permanently occurring mixing events of hot and cold flows at junction locations (t-sections) may induce high cycle fatigue loads. Certain plant conditions may induce temperature stratification events within larger pipes at lower flow rates and an existing temperature difference. These phenomena may equally induce cyclic loads in the pipeline and the attached components. Of course, cyclic mechanical loads such as internal pressure or piping loads have to be considered for the fatigue check as well. Until now, the design and operation of NPPs was concentrated on the purpose of base load generation for the respective electrical network. In the future, NPPs will have to take increasing parts of the average and peak load generation due to the growing utilization of renewable energy sources such as wind and solar energy. This generates permanently changing states of the plant which have to be considered within the fatigue check. All these expected loads are examined in the design phase of a power plant as well as recorded and described in a design transient catalogue. These so called design transients are characterized by the expected temperature ranges, temperature change rates and the expected frequencies of occurrence. Furthermore, the expected internal pressure and - if applicable - stratification states are considered. This transient catalogue constitutes the basis of the design fatigue checks. This first fatigue check is part of the licensing documents and should also indicate fatigue relevant positions and plant processes. If necessary, modifications of the plant processes and/or components are carried out in the design phase with the aim of eliminating potentially critical positions. The task of the fatigue check changes somewhat during the operation of the NPP. In this phase, the fatigue check is primarily used to show that the operation of the plant is within the specified limits. That is to say, the fatigue usage factor for the relevant components has to be reported in a regulated cyclic sequence. As the operational processes differ even for identical technological procedures a simple counting of technological procedures (events) will not necessarily deliver a covering fatigue usage factor. It is also possible to overestimate fatigue usage in case of conservatively specified design transients.

1.4 Local monitoring concept

The complex fluid flow events occurring during the operation of NPPs are influenced by the automatic operational control processes. Nevertheless, as a consequence of the manifold manual intervention opportunities equal technological processes may induce different local loading sequences for the components. In other words, an assessment of components exclusively based on operational measuring instrumentation is insufficient. Local data acquisition and monitoring of local loads at the fatigue relevant components is the better solution. Local effects such as the swapping flow after feeding interruption can only be recorded in the load data set this way. It is to be pointed out that the safety check against cyclic loads of the components has to be a permanent operation accompanying procedure. The German KTA rules regulate this issue as part of the rule for operational monitoring (KTA 3201.4) [3].

1.5 Modified codes

On one hand, the checks have to be harmonized with the valid design code. On the other hand, the state of the art in science and technology has to be considered. Recently, the detrimental influence of the medium (high temperature reactor water) on the fatigue process – which has been examined since the 1980ies – is the subject of code modifications tending towards tightening code rules. The term environmentally assisted fatigue (EAF) is synonymous to the corrosive influence of the medium on the fatigue behavior. The usual way of considering EAF in fatigue analysis is the application of penalty factors F_{en}. The modified code rules mostly based on [2] have to be considered and applied both within the lifetime extension projects and the new built projects of NPPs.

1.6 Fatigue monitoring systems

During the early operation of NPPs in the 1970ies and 1980ies local loads occurred at different locations causing fatigue cracks. These were either due to new loading conditions which were not considered in the design phase (e.g. temperature stratification) or insufficient manufacturing quality (e.g. welded joints). These problems constituted the starting signal for the development of fatigue monitoring systems. Thus, FAMOS was for instance developed by then Siemens KWU at the end of the 1980ies and installed in German NPPs. At that time, this was a very progressive data logging system. Henceforth, it was possible to measure the local loading effects. The fatigue relevance of those effects was analyzed by simple assessment methods. These experiences gave rise to a better understanding of the ongoing loading phenomena. The fatigue assessment induced the necessity of retrofitting of components or the modification of the operating mode. For instance, the feedwater sparger of the steam generator was subsequently designed in a way that the stresses of cyclically occurring stratification transients were minimized. Nevertheless, the technology of the data logging system at that time still had certain limits with respect of the frequency of data logging and the recording and storage. A data logging frequency of 10s (0.1Hz) constituted the upper limit (nowadays, 1s respectively 1Hz is usual). Furthermore, the capacitive effect of the applied measurement sections was underestimated in their transient behavior. Nowadays, this effect is appropriately considered by correction factors specific to the respective measurement section.

1.7 New software, hardware and ageing requirements

Hence, these limits do not exist anymore in the application of up-to-date data logging systems. The application of modern BUS technologies paves the way for an economic and flexible measuring system without long cables. Handling and performance has been significantly improved for the modern systems. AREVA offers FAMOSi as a modern data logging system of that kind. It delivers the data base for the further fatigue monitoring and for the further assessment in a highly reliably way.

1.8 Data post processing and evaluation of cumulative usage factors (CUFs)

As it was mentioned before the evaluation of the recorded measurement data is done by application of a three staged assessment system. Only those analyses are applied that are required for the fatigue assessment of the according component. If the CUF obtained by application of the first analysis stage – the simplified fatigue estimation (SFE) – falls below the defined target value no further evaluation will be required. If it does not the second stage – fast fatigue evaluation (FFE) – will be activated and so on.

All data have to undergo a plausibility check before being further processed. The plausibility check enables a data control based on predefined limiting values. Erroneous data blocks originating for instance from electromagnetic pulses (e.g. switching operations within the main coolant pump) are detected by identification of physically questionable values of the temperatures or their gradients. These data sets are corrected to logical values. The correction is precisely recorded and can be reproduced at any time. Thus, the generated plausible data set constitutes the basis for the further data processing and assessment.

2. Design analysis before operation

Before commissioning and operation of the plant, a catalogue of thermal transients is compiled. These thermal transients are considered as design transients in contrast with the real transients based on temperature measured during operation. In the past, the anticipated transients were covering 40 years of plant operation. Now, the period to be covered is 60 years. Moreover, the specification is done for normal, upset, emergency and testing conditions. The design thermal transients are specified according to different plant models and experiences. They should always be conservative concerning frequency of occurrences, temperature range, rate of temperature change and load type (thermal stratification, thermal shock). Due to this conservatism, the usage factor calculated in the design phase, under normal circumstances, will be more severe than the results of the detailed fatigue calculation performed at a later operation stage taking into account the real operational thermal loads. As a consequence, usage factors around 1.0 are still tolerable in the design phase. They indicate the fatigue sensitive positions. These locations are selected for future instrumentation and non-destructive testing. Design improvements of components, depending on the calculated fatigue usage factors, can also be taken into account at this early stage. Additionally, optimization of operating modes can be considered. Thermal transients with low influence on fatigue behavior are identified as well. Depending on the different design codes [1], [4], [5], some procedures can allow the exemption of non significant loads. In the end, the predicted fatigue usage factors, which were calculated with design transients, shall be verified and the fatigue status shall be updated during lifetime operation.

3. Fatigue monitoring system

The acquisition of realistic operational load data in the power plant is one essential pillar of the AFC. Its function is to determine the realistic thermal loads. FAMOS was developed in the early eighties. At that time, German licensing authorities demanded for the realization of a comprehensive measurement program in one German NPP. This was in order to get detailed information on the real component loadings during plant operation. This proof should give the information that the real operating conditions are not different from the design data. At that occasion, the advantages of monitoring real operating loads and using the measured data as an input for fatigue analyses became obvious. Therefore, a sophisticated fatigue monitoring system was developed. As a consequence, many NPPs in Germany and abroad were equipped with FAMOS (see also [6]).

Depending on each power plant, a fatigue handbook is developed to identify the locations relevant to fatigue in the NPP. The instrumentation of these locations is specific for each plant and depends on system design and further requirements.

For the acquisition of load data FAMOS uses two different methods: the global fatigue monitoring and the local fatigue monitoring [7]. The global monitoring is made by existing operational measurement. The corresponding operational signals could be fluid pressure, fluid temperature, the position of valves etc. measured at different parts of the systems.

Local fatigue monitoring is located at fatigue relevant locations at the outer surface of pipes and is based on additional temperature measurement by means of thermocouples. The thermocouples are manufactured as measurement sections.

Figure 2 shows the typical locations of measurement sections in a pressurized water reactor (PWR). Indeed, FAMOS gathers measurement sections, which are mostly located on the:

- primary loops
- surge line
- spray lines
- volume control system
- feedwater system and further positions.

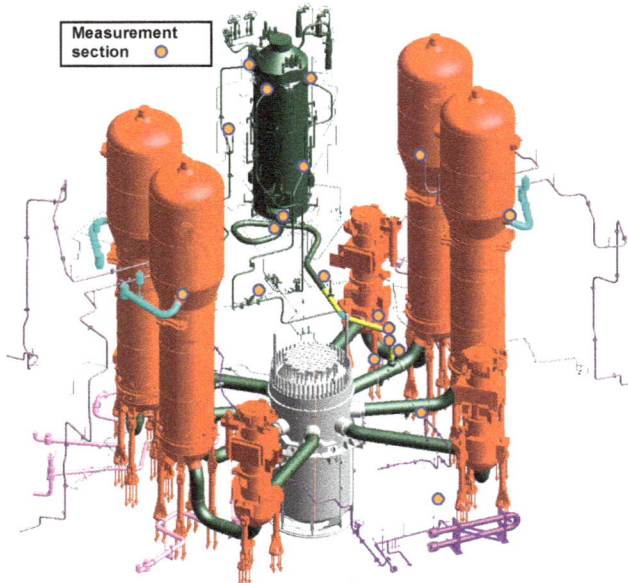

Fig. 2. FAMOS measurement sections in a PWR

The different FAMOS measurement sections can be composed of seven or more thermocouples if some thermal events like stratification are suspected (horizontally installed pipes). However, in case of plug flow the application of only two thermocouples is sufficient (vertically installed pipes).

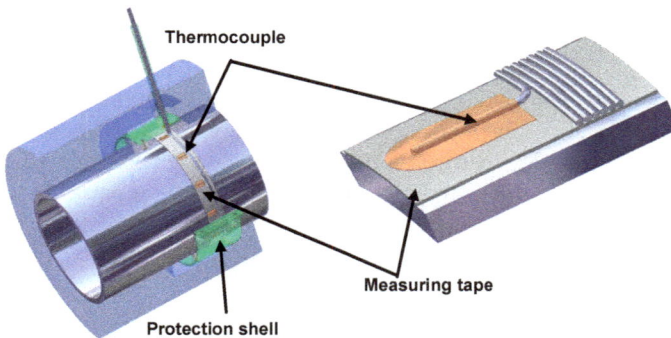

Fig. 3. FAMOS principle

Figure 3 shows how the application of thermocouples at the outer surface of a pipe is performed. More details on the technical bases of FAMOS can be found in [8].

Each measurement section consists of the thermocouples installed on a thin metal tape and a robust protection shell to prevent the thermocouples from being damaged. Both, metal tape and protection shell are installed around the pipe under the piping insulation. The installation takes place at a certain distance to pipe welds. Thus, the dismounting of measurement sections during an ultrasonic testing of the weld is unnecessary. Furthermore, a distance to thick walled components (e.g. nozzles) is needed in order to minimize the thermal impact on the temperature measurement.

The measurement sections are designed for a fast mounting process. A very short installation time is absolutely necessary to ensure low dose rates for the mounting staff. Special manufacturing processes and thermal responses tests of the thermocouples guarantee realistic thermal load data. Actual measurement sections are characterized by a minimization of heat capacity effects and excellent thermal sensitivity.

All thermocouples are wired with extension lines to junction boxes where the cold junction compensation of the thermo voltage signal is performed. For the compensation a board with an isothermal terminal for up to 30 channels is used to connect the wires made of thermocouple material with a trunk cable made of copper. The temperature at the isothermal terminal is measured with a resistance thermometer. The voltage signals from different junction boxes are connected to the information and control system (I&C) of FAMOS. That system consists of two or more information modules with signal processing units and analog-to-digital converters. The information modules are connected by means of a data bus to a processing unit. The data of all thermocouples are analyzed, stored and transferred to a computer network in real time. To avoid a high data volume in the storage system a data reduction method is used for the samples of the signals. By application of the bandwidth method samples will only be stored if signals leave the bandwidth. Thus, it is ensured that all samples of fatigue relevant load data during transients are stored and only a few samples during steady state operating mode of the plant with no fatigue relevant transients.

The information modules and the processing unit are installed within the containment of the power plant. All requirements of the containment environment such as temperature, humidity, atmosphere and material restrictions have been considered during engineering and design of the I&C components. The system is equipped with redundant power supply and a diagnostic system to detect malfunctions in different modules. Thus, a long term monitoring and storage of load data in a containment environment is ensured.

Outside the containment a sever unit is installed in the safeguards building, main control side room or other computer rooms of the power plant. That server stores all load data in a database over years. In addition to the FAMOS data, load data of the global instrumentation are also stored in the sever unit. For the connection of the server unit with the processing unit inside the containment fibre optic cables or normal ethernet connections made of copper are used. These cables run in special penetrations through the pressure-sealed and gas tight containment wall.

By means of a network connection to the server unit all data of FAMOS and the global monitoring can be viewed in real time on any connected computer. The access to the database features the view on the load data of several years of power plant operation. The real time data analyzing process done by the processing unit generates messages about high thermal loads or malfunctions automatically. Thresholds for temperature rate of change or thermal stratification can be set. Is the threshold reached during a transient the system generates a warning about high thermal loads. In combination with valve position signals from the global monitoring, the system detects insufficiently closed valves even if the valve position indicates it.

With the help of automated warnings, unfavourable thermal load events can be analyzed and operating modes optimized. For the purpose of preventive maintenance FAMOS offers all necessary information to prevent the plant's components from avoidable load events.

The FAMOS I&C system operates in real time with signals of 120 or more than 300 thermocouple channels. The number of channels depends on the power plant layout, the number of primary loops and the monitored components. The engineering of FAMOS in a NPP starts with the generation of a FAMOS manual. This process contains a deep analysis to identify components relevant to fatigue in the primary, secondary, auxiliary and safeguards systems. To do so, design documents, operating experience and feedback from similar plants are considered. A measurement point plan is elaborated and all activities are coordinated with the plant operator and, if required, with independent experts.

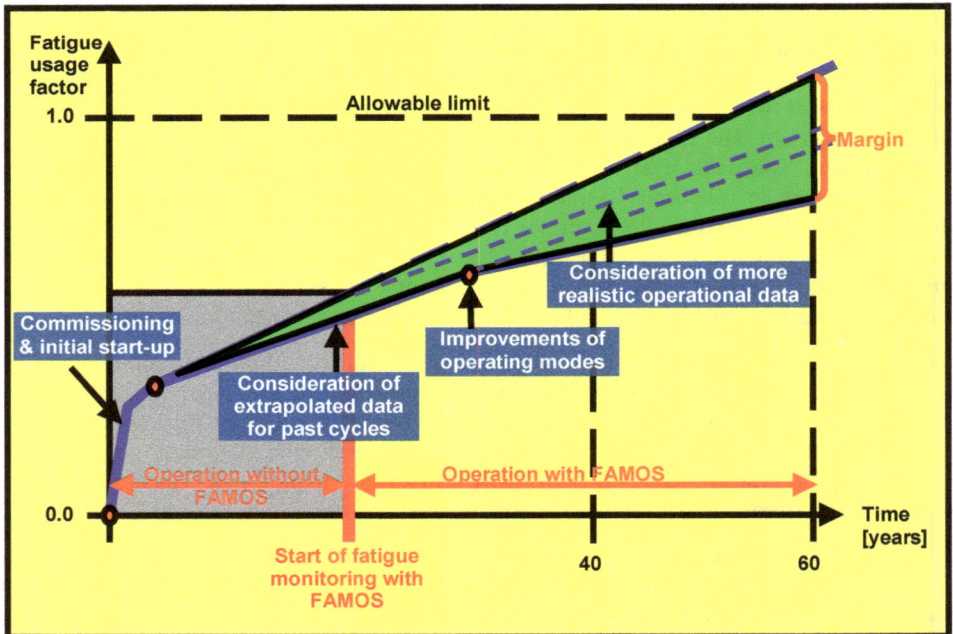

Fig. 4. Development of fatigue usage factor considering local monitoring and improvements

If FAMOS is installed before commissioning of the plant the fatigue status can be calculated over the complete lifetime with realistic load data. Indeed, the commissioning phase is often characterized by the highest loads of the entire lifetime of the power plant. But an installation into a running plant is useful after several years of operation e.g. during lifetime extension.

Getting the real measurements at this stage implies a consequent reduction of the fatigue usage factor when the later detailed fatigue calculation is required as explained in Figure 4.

The objectives of FAMOS are summarized here below:

- to determine the fatigue status of the most highly stressed components
- to identify and optimize the operating modes which are unfavourable to fatigue e.g. valve leakages
- to improve the catalogue of transients used at the design phase
- to establish a basis for fatigue analysis based on realistic operating loads
- to use the results for lifetime management and lifetime extension.

An important connecting link between the recording of measurements and the stress and fatigue analysis is the load data evaluation (see Figure 1) and the specification of thermal loads in transient catalogues. Expert knowledge of the processes and the operation of the systems is essential. The first step to specify thermal loads is the identification of the operational processes leading to relevant transients of temperature and/or pressure. For these events an appropriate number of model transients is selected. Usually, it is necessary to split up these model transients into subclasses which are different e.g. in the temperature difference of the transient. As it is shown in Figure 1 the specified transients are input data for the detailed fatigue check.

Indeed, three graded methods were developed fulfilling the different requests in terms of fatigue. The choice of one method depends on the expected degree of fatigue relevance and the expected grade of details in fatigue calculation.

Step 1: Simplified fatigue estimation

Simple estimations of fatigue relevance of real loads for components are based on thermal mechanical considerations using the equation of completely restrained thermal expansion. A basic decision about fatigue relevance (yes/no) for the monitored position is made. In case of fatigue relevance a further evaluation is done according to step 2.

Step 2: Fast fatigue evaluation

A code-conforming usage factor U is calculated in a highly automated way based on the simplified elasto-plastic fatigue analysis. If $U \leq U_{admissible}$ the fatigue check will be success-fully finished. If $U > U_{admissible}$ further analyses will be based on step 3.

Step 3: Detailed fatigue calculation

Fatigue analysis is based on a detailed catalogue of transients. This catalogue of transients results from the evaluation of the real loads for the monitored component. The detailed fatigue check is based on finite element analyses mostly including elasto-plastic material behavior.

Both step 2 and step 3 allow for the consideration of EAF in the analysis process.

4. Simplified fatigue estimation

The results of the temperature measurement are to be processed quickly in order to get a first estimation of the fatigue state. One important task before the simplified and automated evaluation is the verification of the acquired data. Detection and adjustment of implausible data are parts of this process. In this context, data plausibilization is based on the limits of the measurement range (e.g. 0°C – 400°C) as well as the predefined limiting gradients. Irregularities such as those resulting from the switching of the main coolant pump by electromagnetic pulses are recognized and corrected this way. The original set of data is not modified. All adaptations are reproducibly recorded. These plausibility and quality checks of the measured data have to be done by experienced specialists. In other words, the specialists must be capable of checking the operational events with respect of their plausibility. The result is a preprocessed database for data evaluation and fatigue assessment.

In the very first step of the SFE, the changes of temperatures are subject to a rain-flow cycle counting algorithm (see e.g. [9] and [10]). In this process the temperature ranges at the locations of measurement are identified, counted and classified. The according temperature differences between a subsequent minimum and maximum are inserted into a rain-flow matrix. The temperature difference, the starting value and the temperature change as well as the stratification differences are the parameters of this matrix. An exemplary matrix is shown in Figure 5.

These thermal load cycles are input data for a stress and fatigue assessment of the monitored components based on conservative analytical computation formulae.

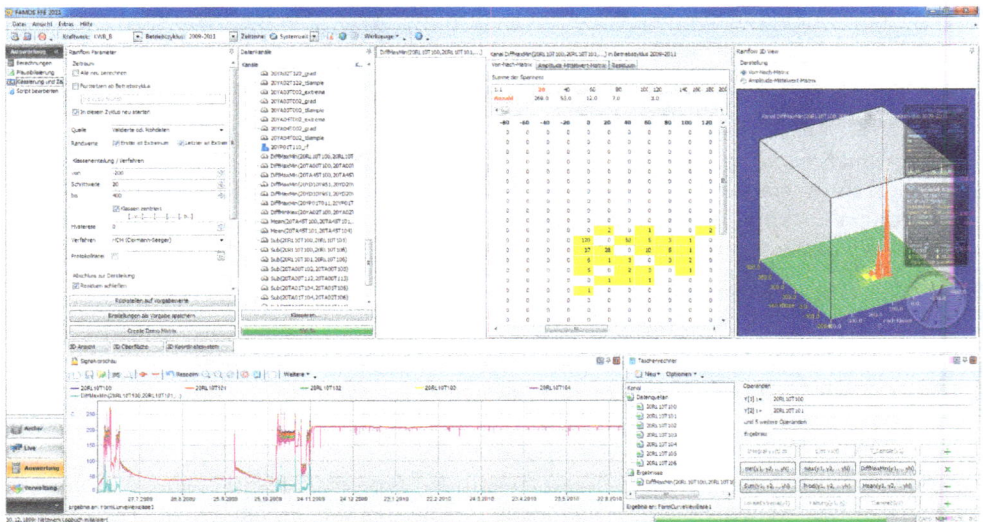

Fig. 5. Exemplary rain-flow matrix for SFE application

The temperature differences are processed to stresses by applying the equation of the completely restrained thermal extension $\sigma = E \cdot \alpha \cdot \Delta T$ (σ ... Stress, E ... Young's modulus, α ... coefficient of thermal extension, ΔT ... temperature difference). The internal pressure induced stresses are added. The stresses calculated in this way are multiplied by stress concentration factors as a function of the component geometry. The well-known stress concentration factors for pipe bends, t-joints or weld seams (see e.g. chapters on piping design in the ASME code [1] or the KTA rules [4]) are applied. Based on these stresses and their frequency of occurrence the partial usage factors are calculated and summed up to the total usage factor. The basic function of this method is a check of fundamental fatigue relevance of the component subjected to the recorded loading. It constitutes a simple qualitative assessment method. In case of calculated CUFs < 5 per cent there is no fatigue relevance of the component. Additionally, SFE allows for a simple qualitative comparison of annual partial usage factors. By means of extrapolation the future fatigue potential can be predicted.

SFE has been successfully applied in many German NPPs for about 20 years. This rough real time fatigue estimation is done after every operational cycle and allows for a direct comparison of thermal loads and an evaluation of the current fatigue usage factor. The result of this SFE provides a qualitative tendency. Although the correlation of the real temperature ranges is fairly simple, it is suitable for a comparison of different real sequences of loads and allows for a qualitative evaluation of the mode of operation and the detection of fatigue critical locations. Furthermore, the investigation of the results allows for the detection of anomalies.

5. Fast fatigue evaluation

5.1 General remarks and context

With the help of the fast running SFE method, an overview of the fatigue level for every monitored component is given. For highly loaded components a more detailed method, the FFE, can be used to calculate the usage factors in a more realistic way as indicated in [11]. This method uses FAMOS measured data from the outside surface of a pipe and can evaluate a fatigue level of the component for different thermal loads (plug flow, stratification).

The measuring location of FAMOS is chosen close to a fatigue relevant component and the measurement sections installed at the outer surface of the pipe. Nevertheless, the points of interest are at the inner surface of the component. Therefore, the calculated temperature at the inner surface of the pipe will be transferred to the inner surface of the component. The thermal load cycles are well known after that step and the stress time history is calculated with the Green's function approach. This approach deals with two unit (elementary) transients of +/- 100 K, which are used to scan the original temperature time history at each time step. By means of elementary transients, stresses are calculated at all fatigue relevant locations, which are monitored with FAMOS. Pressure cycles as well as section moments will also be evaluated based on the Green's function approach.

After the calculation of the stress tensor, the mechanical load cycles can be classified by application of the rain-flow cycle-counting algorithm. Then, comparisons with the fatigue curve result in fatigue levels and are performed for all relevant locations.

Moreover, an enveloping fatigue level can still be calculated. In other words, for highly loaded components, the application of the FFE method can provide a more realistic stress calculation and enveloping fatigue level calculation. Depending on the real number of load cycles, the new and more stringent code requirements can also be complied with.

In the end, if the calculated fatigue usage factor is lower than the allowable limit, the fatigue check will be successfully finished. If not, further analyses will be performed, according to the detailed fatigue calculation (DFC).

5.2 Detailed description

The fast running SFE method draws up an overview of the temperature changes and a qualitative stress estimation for every monitored component. The fatigue handbook and the knowledge of these parameters determine the pertinence of a fatigue analysis for the component. In that way a more detailed and automatic method, FFE, can be used to calculate the CUFs.

The determination of the time-history of loads and the resulting local stresses are the basis of the method. On a measuring section, close to the fatigue relevant location, FAMOS measures the temperature at the outside surface of the pipe. First of all the interpretation of this temperature has to be explained.

In a homogeneous isotropic solid, where the temperature is a function of time and space $T = f(x,y,z,t)$, the equation of heat conduction is as follows:

$$\frac{\partial^2 T}{\partial x^2} + \frac{\partial^2 T}{\partial y^2} + \frac{\partial^2 T}{\partial z^2} - \frac{\rho c}{\lambda}\frac{\partial T}{\partial t} = 0 \tag{1}$$

λ ... thermal conductivity of the solid [W/(m·K)]
ρ ... density of the solid [kg/m³]
c ... specific heat of the solid [J/(kg·K)]

This equation can be written in a cylindrical coordinate system:

$$\frac{\partial^2 T}{\partial r^2} + \frac{1}{r}\frac{\partial T}{\partial r} + \frac{1}{r^2}\frac{\partial^2 T}{\partial \theta^2} + \frac{\partial^2 T}{\partial z^2} - \frac{\rho c}{\lambda}\frac{\partial T}{\partial t} = 0 \tag{2}$$

The stratification effects will not be considered here and the application of the method will be restricted to plug flow events. In this case, the temperature evolution is independent of the circumferential direction $T = f(r,z,t)$. Moreover the measuring section on the pipe is located relatively far away from geometrical discontinuities and $T(z-\delta z) = T(z+\delta z)$ holds true. Following this presumption, the temperature in the assessed section can be written as: $T = f(r,t)$. This involves:

$$\frac{\partial^2 T}{\partial r^2} + \frac{1}{r}\frac{\partial T}{\partial r} = \frac{\rho c}{\lambda}\frac{\partial T}{\partial t} \tag{3}$$

This equation handles the thermal evolution inside the thickness of the pipe. The solution of the equation depends on the applied boundary conditions. As the varying load is the medium temperature flowing throughout the component, the heat transfer between the fluid and the inner surface is governed by a Newton's law of cooling:

$$\lambda\frac{\partial T}{\partial r}\bigg|_{r=ri} = h(T - T_\infty)_{r=ri} \tag{4}$$

h ... heat transfer coefficient [W/(m²·K)]

For further explanations of the mathematical background of the method see e.g. [12].

All the difficulty to apply this equation during unsteady fluid temperature states is due to the determination of the heat transfer coefficient h. Indeed, this parameter depends on the velocity, the thermo-hydraulics conditions and the geometry of the surface. The time dependent knowledge of all these parameters with sufficient accuracy is hardly compatible with a fast determination of the real loads. To solve this problem, the inverse philosophy was developed to calculate the stresses in the structure. Indeed, to perform a structural analysis, the knowledge of the temperature distribution throughout the wall is sufficient. The FFE method is based on the time history determination of the inner wall temperature by solving the inverse problem of conduction of heat. The according flowchart is shown in Figure 6.

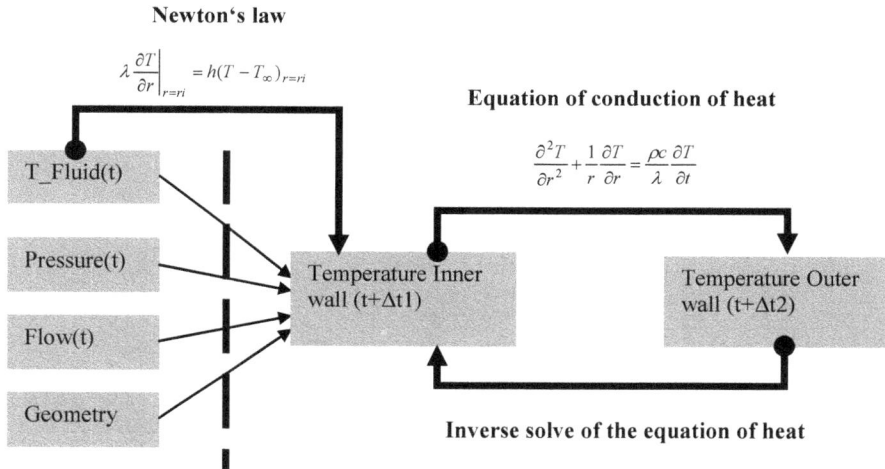

Newton's law

$$\lambda \left. \frac{\partial T}{\partial r} \right|_{r=ri} = h(T - T_\infty)_{r=ri}$$

Equation of conduction of heat

$$\frac{\partial^2 T}{\partial r^2} + \frac{1}{r}\frac{\partial T}{\partial r} = \frac{\rho c}{\lambda}\frac{\partial T}{\partial t}$$

T_Fluid(t)

Pressure(t)

Flow(t)

Geometry

Temperature Inner wall (t+Δt1)

Temperature Outer wall (t+Δt2)

Inverse solve of the equation of heat

Fig. 6. Different way to get inner wall temperature

Solving the inverse conduction equation of heat is done by application of potential functions (unit transients). A unit transient is applied at the inner surface of the pipe (boundary condition), the equation of conduction of heat is solved, the resulting time-history of temperature at the outer wall is observed. The resolution of the equation of heat can be done by means of an analytical method or with the help of a finite element program (ANSYS®). In that case a two-dimensional model of the section of the pipe is generated. The benefit of this last choice is the opportunity to integrate the thermal influences of the thermocouple installation at the outer surface of the pipe in the solution (see Figure 7).

The determined temperature response calculated in the thermocouple (outer wall) will be considered as a reference. Its evolution is characteristic by the applied unit transient at the inside surface of the pipe (characterized by a temperature rate of changes and a thermal amplitude ΔT_{ref}). Thus the FAMOS measured outside temperature will be scanned step by step (typically every second). The temperature difference at the outer wall between two time steps is compared with the simulated outer wall temperature (reference). The factor resulting from this comparison is through linearity properties also available at the inner side of the structure. Thus, step by step the inside temperature of the pipe can be restituted. A computation algorithm of this process was developed. The acquired measured data of FAMOS are read into the FFE program. A preparatory work consists of calculating, for the different observed piping sections, the thermal references. These last ones depend on the material, pipe thickness and measurement thermocouple. After this pre-processing work, the computation of the transient inner wall temperatures is completely automated.

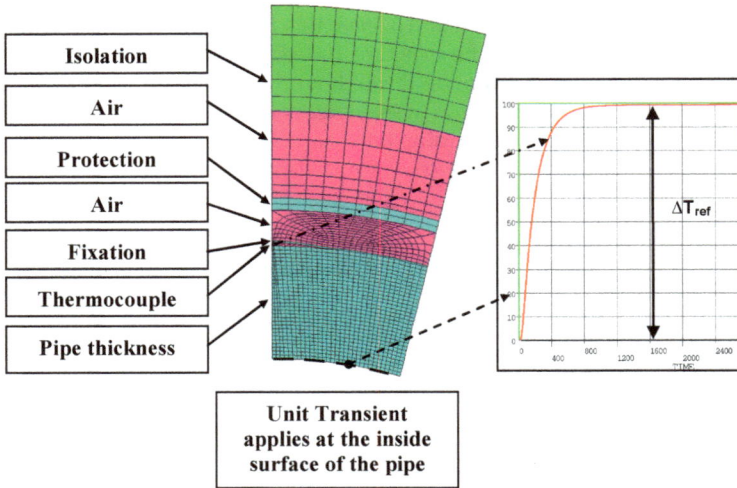

Fig. 7. FE calculation of the temperature response at the outside of the pipe

The determined inner wall temperature will be used to calculate the thermal stress at the fatigue relevant locations. An appropriate temperature transfer function can readily be used for correction of the axial dependency of the temperature if the FAMOS section is far away from the stress calculation locations $T(z - \delta z) \neq T(z + \delta z)$. The procedure is shown schematically in Figure 8.

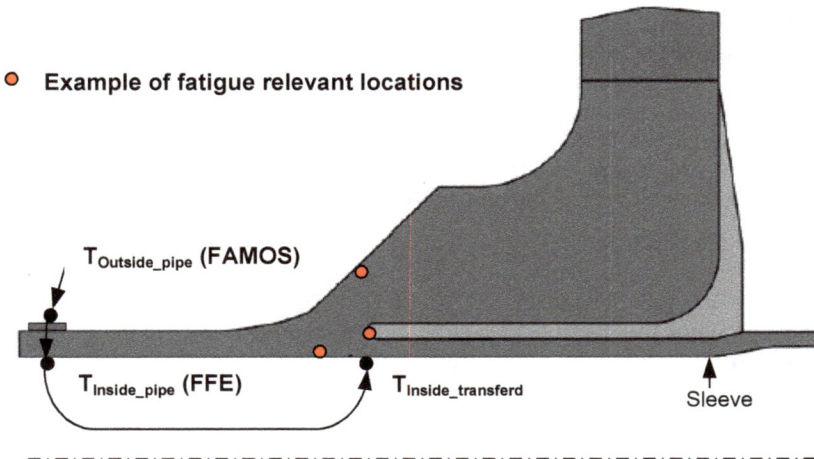

Fig. 8. Inner wall calculation and transfer of thermal loads to the fatigue relevant locations

The thermal stress determination is done according to a similar process as previously explained. A two- or three- dimensional finite element model of the monitored component is generated (nozzle, heat exchanger,...). A unit (elementary) transient is used as a reference load of a thermal calculation. Thus, the thermal field in the structure is calculated. Subsequently, the thermal stresses are calculated by a linearly elastic structural analysis.

The resulting thermal stresses are determined for typical fatigue relevant locations. The calculated stress components are the response to a reference load characterized by a temperature rate of change and a thermal amplitude ΔT_{ref}. The exemplary procedure is shown in Figure 9.

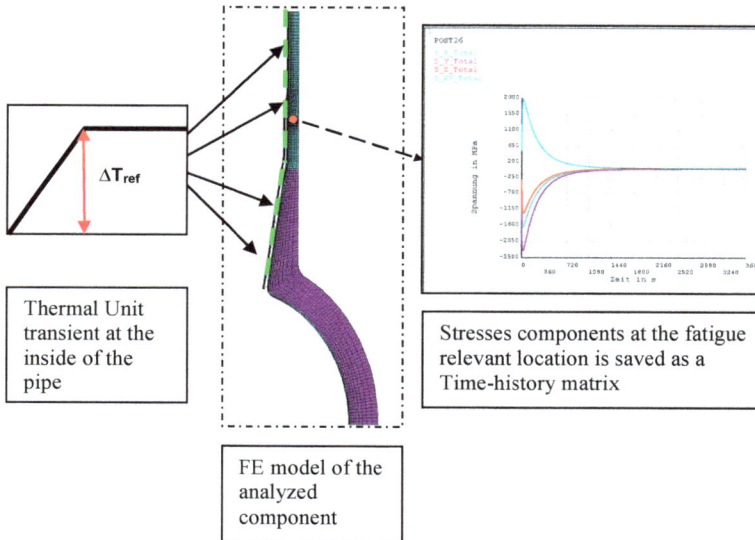

Thermal Unit transient at the inside of the pipe

Stresses components at the fatigue relevant location is saved as a Time-history matrix

FE model of the analyzed component

Fig. 9. FE calculation of the stress responses at a fatigue relevant location

The inside temperature calculated in the previous step by means of FFE, is scanned step by step. Between two time steps, the temperature difference is interpreted as a unit transient the amplitude of which is compared to the reference unit transient of amplitude ΔT_{ref}. Because of linearity in the thermal stress calculation, the comparison between the measured amplitude and the reference gives a coefficient to be applied to the reference stress matrix in order to obtain the stress contribution resulting from the thermal load at the calculated time. The time-dependent stress components are then obtained by the summation of all these single contributions. The process is also completely automated within the FFE program. The stress matrix references have to be calculated previously in an FE program. The results are then added to the database of FFE: it is the pre-processing work. Subsequently, the calculation at the selected locations can be processed. Within a few minutes, thermal loads and stress components of the entire operating cycles are calculated (see e.g. Figure 10).

If information on the time dependent pressure or piping section forces and moments are available based on operational instrumentation, the resulting mechanical stress components can be calculated equally by means of FFE (scaling of unit loads). Thermal and mechanical stress components are added and the equivalent stress is calculated. The use of a rain-flow algorithm will classify the stress ranges, a standard conform comparison with the fatigue curves will give the fatigue level of the selected locations.

Finally, if the calculated fatigue usage factor is lower than the allowable limit, the fatigue check will be successfully finished. If not, further analyses according to the detailed code based fatigue check will be performed.

In order to optimize the costs and user flexibility, the FFE program was based on a modular architecture. Thus, only information required by the customer/user is calculated. This architecture also permits an easy upgrade of the program to implement new modules e.g. as a consequence of changes of nuclear standards (new fatigue curves, environmental factor integration,…) or further calculation methods (automated stratification consideration).

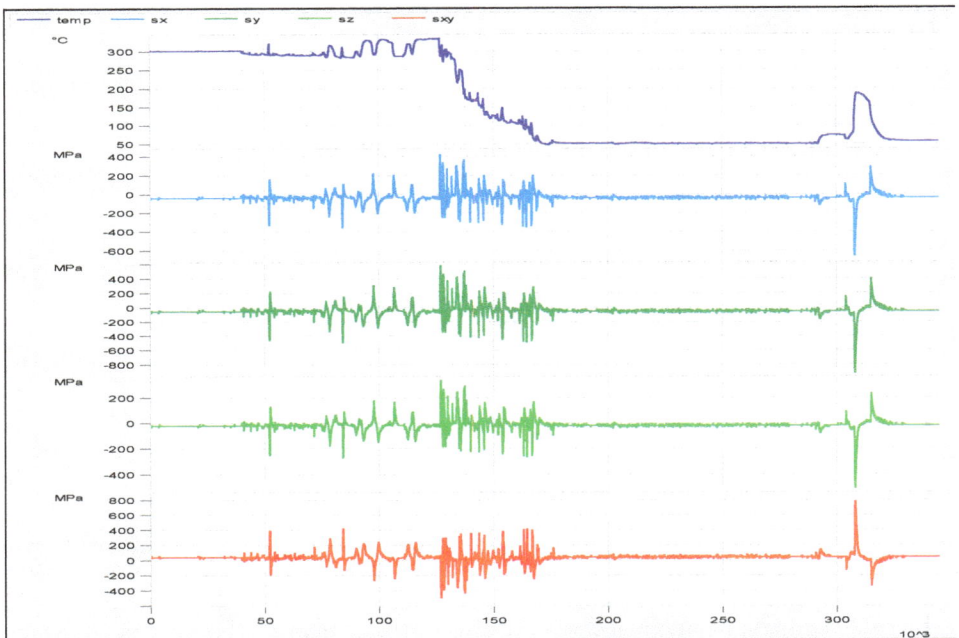

Fig. 10. FFE temperature and thermal stresses calculation for shut down event

6. Detailed fatigue calculation

6.1 General remarks and context

The detailed fatigue calculation (DFC) is usually carried out after a certain time period of plant operation, every ten years for instance. These analyses are often performed in the framework of the periodic safety inspection (PSI). Loading data of the operational period as well as anticipated loads of future operation are used as essential input parameters. Hence, usage factors are calculated for the current state of the plant and some prognoses are taken into account to get results until the end of life.

The simplified elasto-plastic fatigue analysis based on elastic FE analyses and plasticity correction (fatigue penalty or strain concentration factors K_e) e.g. according to paragraph 7.8.4 of [4] or equally NB 3228.5 of [1] is known to yield often overly conservative results.

In the practical application this may yield high calculated usage factors. As a consequence, the less conservative elasto-plastic fatigue analysis method based on non-linear FE analyses will often be used for fatigue design. This is associated with an increased calculation effort. Computing times for complex geometries and numerous transients may be significant. Under these circumstances the specified transients have to be rearranged in a small set of covering transients, approximately ten, for calculation purposes.

The possible modification of design codes in respect of more severe fatigue curves and particularly the consideration of EAF will significantly influence the code based fatigue design. Of course, these developments are attentively followed and actively accompanied; see "supporting functions" in Figure 1.

The usual workflow of the fatigue analysis of NPP components is shown in Figure 11. The structural analysis might be simplified elasto-plastic or fully elasto-plastic. The transient temperature fields are analyzed for all relevant N model transients according to Figure 11. These transient temperature fields are themselves the input data for the subsequent transient (linear or non-linear) structural mechanical analyses yielding the local stresses and strains required for code-conforming fatigue assessment. Cycle counting is done in accordance with the requirements of the ASME code as implemented in the ANSYS® Classic Post 1 Fatigue module. It is explained in more detail in the following section.

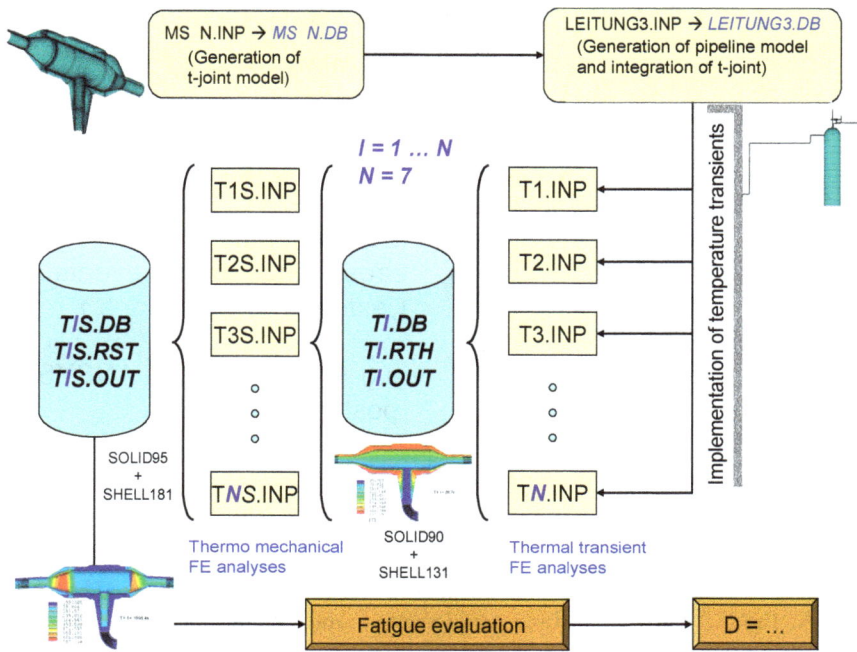

Fig. 11. Workflow of detailed fatigue analyses

6.2 Cycle counting

Cycle counting is the prerequisite for any fatigue or service durability assessment method dealing with arbitrary operational load sequences. Consequently, an appropriate cycle counting algorithm is required.

Cycle counting methods in general are characterized by the following features [10]:

- decomposition of a given course of load (stress) – time history into a sequence of reversal points
- definition of a relevant elementary event (e.g. hysteresis)
- formulation of an algorithm for the detection and processing of elementary events.

The superposition of transients according to the design code (ASME Code, NB 3222.4, see Figure 12) is based on the peaks and valleys method. The largest stress ranges are usually determined from "outer combinations" (e.g. load steps across different transients respectively events). The associated frequency of occurrence results from the actual number of cycles of the participating two events with the smaller number of cycles. This event provides the associated contribution to the partial usage factor U_i. The summing up of all partial usage factors according to Miner's rule delivers the accumulated damage (usage factor U) or cumulated usage factor CUF.

Extreme value method

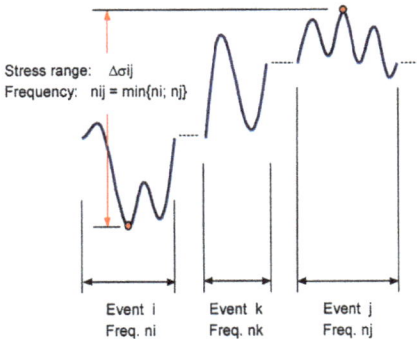

Stress range: $\Delta\sigma_{ij}$
Frequency: $n_{ij} = \min\{n_i;\ n_j\}$

Event i
Freq. ni

Event k
Freq. nk

Event j
Freq. nj

2007 SECTION III, DIVISION 1 — NB

NB-3222.4 Analysis for Cyclic Operation

(5) Cumulative Damage.

- Usual counting method according to ASME-Code

- Search for the largest stress range across events ("external combinations")

- e.g. implemented in ANSYS®- postprocessing

Fig. 12. Cycle-counting method according to [1]

Additionally, a counting of sub cycles within the events should be carried out according to the rain-flow cycle-counting method [e.g. 10] although it is not explicitly addressed by the design code [1]. This is standard practice in the framework of the AFC. The Hysteresis Counting Method (HCM) according to Clormann and Seeger [10] is applied for this purpose. Additionally, the introduction of so called basic events allows a more realistic consideration of the load time sequence [13].

6.3 Application example

As an application example, the spray line of a PWR system is subject of a detailed code-based fatigue analysis. The load input covers the temperature transients measured during operation. Figure 13 gives an example of a specified temperature transient. This specification of the plant-specific thermo-hydraulic loads in official reports and transient handbooks is realistic, but still conservative. It is obvious that the load specification takes a decisive influence on the fatigue usage calculation and is a source of conservatism.

In the subsequent calculation process all relevant loads and relevant components have to be considered. Furthermore the interaction between the components and the adjacent piping system (supplementary loads resulting from the deformation of the piping sections) cannot be neglected. Based on this requirement, a decision was taken to model the complete spray line system by means of shell type elements. This model containing the spray line, the auxiliary spray line, and the pressurizer is shown in Figure 14. It allows for the identification of realistic transient piping loads on the fatigue relevant components which are modeled in detail based on brick type elements.

The spray line nozzle, the spool, and the t-section (see Figure 15) were identified as fatigue-relevant components within the spray line system. In a next step, detailed brick type FE models of these three components were generated and integrated into the overall shell type model of the spray line. It is pointed out that the detailed models were considered subsequently and not simultaneously due to model and file sizes as well as analysis time. Some model statistics are given in Figure 16 for the example of the t-section.

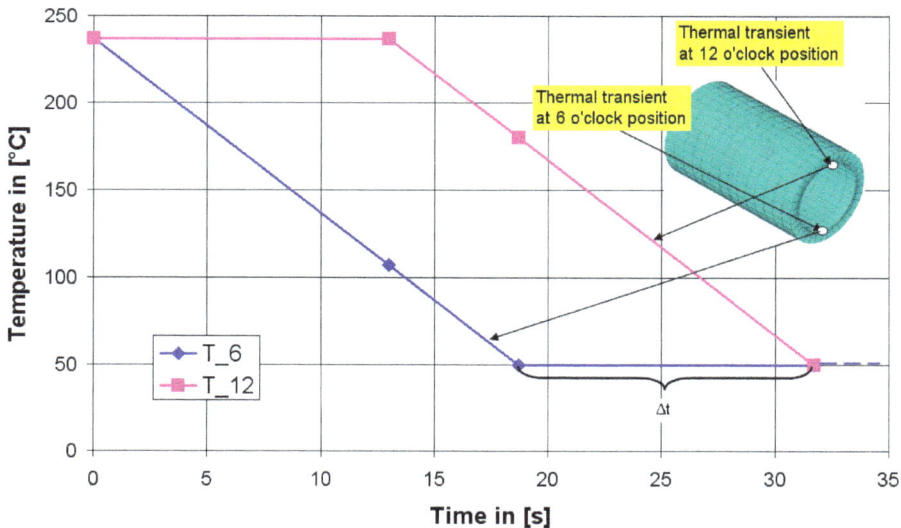

Fig. 13. Example of a specified temperature transient

The connection between the shell type and the brick type part of the complete model is achieved by constraint equations (CE) in the case of the transient temperature field analyses and by means of multiple point constraints (MPC) in the case of the structural mechanical analyses. Nearly one million nodes yield considerable computing times in the transient nonlinear analyses. The complete workflow of the fatigue analysis is shown in Figure 11. A simplified elasto-plastic fatigue analysis was not applicable in this application example. All transient analyses were based on a nonlinear material law with kinematic hardening.

The transient temperature fields were analyzed for all relevant N model transients according to Figure 11. These transient temperature fields were the input data for the subsequent transient nonlinear structural mechanical analyses yielding the local strains required for code-conforming fatigue assessment. Note that the code-conforming damage accumulation algorithm is not trivial in the case of elasto-plastic analyses. More details on the implementation can be found in [14]. Cycle counting was done in accordance with the requirements of the ASME code as implemented in the ANSYS® Classic Post 1 Fatigue module.

An example of the temperature distribution in the t-joint is shown in Figure 17. It represents one point of time of one model transient. Note that the temperature distribution remains continuous at the border between the solid type and the shell type part of the model. The resulting von-Mises stress distribution for the same point of time is shown qualitatively in Figure 18. Again, the stress distribution remains continuous at the border between the solid type and the shell type part of the model. The MPC approach works very well. It is clearly shown that the thick walled transition regions and thermal sleeve connections are particularly prone to fatigue damage.

Consequently, the fatigue usage analysis revealed the thick walled transition region to be the most relevant location for the fatigue check. Only the elasto-plastic fatigue analysis assured fatigue usage below the admissible value of 1.0.

For more details on the code-conforming fatigue check and associated further developed methods see e.g [15].

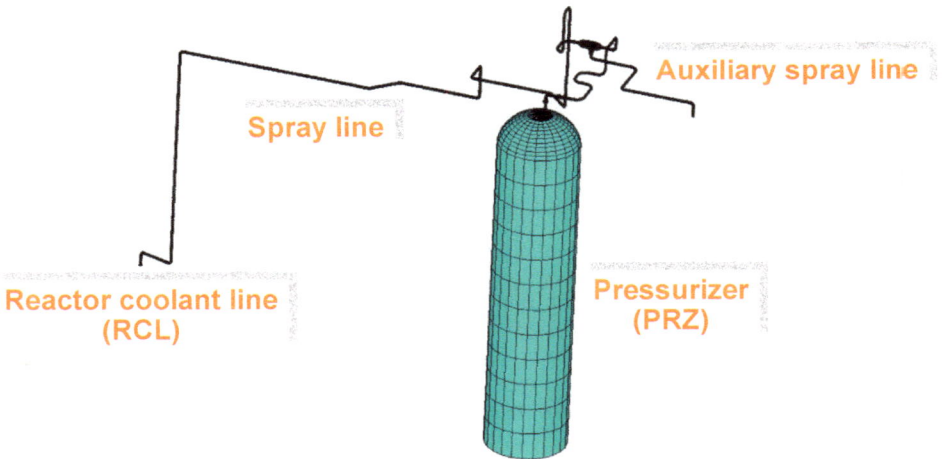

Fig. 14. FE model of spray line and pressurizer (shell type elements)

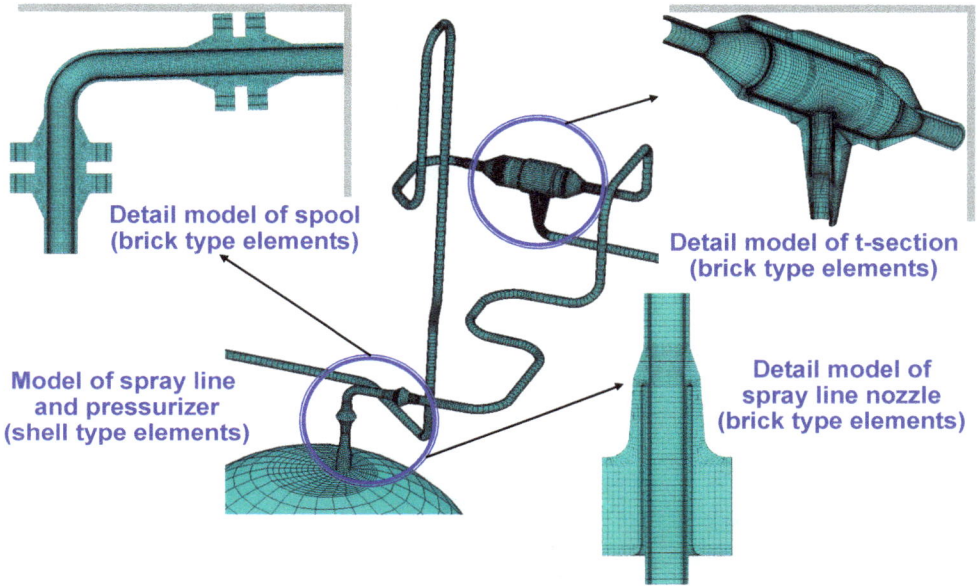

Fig. 15. Detail models of spray line nozzle, spool and t-section

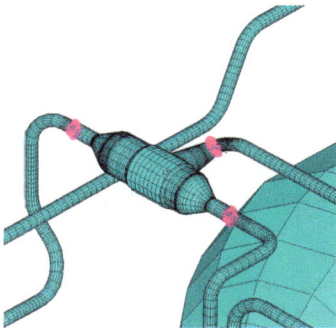

Transient temperature field analyses:
• *SHELL131 + SOLID90*
• *Constraint equations (CE)*

Structural mechanical analyses:
• *SHELL181 + SOLID95*
• *„Multiple Point Constraints" (MPC)*

Model statistics:
• *912281 nodes, 243120 elements*
• *9 supports*
• *Pipeline (SHELL):*
     35700 nodes, 35568 elements
• *T-joint (SOLID):*
     877331 nodes, 207552 elements

Fig. 16. Model statistics for the t-section

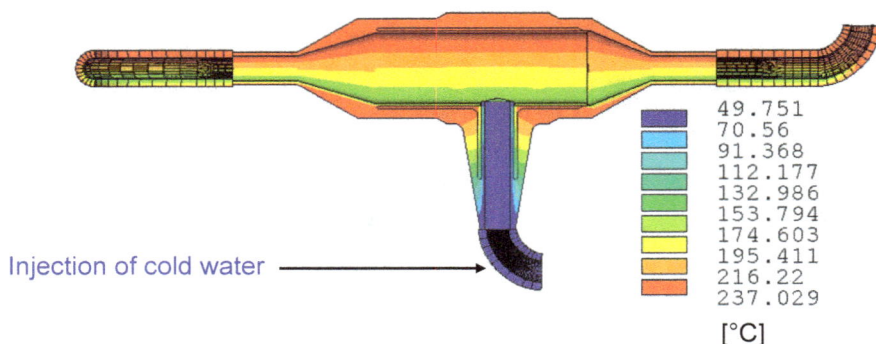

Injection of cold water

```
49.751
70.56
91.368
112.177
132.986
153.794
174.603
195.411
216.22
237.029
```
[°C]

Fig. 17. Exemplary temperature distribution in the t-joint

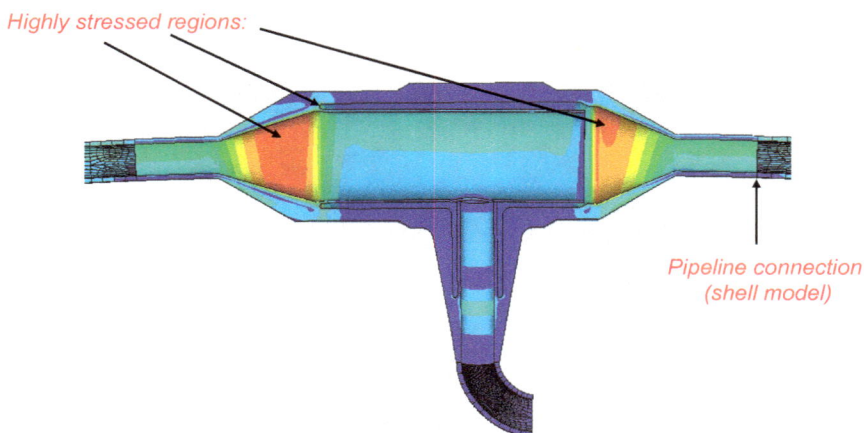

Highly stressed regions:

Pipeline connection
(shell model)

Fig. 18. Exemplary von-Mises stress distribution in t-section due to thermal loading

7. Conclusions

The AREVA integrated and sustainable concept of fatigue design expresses the importance of design against fatigue in NPPs. Actually, new plants with scheduled operating periods of 60 years, lifetime extension, the modification of the code based approaches and the improvement of operational availability are driving forces in this process. Therefore, applying the AFC is an expression of responsibility sense, as well as an economic requirement. Moreover, the fatigue concept is widely supported by measured data. Indeed, the results of the fatigue monitoring can be the basis for decisions of optimized operating modes and thus influence the fatigue usage factors.

The main modules are FAMOS, the first design analysis before operation, and the three stages of fatigue data evaluation. As all modules are closely connected, it is reasonable to apply the approach as a whole, with an additional cost reduction effect, compared to separate solutions. An integrated software ensures the effective data processing from measurement to fatigue data calculation and offers the user an easy to use interface to NPP's loading data. Thus, the integrated fatigue approach makes a significant contribution to the safety margins monitoring, the operational availability and the protection of investment.

8. Acknowledgment

The authors wish to express special thanks to all contributors to the AFC within AREVA.

9. References

[1] ASME Boiler and Pressure Vessel Code, 2007 Addendum 2009b Section III, Division 1 – Subsection NB: Class 1 Components. "Rules for Construction of NPP Components"

[2] Chopra, O. K.; Shack, W. J.: "Effects of LWR Coolant Environments on the Fatigue Life of Reactor Material" NUREG/CR-6909 ANL-06/08, Argonne National Laboratory for U.S. Nuclear Regulatory Commission

[3] KTA rules (Safety Standards of the Nuclear Safety Standards Commission). KTA 3201.4 (06/96): Components of the Reactor Coolant Pressure Boundary of Light Water Reactors. Part 4: In-service Inspections and Operational Monitoring. (Komponenten des Primärkreises von Leichtwasserreaktoren; Teil 4: Wiederkehrende Prüfungen und Betriebsüberwachung)

[4] KTA rules (Safety Standards of the Nuclear Safety Standards Commission). KTA 3201.2 (06/96), 1996: Components of the Reactor Coolant Pressure Boundary of Light Water Reactors. Part 2: Design and Analysis (Komponenten des Primärkreises von Leichtwasserreaktoren; Teil 2: Auslegung, Konstruktion und Berechnung)

[5] RCC-M, Edition 2007: "Design and Construction Rules for Mechanical Components of PWR Nuclear Islands"Section I, Subsection B: Class 1 components.

[6] Kleinöder, W.; Poeckl, C.: "Developing and implementation of a fatigue monitoring system for the new European pressurized water reactor EPR"Proceedings of the International Conference "Nuclear Energy for New Europe 2007", September 10-13, 2007, Portoroz, Slovenia

[7] Miksch, M.; Schön, G.; Thomas, B.: "FAMOS – a tool for transient recording and fatigue monitoring"PVP-Vol. 138/NDE-Vol. 4 "Life extension and assessment: nuclear and fossil power plant components", presented at the 1988 ASME Pressure Vessels and Piping Conference, Pittsburgh, Pennsylvania, June 19-23, 1988

[8] Abib, E.; Bergholz, S.; Pöckl, C.; Rudolph, J.; Bergholz, S.; Wirtz, N.: "AREVA Fatigue Concept (AFC) – an integrated and multi-disciplinary approach to the fatigue assessment of NPP components."Proceedings of ICAPP 2011, Nice, France, May 2-5, 2011, Paper 11027

[9] Matsuishi, M.; Endo, T.: "Fatigue of metals subjected to varying stresses"Proceedings of the Kyushu branch of the Japanese Society of Mechanical Engineers, pp 37/40; March 1968

[10] Clormann, U.H.; Seeger, T.: "Rainflow – HCM, ein Zählverfahren für Betriebsfestigkeitsnachweise auf werkstoffmechanischer Grundlage". Stahlbau 55 (1986), Nr. 3, S. 65/71

[11] Heinz, B., 2010: "AREVA Fatigue Concept – a new method for fast fatigue evaluation"PVP2010-25935; Proceedings of the ASME 2010 Pressure Vessels & Piping Division Conference; July 18-22, 2010, Bellevue, Washington, USA

[12] Carslaw H.S-; Jaeger, J.C.: „Conduction of heat in solids"Oxford University Press, 2nd edition. 1959

[13] Lang, H.: "Fatigue – Determination of a more realistic usage factor" Nuclear Engineering and Design 206 (2001), S. 221/234

[14] Rudolph, J.; Götz, A.; Hilpert, R.: "Code-conforming determination of cumulative usage factors (CUF) for general elasto-plastic finite element analyses"Proceedings of ANSYS Conference & 29th CADFEM Users' Meeting 2011, October 19-21, 2011 Stuttgart, Germany

[15] Rudolph, J.; Bergholz, S.; Willuweit, A.; Vormwald, M.; Bauerbach, K.: "Methods of detailed thermal fatigue evaluation of NPP components". Proceedings of SoSDiD 2011, May 26th-27th, 2011 Darmstadt, Germany

12

Phase Composition Study of
Corrosion Products at NPP

V. Slugen, J. Lipka, J. Dekan, J. Degmova and I. Toth
Institute of Nuclear and Physical Engineering
Slovak University of Technology Bratislava, Bratislava
Slovakia

1. Introduction

Corrosion at nuclear power plants (NPP) is a problem which is expected. If it is managed properly during the whole NPP lifetime, consequences of corrosion processes are not dramatic. For adequate protection against corrosion it is important to collect all relevant parameters including exact phase composition of registered corrosion products.

Corrosion is more frequent and stronger in secondary circuit of NPP. Steam generator (SG) is generally one of the most important components from the corrosion point of view at all NPP with close impact to safe and long-term operation. Various designs were developed at different NPPs during last 50 years. Wide type of steels was used in respect of specific operational conditions and expected corrosion processes. In our study we were focused on the Russian water cooled and water moderated reactors (VVER). These reactors are unique because of horizontal position of SGs. It takes several advantages (large amount of cooling water in case of loss of coolant accident, good accessibility, large heat exchange surface, etc. ...) but also some disadvantages, which are important to take into account during the operation and maintenance. Material degradation and corrosion/erosion processes are serious risks for long-term reliable operation. In the period of about 10-15 year ago, the feed water pipelines were changed at all SG in all 4 Bohunice units (V-1 and V-2, in total at 24 SGs). Also, a new design of this pipeline system was performed. Actually, there is a time to evaluate the benefit of these changes.

The variability of the properties and the composition of the corrosion products of the stainless Cr-Ni and mild steels in dependence on the NPP operating conditions (temperature, acidity, etc.) is of such range that, in practice, it is impossible to determine the properties of the corrosion products for an actual case from the theoretical data only. Since the decontamination processes for the materials of the VVER-440 secondary circuits are in the progress of development, it is necessary to draw the needed information by the measurement and analysis of the real specimens [1].

2. Mössbauer spectroscopy advantages

The phenomenon of the emission and absorption of a γ-ray photon without energy losses due to recoil of the nucleus and without thermal broadening is known as the Mössbauer effect. Its unique feature is in the production of monochromatic electromagnetic radiation with a very narrowly defined energy spectrum that allows resolving minute energy differences [2,3].

Mössbauer spectroscopy (MS) is a powerful analytical technique because of its specificity for one single element and because of its extremely high sensitivity to changes in the atomic configuration in the near vicinity of the probe isotopes (in this case ^{57}Fe). MS measures hyperfine interactions and these provide valuable and often unique information about the magnetic and electronic state of the iron samples, their chemical bonding to co-ordinating ligands, the local crystal symmetry at the iron sites, structural defects, lattice-dynamical properties, elastic stresses, etc. [1,4]. Hyperfine interactions include the electric monopole interaction, i.e., the isomer shift, the electric quadrupole interaction, i.e., the quadrupole splitting, and the magnetic dipole (or nuclear Zeeman) interaction, i.e., hyperfine magnetic splitting. These interactions often enable us detailed insight into the structural and magnetic environment of the Mössbauer isotope. Indeed, more than four decades after its discovery (1958), Mössbauer spectroscopy still continues to develop as a sophistical scientific technique and it is often the most effective way of characterizing the range of structures, phases, and metastable states.

In general, a Mössbauer spectrum shows different components if the probe atoms are located at lattice positions, which are chemically or crystalographically unequivalent. From the parameters that characterise a particular Mössbauer sub-spectrum it can, for instance, be established whether the corresponding probe atoms reside in sites which are not affected by structural lattice defects, or whether they are located at defect-correlated positions. Each compound or phase, which contains iron, has typical parameters of its Mössabuer spectrum. It means, the method is suitable for quantitative as well as qualitative analysis. Mössbauer spectroscopy is non-destructive and requires relative small quantities of samples (~100 mg) [5-8].

Application of Mössbauer spectroscopy for precise analysis of phase composition of corrosion products was performed from selected areas of primary and secondary circuit and SG. Interpretation of measured results, having in vision the long-term operation and nuclear safety, is not easy, nor straightforward. Thanks to our more than 25 years of experiences in this area, there exists already a base for the relevant evaluation of results. Optimisation of operating chemical regimes as well as regimes at decontamination and passivation seems to be an excellent output.

3. Safety analyses of Slovak steam generators and latest upgrades

The safe and reliable operation of steam generators is the essential pre-condition for the safe operation of the whole NPP, but also for all economical parameters at the unit. The steam generator has to be able to transfer the heat from the reactor in all operating or accidental regimes.

It is well-known that VVER-440 units have the horizontal steam generators with much higher capacity of cooling water in tank than in vertical steam generators, which are normally used in western NPPs. Undoubtedly; these horizontal steam generators are safer. On the other hand, due to horizontal design as well as in total 6 loops with additional welds and pumps cover a huge area, which is a limiting factor for containment construction of the unit. The exchange of steam generators is extremely difficult (in some reactor types almost impossible), therefore their optimal operation and clever maintenance (upgrades) is one of essential duty of NPP staff.

Based on operational experiences, the mitigation of damages and leak tightness defects in pipelines or collectors require much more time and money, than prevention measures. It is necessary to keep in mind the actual development in nuclear industry towards NPP lifetime prolongation and power increase (one of essential goals of European Commission 7FP-NULIFE). Fortunately, VVER-440 steam generators were designed with the huge power reserve (possible overloading of 20%). Beside several leakages in primary pipelines (ø16 mm), which can be in case of VVER-440 SGs relatively easy solved (blended), the corrosion deposits in feed water pipeline system occurred at many VVER-440 units [9]. The identified damages were caused mostly due to corrosion/erosion processes attacking materials familiarly called "black steels" with insufficient resistance against corrosion.

Based on experience from Finland, also other countries including Slovakia changed the old feed water pipeline system. At this moment we would like to mention that the incident at the 2nd unit of NPP Paks (Hungary, 2003) connected to cleaning of fuel assemblies in special tank had the root causes in insufficient passivation of pipelines in SG after the steam water pipeline system exchange in 1997.

All steam generators at four VVER-440 units in Bohunice were gradually changed. At that time, there were two possibilities for the new feed water pipeline system. Out from two conceptions: Vítkovice a.s. design and OKB Gidropress design. The first solution was selected and lately improved to so called "Bohunice solution". Actually, experience from the last 10 years after upgrade was utilized.

A detailed description of VVER-440 steam generators delivered to NPP Bohunice is in [10-12]. The safety analyses were performed in 1977 by OKB Gidropress according to the Russian norms. The Russian designer and producer made the feed water pipelines (secondary side) from the carbon steel (GOST norm 20K and 22K). Water inlet pipeline was connected to the T-junction. From this point, 2 lines of the pipe with nozzles distributed the cooling secondary water in the space between primary pipes. Several problems having occurred in other NPPs were published in [13-16]. A disadvantage of such steam generators are difficult accessibility to the T-junction and next pipelines in the bundle.

3.1 Design changes at VVER-440 steam generators

The steam generator with technical mark RGV-4E is one body SG [10-12]. The heat-exchange area is incorporated inside as surface of primary pipelines bundle with U-shape. The ends of these pipelines are fixed to the walls of the primary collector. Inside of SG body several separators and system of the steam water distribution are placed. The PGV-4E steam generator is foreseen for dry steam production with the pressure of about 4,61 MPa at a temperature of about 256°C.

The basic 1977 design from was improved after 1994 by new feed water pipeline system. There was also change in the type of steel of these pipelines. Instead conventional carbon steel, the austenite steel was used in distribution boxes as well as feed water pipelines.

All components in the Bohunice innovated feed water pipeline system were made of austenitic steel according to the Czechoslovak norm ČSN, class 17. Advantages of the new construction are not only higher resistance against corrosion, but also much more comfortable visual inspection. The innovations can be seen in Figs. 1, 2.

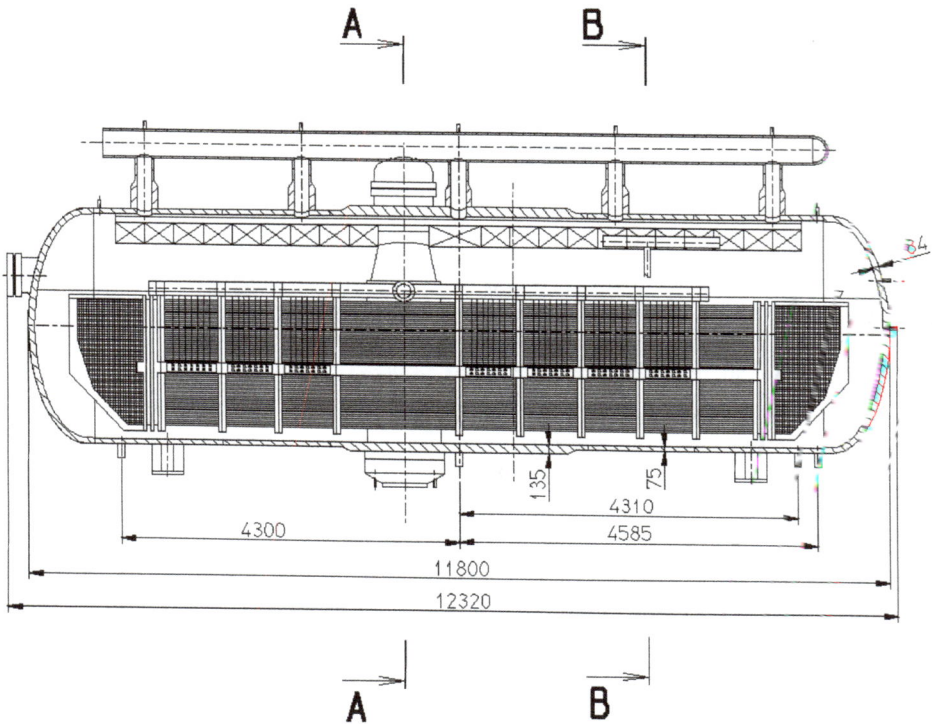

Fig. 1. VVER-440 (Bohunice) steam generator – cross section. NPP Bohunice innovation

The feed water comes via nozzle to distribution pipeline system and gets inside to left and right incoming line. From this place, water flows via pipelines ϕ 44,5×4mm into chambers and gets out via ejectors. This flow is mixed together with boiler water, so the final flow on the small primary pipelines is not extremely hot and does not cause a disturbing thermal load. Simultaneously, the circulation in SG tank was improved and places with increased salt concentration are reduced.

The main advantage is that the visual inspection of the feed water pipeline can be performed immediately due to placement of the whole water distribution system over the primary pipelines bundle. Using this system, the possible defects are easier observable. The next advantage is connected to 7 boxes with ejectors which mix properly the feed water with boiler water and the thermal load decreases in this way. An additional advantage is the checking and exchange possibility of distribution boxes in case of their damage.

A-A

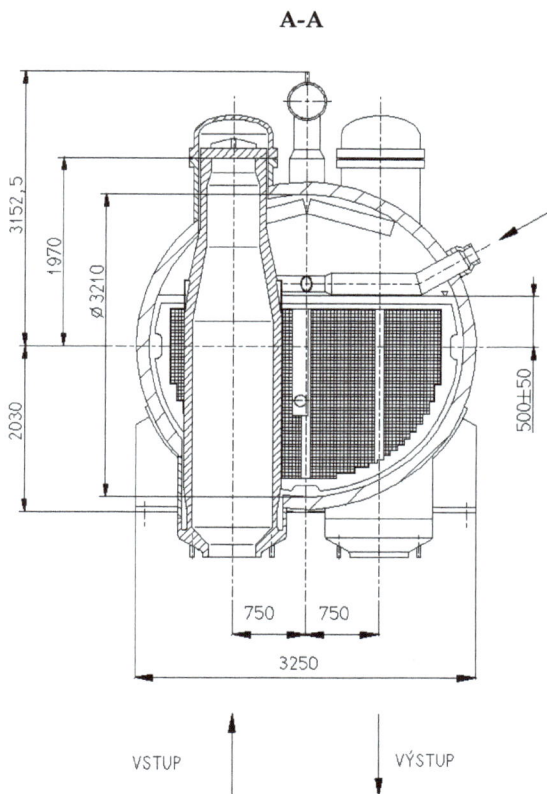

Fig. 2. VVER-440 (Bohunice) steam generator – cross section A-A, NPP Bohunice innovation

4. Experimental

For the experimental measurements, several specimens containing corrosion products were taken from different parts of all of 4 NPP Bohunice units. In the first step, corrosion process at the steam generators was studied. The corrosion layers were separated by scraping the rust off the surface and the powder samples were studied by transmission Mössbauer spectroscopy. It should be noted that the gamma spectroscopic measurements gave no evidence of the presence of low-energy gamma radiation emitted from the samples. Later, the corrosion products were collected also from different parts of secondary circuit components and several filter deposits were analysed as well.

The room temperature Mössabuer study was performed on two different steam generator materials using conventional transmission Mössbauer spectrometer with the source ^{57}Co in Rh matrix. The spectra were fitted using NORMOS program.

The original STN 12022 material used at the 4th (SG46) over 13 years was compared to STN 17247 steel used at the 3rd unit (SG35) for about 5 years (1994-1998). The chemical compositions of both materials are shown in Table 1.

Steam-generator	Type of the steel	Chemical composition [weight %]							
		C	Mn	Si	Cr	Ni	Ti	P	Cu
SG35	STN 17247	max. 0,08	max. 0,08	max. 1,0	17,0 - 19,0	9,5 - 12,0	Min. 5x%C	max. 0,045	-
SG46	STN 12022 (GOST 20K)	0,16- 0,24	0,35- 0,65	0,15-0,30	max. 0,25	max. 0,25	-	0,04	Max. 0,3

Table 1. Chemical composition of investigated base material

Samples of corrosion products scrapped from different parts of the steam generators SG 35 and SG46 were analysed. The scrapped corrosion particles were homogenised by granulation and sieved through a of 50μm wire sieve.

5. Results from Mössbauer spectroscopy analyses

The advanced evaluation of phase analyses of corrosion products from different parts of VVER-440 steam generators via Mössbauer spectroscopy is our active and unique contribution in this area. The scientific works go over 25 years. The first period (mostly 80-ties) was important for improving Mössbauer technique. The benefit from this period is mostly in experience collection, optimization of measurement condition and evaluation programs improvement [5]. Unfortunately, not all specimens were well defined. Having in mind also different level of technique and evaluation procedures, it would be not serious to compare results from that period with the results obtained from measurement after 1998.

5.1 Comparison of the corrosion products before and after SG design changes

In the period 1994-1999 we focused our study on the comparison of the phase composition of corrosion products taken from the NPP Bohunice before and after changes in the feed water pipeline system.

Schematic drawings of VVER steam generators (SG) with indicated places of scrapped corrosion specimens are presented in Fig.3.

Serious damages were observed in the region of T-junction (position 4 in Fig.3) as well as of pipe-collector and outlet nozzles on many VVER440 SGs after approximately ten years of operation [9,17]. Therefore, the former feed-water distributing system has been replaced by an advanced feed-water distributing system of EBO design at SGs of NPP Jaslovske Bohunice [18, 19]. The advanced system consists of a V-shaped junction connected to the left – and the right part water distributing chambers both located above the tube bundle and few feed water boxes with water ejectors inserted into the tube bundle and connected to the distributing chamber by distributing pipelines.

After five year´s operation in the SG No. 35 in the NPP outage one feed water box and corresponding distributing pipelines were replaced by new ones with the aim to analyse their overall stage and corrosion products on walls. For comparison, some parts of the former feed-water-distributing system from the SG Number 46 were cut out and analysed.

More than 50 specimens were collected from the NPP Bohunice secondary circuit in 1998-2000. The investigation was focused mainly on the corrosion process going on in steam generators SG35 with new design and SG46 with old design. Nevertheless, additional measurements performed on the corrosion products from SG31 and SG32 confirmed that corrosion process in all 6 steam generators of one reactor unit is the same and corrosion layers are on the some places altogether identical.

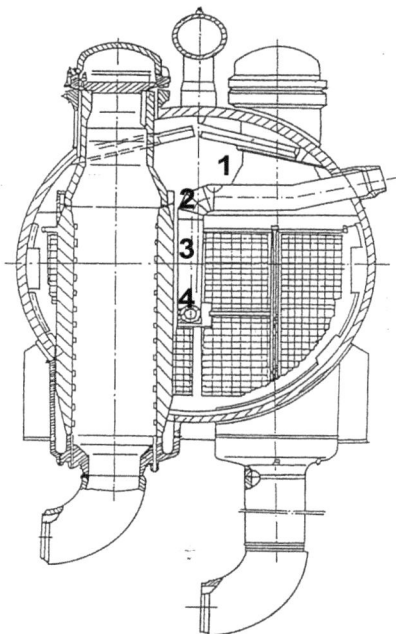

Fig. 3. Cross section of SG46 (Numbers indicate the places, where the specimens were scrapped)

All measured specimens contain iron in magnetic and many times also in paramagnetic phases. Magnetic phases consist in form of nearly stoichiometric magnetite (γ-Fe$_3$O$_4$), hematite (α-Fe$_2$O$_3$), and in some case also iron carbides. The paramagnetic fractions are presented in Mössbauer spectra by a doublet and a singlet. Its parameters are close to hydro-oxide (FeOOH) parameters or to parameters of small, so called superparamagnetic particles of iron oxides (hydrooxides) with the mean diameter of about 10 nm (see Table 2 and Table 3).

MS confirmed its excellent ability to identify steel samples phase composition although its sawdust form and relative small amount (~ 100 mg). Our experiences with such measurements applied on different VVER-440 construction materials were published in [15, 20-21]. From other works it is possible to mention [22]. MS confirmed an austenitic structure of STN 17247 steel and ferrite structure of STN 12022 steel. Differences between these two materials are well observable (see Table 2 and Fig. 4 and Fig. 5). According to the *in-situ* visual inspections performed at SG35 (1998) and SG46 (1999) as well as MS results, significant differences in corrosion layers and material quality were observed. The feed water tubes in SG46 were significantly corroded after 13 years operation.

Sextets						Singlet		
Sample	H_1	A_{rel}	H_2	A_{rel}	H_3	A_{rel}	IS	A_{rel}	Fig.
(sawdust)	(T)	(%)	(T)	(%)	(T)	(%)	(mm/s)	(%)	
SG35	29.3	18.7	26.3	18.5	22.8	27.4	-0.12	35.4	4
SG46	33.3	80.5	31.2	19.5		-	-	-	5
Accuracy	±0,1	±0,5	±0,1	±0,5	±0,1	±0,5	±0,04	±0,04	

Table 2. MS parameters of the steam generators base material

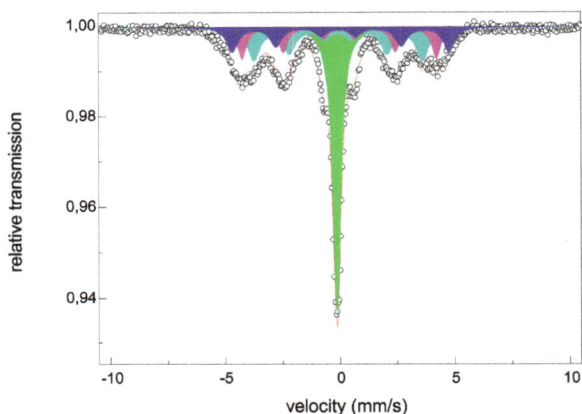

Fig. 4. MS spectrum of SG35 base material

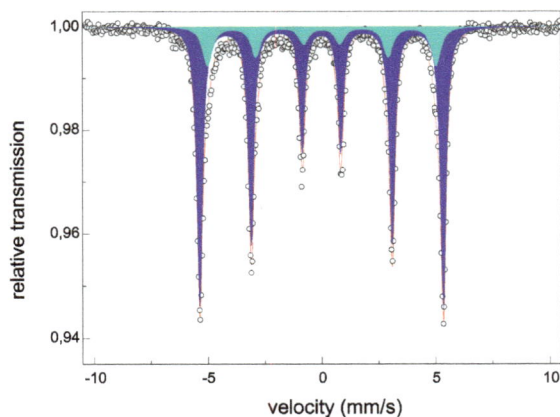

Fig. 5. MS spectrum of SG46 base material

Our results confirmed that during operation time a faint oxidation surroundings was in the observed steam generator SG35 after 5 years of operation time and the corrosion samples were fully without base material particles.

Magnetite was identified as dominant component in all studied samples (see Table 3). Mössbauer spectrum of the steam generators (both SG35 and SG46) surface layer is the superposition of two sextets with hyperfine magnetic field H_{efA} = 49,4T and H_{efB} = 45.8T. Sextet H_{efA} corresponds to the Fe^{3+} ions in tetrahedral (A) sites and sextet H_{efB} corresponds to Fe^{2+} and Fe^{3+} ions in octahedral (B) sites in magnetite spinel structure (Fe_3O_4).

Sample	Magnetite					Doublet			Singlet	
	H_A	A_{rel}	H_B	A_{rel}	r_{AB}	IS	QS	A_{rel}	IS	A_{rel}
	(T)	(%)	(T)	(%)	(a.u.)	(mm/s)	(mm/s)	(%)	(mm/s)	(%)
L754	49.0	35.0	45.9	65.0	0.538					
L755	49.0	35.3	45.9	64.7	0.546					
L757	49.0	34.8	45.9	65.2	0.534					
L758	49.0	34.1	45.9	62.3	0.547				-0.20	3.6
L789	49.0	34.4	46.0	65.6	0.535					
L790	49.0	34.9	46.0	64.3	0.543				-0.18	0.8
L759	49.0	35.4	45.9	63.9	0.534				-0.20	0.7
L777	49.0	35.0	45.9	65.0	0.538					
L786	49.0	35.2	46.0	64.8	0.545					
L787	49.1	36.5	46.0	56.1	0.651	0.22	0.67	2.0	-0.20	5.4
L760	49.0	34.1	45.9	64.3	0.530				-0.17	1.6
L761	49.0	35.0	45.9	63.9	0.547				-0.23	1.1
L762	49.1	34.8	46.0	56.4	0.617				-0.20	8.8
L779	49.0	33.4	45.9	62.9	0.531				0.10	3.7
Accuracy	±0,1	±0,5	±0,1	±0,5		±0,04	±0,04	±0,5	±0,04	±0,5

Table 3. MS parameters of corrosion products taken from the steam generator SG35[1]

In contrast to magnetite, whose spectrum is characterised by two sextets, the hematite phase present in the powders gives a single sextet. The relatively narrow line width (Γ) of the α-Fe_2O_3 (mainly 0,24 ÷ 0,26 mm/s) indicates presence of a well-crystallised phase with few, if any, substitutions of other elements for Fe. However, in some spectra (mainly from filter deposits studied later), both the lower hyperfine field and the larger width (about 0.33 – 0.34 mm/s) could indicate a poorer crystallinity and/or a higher degree of substitution. These findings are in good agreement with those obtained by E. De Grave [23]. Similar inspirative results (focused also on corrosion products from VVER-440 construction materials) were published in [24-26].

For the ideal stoichiometric Fe_3O_4 the quantity r_{AB} (ratio between A and B sub-component areas) is equal to 0.535. In the case that magnetite is the dominant (sole) phase in the sample, the deviation from the ideal value of r_{AB} is minimal (see Table 3). Significant deviations could be explained by a small degree of oxidation of magnetite, resulting in presence of vacancies or substitution by non/magnetic irons in the octahedral sub-lattice. Slight substitution of other elements (Mg, Ni, Cu, …) for Fe in the magnetite lattice is not unlikely, and this has a similar effect on the A- to B-site area ratio. Therefore, it is not feasible to conclude anything quantitatively about the degree of oxidation. Qualitatively, it can be inferred that this degree must be very low.

---

[1] Samples 1754-1757 were taken from the feed water pipelines *in situ* during the reactor shut down. Samples 1758-1790 were taken from the same steam generator from selected parts of feed water dispersion box (see Table 3 and Fig. 6, positions 1-14)

During visual inspection of removed feed water dispersion box (1998), 2 disturbing undefined metallic particles, fixed in one of outlet nozzle, were found. Both were homogenised and analysed by MS. It has been shown that these high-corroded parts ("loose parts" found in outlet nozzle of ejector) originate not from the 17247 steel but high probably from GOST 20K steel (probably some particles from the corrosion deposit from the bottom part of the steam generator moved by flow and ejection effect into the nozzle).

Fig. 6. Position of corrosion product scraps from the feed water dispersion box (SG35)

Code	Hematite		Magnetite				Base material				Doublet 1		Doublet 2	
	H1	A_{rel}	H_A	A_{rel}	H_B	A_{rel}	H4	A_{rel}	H5	A_{rel}	IS1	A_{rel}	IS2	A_{rel}
	(T)	%	(T)	%	(T)	%	(T)	%	(T)	%	(mm /s)	%	(mm /s)	%
M005			49.0	35.4	45.8	64.6								
M006			49.1	36.5	45.9	63.5								
M007	50.0	16.9	49.2	25.6	45.8	38.2	33.0	1.6					0.84	17.7
M008			49.0	35.6	45.9	64.1								
M009	51.5	13.4	49.1	32.1	45.9	54.5								
M010			49.1	36.5	45.8	63.5								
M012	51.5	12.5	49.2	31.9	46.0	55.6								
M013			48.8	25.3	45.7	40.5	33.0	30.2	30.8	4.0				
M014			49.0	9.9	45.8	13.6	33.0	66.6	30.7	9.9				
M015			48.5	6.0	45.6	8.6	33.0	73.1	30.6	12.3				
Accuracy	±0,1	±0,5	±0,1	±0,5	±0,1	±0,5	±0,1	±0,5	±0,1	±0,5	±0,1	±0,5	±0,1	±0,5

Table 4. MS parameters of corrosion products taken from the steam generator SG46[2]

---

[2] Samples m006, m008, m010 were taken from outside surface, samples M007, M009, M012 from inside surface of the feed water pipeline according to the same positions 1, 2 and 3, respectively. Sample M15 - see Fig. 7, position 7).

Mössbauer measurements on the corrosion specimens scrapped from different position of the feed water distributing system show that the outside layer consists exclusively from magnetite but the inside layer contains also hematite. Its amount decreases in successive steps towards the steam generator. The cause of this result is probably in fact that outside the system there is boiling water at the temperature of approximately 260 °C with higher salt concentrations and inside there is the feed water at the temperature up to 225 °C. Changes in the inside temperature in region (158-225 °C) can occur in dependence on the operation regime of high-pressure pumps in NPP secondary circuit.

The most corroded areas of the former feed water distributing system are the welds in the T-junction (see Fig. 7). Due to dynamic effects of the feed water flow with local dynamic overpressures of 20 to 30 kPa or local dynamic forces up to 1000 N (in the water at the pressure of about 4,4 MPa) on the inner pipe wall in the region of T-junction, the content of corrosion products was reduced and moved into whole secondary circuit. Particles of the feed water tube of SG46 base material were identified also in sediments.

Fig. 7. Position of corrosion product scraps from the feed water dispersion tube (SG46)

5.2 Results from visual inspection of heterogenic weld at SG16 from April 2002

In the period 2002-2003 we focused on the „Phase analyses of corrosion induced damage of feed water pipelines of SG 16 near the heterogenic weld". In frame of this study visual inspections as well as original "in situ" specimens scrapping was performed. Conclusions from visual inspections (performed at 19.4.2002 and 29.4.2002 at SG16) were the following:

SG16 was dried under the level of primary pipelines bundle and decontaminated. During the visual inspection of SG16 internal surface as well as hot and cold collectors (after 23 years of operation) no defects or cracks were identified. The SG16 was in excellent status with minimal thickness of corrosion layer or other deposits. For comparison to our previous experience from visual inspections from 1998, the SG16 was in better condition than SG35 or SG46 (14 and 13 years in operation, respectively). Moreover, the radiation situation after decommissioning procedures was two times better.

Visual inspection on 29.4.2002 was focused on heterogenic weld, which connects the feed water pipeline of carbon steel (GOST 20K) to a new feed water pipeline system designed from austenitic steel (CSN17248). Several samplese were taken for MS analyse from the weld as well as surrounding area in form of powder or small particles (samples description is in Table 5 and in Figs 8, 9). The heterogenic weld was well polished.

After visual inspection, the evaluation of corrosion phase composition of samples closed to heterogenic weld was performed. MS results are summarized in Table 6.

Number of samples	Samples description
2.11	Heterogenic weld
2.12	Feed water pipeline (GOST 20), 10 cm from heterogenic weld
2.13	Feed water pipeline (GOST 20), about 40 cm from heterogenic weld, just closed to the SG16 internal body surface.
2.14	Internal body surface, about 1 m under the place of feed water pipeline inlet
2.15	Internal body surface, about 50 cm over the place of feed water pipeline inlet

Table 5. Specimens description

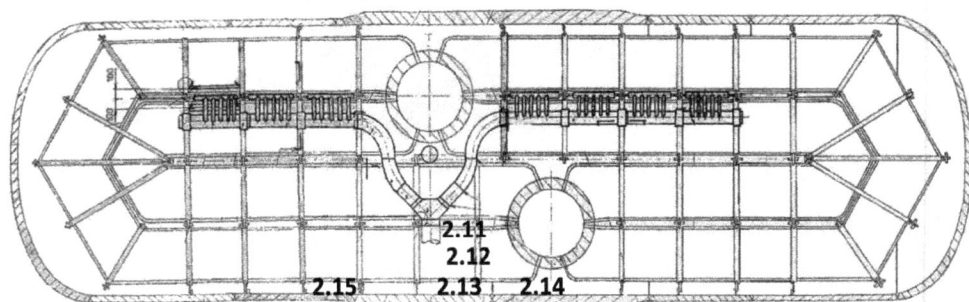

Fig. 8. SG16 with marks and description of places, where MS specimens were taken

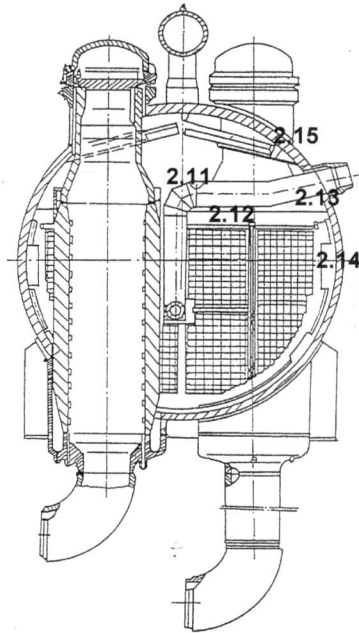

Fig. 9. SG16 cross section with indicated places where specimens were taken

Sample	Haematit				Magnetite						Mag. total	Metalic iron			Dublet/singlet		
	H_1 (T)	QS_1 (mm/s)	IS_1 (mm/s)	A_{rel1} (%)	H_2 (T)	IS_2 (mm/s)	A_{rel2} (%)	H_3 (T)	IS_3 (mm/s)	A_{rel3} (%)	(%)	H_4 (T)	IS_4 (mm/s)	A_{rel4} (%)	QS_4 (mm/s)	IS_4 (mm/s)	A_{rel4} (%)
2.11	51,9	-0,18	0,25	8,0	49,0	0,17	3,8	45,9	0,57	8,0	11,8	33,0	-0,11	12,8	-	-0,19	67,4
2.12	51,6	-0,21	0,26	75,9	49,0	0,16	9,1	45,8	0,56	14,1	23,2	-	-	-	0,40	0,21	0,9
2.13	51,6	-0,21	0,26	77,2	49,0	0,17	9,2	45,9	0,57	12,9	22,1	-	-	-	0,40	0,21	0,7
2.14	51,6	-0,21	0,26	41,1	49,0	0,16	22,0	45,8	0,55	36,9	58,9	-	-	-	-	-	-
2.15	51,8	-0,21	0,26	51,7	49,1	0,17	18,3	45,9	0,54	29,2	47,5	-	-	-	0,40	0,21	0,8
Accuracy	±0,1	±0,04	±0,04	±2	±0,1	±0,04	±0,1	±0,1	±0,04	±2	±2	±0,1	±0,04	±2	±0,04	±0,04	±2

Table 6. Mössbauer spectra parameters

Sample	Hematite				Magnetite						Spolu:	Dublet		
	H_1 (T)	QS_1 (mm/s)	IS_1 (mm/s)	A_{rel1} (%)	H_2 (T)	IS_2 (mm/s)	A_{rel2} (%)	H_3 (T)	IS_3 (mm/s)	A_{rel3} (%)	(%)	QS_1 (mm/s)	IS_1 (mm/s)	A_{rel1} (%)
2.16	51.6	-0.21	0.26	66.4	49.1	0.17	12.1	45.9	0.56	19.6	31.7	0.53	0.23	1.9
2.17	51.6	-0.21	0.26	80.8	49.1	0.16	6.6	46.0	0.55	11.5	18.1	0.47	0.21	1.1
2.18	51.6	-0.21	0.26	33.4	49.0	0.16	22.6	45.9	0.55	42.9	65.5	0.52	0.09	1.1
2.19	51.6	-0.21	0.26	40.3	49.0	0.16	20.5	45.9	0.56	38.0	58.5	0.52	0.13	1.2

Table 8. MS results of specimens taken in 2004 in Bohunice V1 from SG11.

5.3 Results from SG11 (2004)

Four powder specimens were delivered from SG11 to MS analyses. Description is shown in Table 7 and results in Table 8.

Sample	Description of origin	Date of extraction
2.16	Hot collector, HC-SG-11	15.03.04 9:00 h.
2.17	Cold collector, SC-SG-11.	15.03.04 9:00 h.
2.18	SG11 sediments	16. 03.04 10:00 h.
2.19	SG11 sediments cooler (surface of pipelines)	16. 03.04 10:00 h.

Table 7. Specimens from SG11 analysed in 2004

The dominant phase composition of the studied corrosion products taken from SG11 was hematite Fe_2O_3 (66,4% at hot collector, 80,8% at cold collector). The rest is from magnetite Fe_3O_4, presented by two sextets H_2 a H_3 with 31,7%, resp.18,1% contribution. The last component is paramagnetic doublet D_1, which is assumed to be iron hydrooxides – high probably lepidocrockite (gamma FeOOH) presented by 1,9% and 1,1%, respectively.

The magnetite presence in all samples is almost stoichiometric (see the ratio Fe^{3+}/Fe^{2+} which tends to 2,0).

A significantly lower presence of magnetite in case of hot collector can be devoted from 2 parallel factors:

1. Difference in temperature (about 298°C at HC) and (about 223°C at CC) and mostly due to
2. Higher dynamic of secondary water flowing in the vicinity of hot collector, which high probably removed the corrosion layer from the collector surface.

5.4 Period 2006-2008 – The newest measurement of corrosion products at NPP Jaslovske Bohunice

Six samples for Mössbauer effect experiments collected from different parts of NPP Bohunice unit were prepared by crushing to powder pieces (Table 9). These samples consisted of corrosion products taken from small coolant circuit of pumps (sample No. 3.1), deposits scraped from filters after filtration of SG - feed water during operation (sample No. 3.2), corrosion products taken from SG42 pipelines - low level (sample No. 3.3), mixture of corrosion products, ionex, sand taken from filter of condenser to TG 42 (sample No. 3.4), deposit from filters after refiltering 340 l of feed water of SG S3-09 during passivation 27. and 28. 5. 08 (sample No. 3.5) and finally deposit from filters after 367 l of feed water of SG S4-09 during passivation 27. and 28. 5. 08 (sample No. 3.6). All samples were measured at room temperature in transmission geometry using a $^{57}Co(Rh)$ source. Calibration was performed with α-Fe. Hyperfine parameters of the spectra including spectral area (A_{rel}), isomer shift (IS), quadrupole splitting (QS), as well as hyperfine magnetic field (B_{hf}), were refined using the CONFIT fitting software [27], the accuracy in their determination are of ±0.5 % for relative area A_{rel}, ±0.04 mm/s for Isomer Shift and Quadrupole splitting and ±0.5 T for hyperfine field correspondingly. Hyperfine parameters of identified components (hematite, magnetite, goethite, lepidocrocite, feroxyhyte) were taken from [28].

All measured spectra contained iron in magnetic and many times also in paramagnetic phases. Magnetic phases contained iron in nonstoichiometric magnetite $Fe_{3-x}M_xO_4$ where M_x are impurities and vacancies which substitute iron in octahedral (B) sites. Another magnetic fraction is hematite, $\alpha\text{-}Fe_2O_3$. In one sample also the magnetic hydroxide (goethite α-FeOOH) was identified.

Paramagnetic fractions are presented in the spectra by quadrupole doublets (QS). Their parameters are close to those of hydroxides e.g. lepidocrocite γ –FeOOH or to small, so called superparamagnetic particles of iron oxides or hydroxides with the mean diameter of about 10 nm. It should be noted that there is no problem to distinguish among different magnetically ordered phases when they are present in a well crystalline form with low degree (or without) substitution. Both the substitutions and the presence of small superparamagnetic particles make the situation more complicated [29]. In such cases, it is necessary to perform other supplementary measurements at different temperatures down to liquid nitrogen or liquid helium temperatures without and with external magnetic field [30].

Mössbauer spectrum (Fig. 10) of sample no. 3.1 (corrosion products taken from small coolant circuit of pumps) consist of three magnetically split components, where the component with hyperfine field B_{hf} = 35.8 T was identified as goethite (α-FeOOH). Hyperfine parameters of remaining two magnetically split components are assigned to A – sites and B – sites of magnetite (Fe_3O_4). One paramagnetic spectral component has appeared. According to water environment and pH [31], this component should be assigned to hydrooxide (feroxyhyte δ-FeOOH).

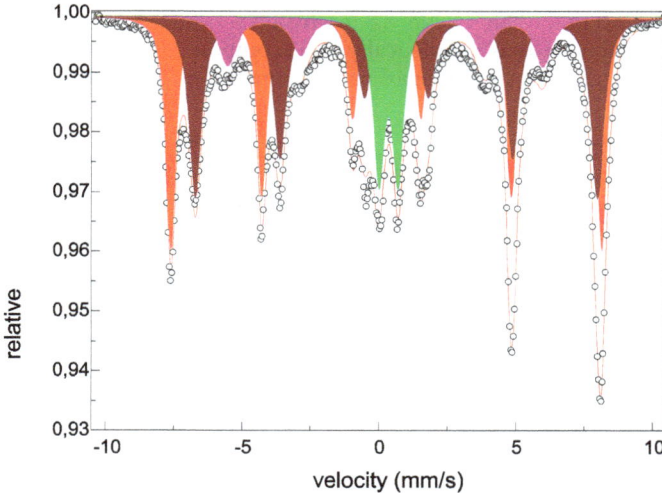

Fig. 10. Mössbauer spectrum of sample no. 3.1. A-site (red), B-site (dark red) magnetite, goethite (pink) and hydroxide (green) was identified

The sample No. 3.2 (deposits scraped from filters after filtration of SG - feed water during operation) also consists of three magnetically split components, where two of them were assigned to magnetite (Fe_3O_4) as in previous spectra, and the remaining magnetically split component was identified as hematite (α-Fe_2O_3). Paramagnetic part of the spectra was formed by one doublet, whose hyperfine parameters were assigned to hydroxide (lepidocrocite, γ-FeOOH). The spectrum is shown in Fig. 11.

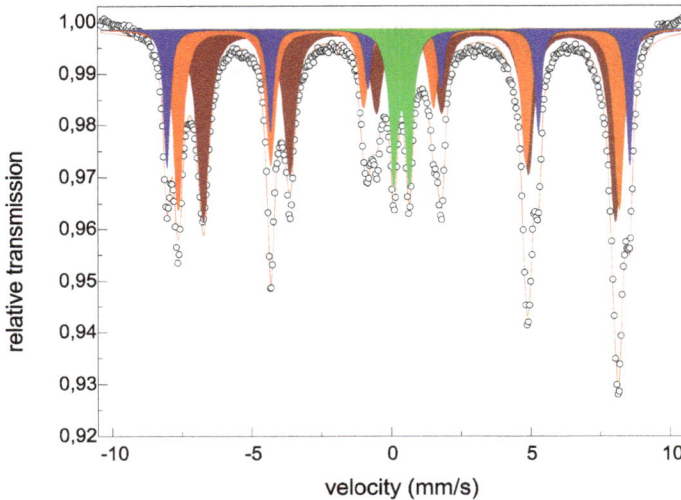

Fig. 11. Mössbauer spectrum of sample no.3. 2. A-site (red), B-site (dark red) magnetite, hematite (blue) and hydroxide (green) was identified

The spectrum (Fig. 12) of the sample No. 3.3 (corrosion products taken from SG42 pipelines - low level) consists only of two magnetically split components with hyperfine parameters assigned to A – sites and B – sites of nearly stoichiometric magnetite (Fe_3O_4) with a relative area ratio $\beta = 1.85$.

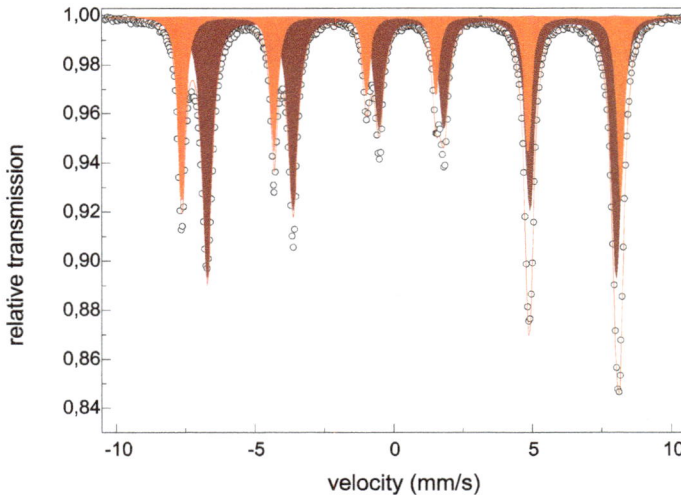

Fig. 12. Mössbauer spectrum of sample no. 3.3. A-site (red) , B-site (dark red) magnetite was identified

The sample No. 3.4 (mixture of corrosion products, ionex, sand taken from filter of condenser to TG 42) also consists of a magnetically split component which corresponds to hematite (α-Fe$_2$O$_3$) and two magnetically split components were assigned to magnetite (Fe$_3$O$_4$) as in previous spectra, and the remaining paramagnetic component was identified as hydroxide. The spectrum of the sample No. 3.4 is shown in Fig. 13.

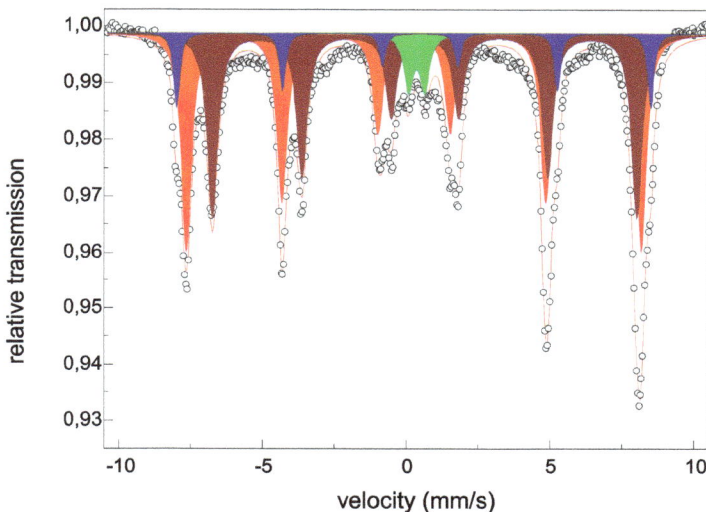

Fig. 13. Mössbauer spectrum of sample no. 3.4. Haematite (blue), A-site (red) , B-site (dark red) magnetite and hydroxide (green) was identified

Both the sample No. 3.5 (deposit from filters after 340 l of feed water of SG S3-09 during passivation 27. and 28. 5. 08) and the sample No. 3.6 (deposit from filters after 367 l of feed water of SG S4-09 during passivation 27. and 28. 5. 08) consist of three magnetically split components, identified as hematite (α-Fe$_2$O$_3$) and magnetite (Fe$_3$O$_4$) and the remaining paramagnetic component in both spectra was assigned to hydrooxide (lepidocrocite γ-FeOOH). The spectra of the samples No. 3.5 and 3.6 are shown in Figs. 14 and 15. Based on comparison of results from samples 3.5 and 3.6 it can be concluded that the longer passivation leads more to magnetite fraction (from 88% to 91%) in the corrosion products composition.

As it was mentioned, above all hydroxides could be also small superparamagnetic particles.

The refined spectral parameters of individual components including spectral area (A_{rel}), isomer shift (IS), quadrupole splitting (QS), as well as hyperfine magnetic field (B_{hf}) are listed in Table 9 for room (300 K) temperature Mössbauer effect experiments. The hyperfine parameters for identified components (hematite, magnetite, goethite, lepidocrocite, feroxyhyte) are listed in [28].

Major fraction in all samples consists of magnetically ordered iron oxides, mainly magnetite (apart from the sample No. 3.1 and 3.2, where also goethite and hematite has appeared, respectively). Magnetite crystallizes in the cubic inverse spinel structure. The oxygen ions form

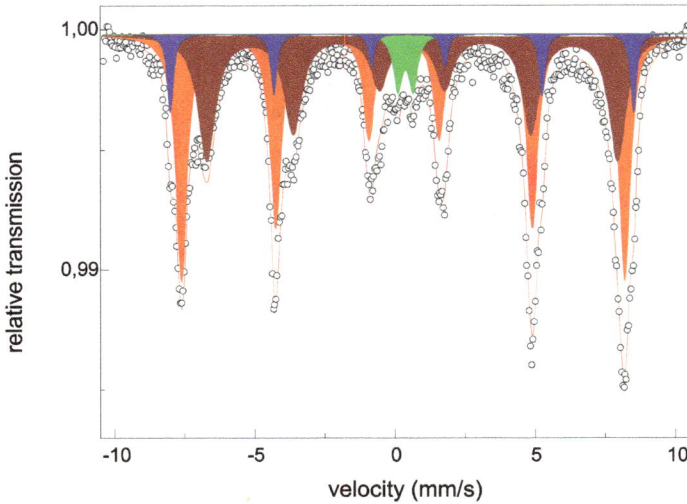

Fig. 14. Mössbauer spectrum of sample no.3.5. Hematite (blue), A-site (red) , B-site (dark red) magnetite and hydroxide (green) was identified

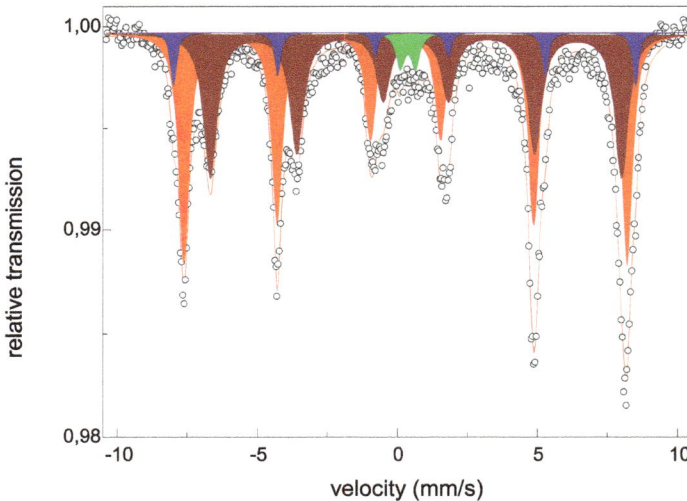

Fig. 15. Mössbauer spectrum of sample no. 3.6. Hematite (blue), A-site (red) , B-site (dark red) magnetite and hydroxide (green) was identified

a closed packed cubic structure with Fe ions localized in two different sites, octahedral and tetrahedral. The tetrahedral sites (A) are occupied by trivalent Fe ions. Tri- and divalent Fe ions occupying the octahedral sites (B) are randomly arranged at room temperature because of electron hopping. At room temperature, when the electron hopping process is fast, the Mössbauer spectrum is characterized by two sextets. The one with the hyperfine magnetic field B_{hf} = 48.8 T and the isomer shift IS = 0.27 mm/s relative to α-Fe corresponds to the Fe^{3+}

sample	Component	Area [%]	Isomer shift [mm/s]	Quadrupole shift/splitting [mm/s]	Hyperfine field [T]
Sample no. 3.1 Small coolant circuit of pumps 17. 10. 2007	magnetite A-site	36.3	0.28	0.00	48.90
	magnetite B-site	37.2	0.64	0.00	45.60
	goethite	14.4	0.36	-0.25	35.80
	hydrooxide	12.1	0.36	0.70	-
Sample no. 3.2. Deposites scraped from filters after filtration of SG - feed water during operation	hematite	15.8	0.38	-0.23	51.56
	magnetite A-site	32.6	0.28	0.00	49.14
	magnetite B-site	41.8	0.65	0.00	45.91
	hydrooxide	9.7	0.38	0.56	-
Sample no. 3.3. SG42 pipelines - low level	magnetite A-site	34.6	0.28	0.00	49.14
	magnetite B-site	65.4	0.65	0.00	45.83
Sample no. 3.4. Mixture of corrosion products, ionex, sand taken from filter of condenser to TG 42	hematite	9.2	0.38	-0.22	51.29
	magnetite A-site	45.4	0.28	0.00	49.20
	magnetite B-site	40.7	0.66	0.00	45.87
	hydrooxide	4.7	0.37	0.56	-
Sample no. 3.5. Deposit from filters after 340 l of feed water of SG S3-09 during pasivation 27. and 28. 5. 08	hematite	8.3	0.36	-0.22	51.33
	magnetite A-site	49.3	0.30	0.00	49.11
	magnetite B-site	38.5	0.61	0.00	45.51
	hydrooxide	3.9	0.37	0.55	-
Sample no. 3.6. Deposit from filters after 367 l of feed water of SG S4-09 during pasivation 27. and 28. 5. 08	hematite	6.4	0.38	-0.25	51.26
	magnetite A-site	50.3	0.29	0.00	49.14
	magnetite B-site	40.7	0.66	0.00	45.61
	hydrooxide	2.6	0.37	0.54	-

Table 9. Spectral parameters of individual components including spectral area (A_{rel}), isomer shift (IS), quadrupole splitting (QS), as well as hyperfine magnetic field (B_{hf}) for each sample with according components

ions at the tetrahedral A - sites. The second one with B_{hf} = 45.7 T and IS = 0.65 mm/s is the $Fe^{2.5+}$ - like average signal from the cations at octahedral B sites. Fe^{2+} and Fe^{3+} are indistinguishable due to fast electron transfer (electron hopping), which is faster (~1 ns) than the ^{57}Fe excited state lifetime (98 ns). The magnetite unit cell contains eight Fe^{3+} ions and eight Fe^{2+} and Fe^{3+} ions, 16 in total at the B sites, therefore, the intensity ratio β = I(B)/I(A) of the two spectral components is a sensitive measure of the stoichiometry. Assuming that the room temperature ratio of the recoil-free fractions f_B/ f_A for the B and A sites is 0.97 [32], the intensity ratio β for a perfect stoichiometry should be 1.94. In non-stoichiometric magnetite, under an excess of oxygen, cation vacancies and substitutions at the B sites are created. The vacancies screen the charge transfer and isolate the hopping process. For each vacancy, five Fe^{3+} ions in octahedral sites become trapped. In the Mössbauer spectrum these trapped Fe^{3+} ions at the octahedral sites and Fe^{3+} ions at tetrahedral sites are indistinguishable without applying an external magnetic field. Therefore, in the spectrum of non-stoichiometric magnetite, intensity transfer from the $Fe^{2.5+}$ to Fe^{3+}-like components is observed. Therefore, the intensity ratio β decreases markedly with the oxidation process, until the stoichiometry reaches the γ-Fe_2O_3 phase. It should be noted that in our samples the intensity ratio β is far from 1.94 (for perfect stoichiometry), varies from 0.97 up to 1.85.

6. Conclusions

Material degradation and corrosion are serious risks for long-term and reliable operation of NPP. The paper summarises results of long-term measurements (1984-2008) of corrosion products phase composition using Mössbauer spectroscopy.

The first period (mostly results achieved in 80-ties) was important for improving proper Mössbauer technique [5]. The benefit from this period came via experience collection, optimization of measurement condition and evaluation programs improvement. Unfortunately, the samples were not well defined and having in mind also different level of technique and evaluation procedures, it would be not serious to compare results from this period to results obtained from measurement after 1998.

The replacement of STN 12022 steel (in Russian NPP marked as GOST 20K) used in the steam generator feed water systems is necessary and very important from the operational as well as nuclear safety point of view. Steel STN 17 247 proved 5 years in operation at SG35 seems to be optimal solution of this problem. Nevertheless, periodical inspection of the feed water tubes corrosion (after 10, 15 and 20 years) was recommended.

Based on results of visual inspection performed at April 19, 2002 at SG16 (NPP V1) it was confirmed, that the steam generator was in good condition also after 23 years of operation. Samples taken from the internal body surface of PG16 confirmed that the hematite concentration increases in the vertical direction (from bottom part to the top).

The newest results from 2008 confirm good operational experiences and suitable chemical regimes (reduction environment) which results mostly in creation of magnetite (on the level 70% or higher) and small portions of hematite, goethite or hydrooxides.

Regular observation of corrosion/erosion processes is essential for keeping NPP operation on high safety level. The output from performed material analyses influences the optimisation of operating chemical regimes and it can be used in optimisation of regimes at

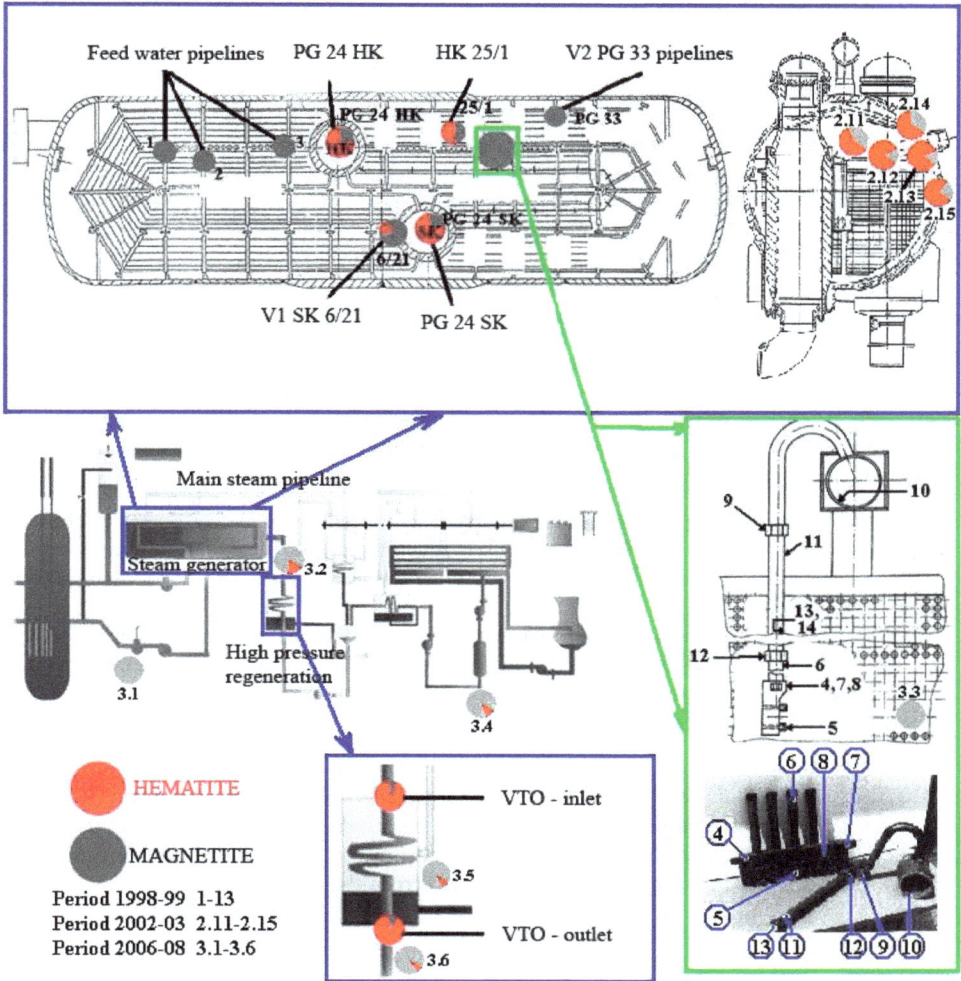

Fig. 16. Summarized figure of corrosion products phase composition at NPP V-2 Bohunice (Slovakia) performed according to results from period 1998-2008

decontamination and passivation of pipelines or secondary circuit components. It can be concluded that a longer passivation time leads more to magnetite fraction in the corrosion products composition.

Differences in hematite and magnetite content in corrosion layers taken from hot and cold collectors at SG11 in 2004 show, that there is a significantly lower presence of magnetite in case of hot collector. This fact can be derived from 2 parallel factors: (i) difference in temperature (about 298°C – HC) and (about 223°C - CC) and mostly due to (ii) higher dynamic of secondary water flowing in the vicinity of hot collector, which high probably removes the corrosion layer away from the collector surface.

With the aim to summarize our results in the form suitable for daily use in the operational conditions a summarized figure was created (see Fig. 16). Corrosion products phase composition (limited on magnetite and hematite only) is presented in form of circular diagrams.

Basically, the corrosion of new feed water pipelines system (from austenitic steel) in combination with operation regimes (as it was at SG35 since 1998) goes to magnetite. In samples taken from positions 5 to 14 (see Fig. 16 – right corner). The hematite presence is mostly on the internal surface of SG body (constructed from "carbon steel" according to GOST20K). Its concentration increases towards the top of the body and is much significant in the seam part of SG where flowing water removes the corrosion layer via erosion better than from the dry part of the internal surface or upper part of pipeline.

The long-term study of phase composition of corrosion products at VVER reactors is one of precondition to the safe operation over the projected NPP lifetime. The long-term observation of corrosion situation by Mössbauer spectroscopy is in favour of utility and is not costly. Based on the achieved results, the following points could be established as an outlook for the next period:

1. In collaboration with NPP-Bohunice experts for operation as well as for chemical regimes, several new additional samples from not studied places should be extracted and measured by Mössbauer spectroscopy with the aim to complete the existing results database.
2. Optimisation of chemical regimes (having in mind the measured phase composition of measured corrosion specimens from past) could be discussed and perhaps improved.
3. Optimisation and re-evaluation of chemical solutions used in cleaning and/or decommissioning processes during NPP operation can be considered.

In connection to the planned NPP Mochovce 3, 4 commissioning (announced officially at 3.10.2008) it is recommended that all feed water pipelines and water distribution systems in steam generators should be replaced immediately before putting in operation by new ones constructed from austenitic steels. The Bohunice design with feed water distribution boxes is highly recommended and it seems to be accepted from the utility side.

7. Acknowledgement

This work was supported by company ENEL Produzione, Pisa and by VEGA 1/0129/09.

8. References

[1] L. Cohen, in: Application of Mössbauer spectroscopy. Volume II. ed. Academic Press, (USA, New York, 1980).
[2] T.C. Gibb, Principles in Mössbauer Spectroscopy, Chapman and Hall, London, (1971)
[3] N.N. Greenwood, T.C. Gibb, Mössbauer Spectroscopy, *Chapman and Hall*, London, (1971)
[4] G. Brauer, W. Matz and Cs. Fetzer, Hyperfine Interaction 56 (1990) 1563.
[5] J. Lipka, J. Blazek, D.Majersky, M. Miglierini, M. Seberini, J. Cirak, I. Toth and R. Gröne, Hyperfine Interactions 57, (1990) 1969.

[6] W.J. Phythian and C.A. English, J. Nucl. Mater. 205 (1993) 162.
[7] G.N. Belozerski, In: Mössbauer studies of surface layers, ed. Elsevier, (North Holland, Amsterdam 1993).
[8] V. Slugen, In: Mössbauer spectroscopy in material science, ed. Kluwer Academic Publishers, Netherlands (1999) 119-130.
[9] S. Savolainen, B. Elsing, Exchange of feed water pipeline at NPP Loviisa. In: Proceedings from the 3rd seminar about horizontal steam generators, Lappeenranta, Finland, 18.-20.10.1994)
[10] Technical descripcion of SG PGV-4E, T-1e, (B-9e/241/), apríl 1978 (in Slovak)
[11] Safety report V-1, chapter IV.3 Primary circuit, Normative documentation A-01/1,2, december 1978 (in Slovak)
[12] Steamgenerator, technical report DTC 1.01.2 - 1.unit V1, Documentation to real status to 30.4.1994 (in Slovak)
[13] G. Brauer, W. Matz and Cs. Fetzer, Hyperfine Interaction 56 (1990) 1563.
[14] G.N. Belozerski, In: Mössbauer studies of surface layers, ed. Elsevier, (North Holland, Amsterdam 1993).
[15] V. Slugen, In: Mössbauer spectroscopy in material science, ed. Kluwer Academic Publishers, Netherlands (1999) 119-130.
[16] J. A. Savicki and M. E. Brett, Nucl. Instrum. Meth. In Phys. Res. B76 (1993), 254.
[17] J. Cech and P. Baumeister, In: Proc. from 2nd International Symposium on Safety and Reliability Systems of PWRs and VVERs, Brno, ed. O. Matal (Energovyskum, Brno1997) 248.
[18] O. Matal, K. Gratzl, J. Klinga, J. Tischler and M. Mihálik, In: Proc. of the 3rd International Symposium on Horizontal Steam Generators, Lappeenranta, Finland (1994).
[19] O. Matal, T. Simo, P. Sousek, In: Proc. from 3nd International Symposium on Safety and Reliability Systems of PWRs and VVERs, Brno, ed. O. Matal (Energovyskum, Brno1999).
[20] V. Slugen, D. Segers, P. de Bakker, E. DeGrave, V. Magula, T. Van Hoecke, B. Van Vayenberge, Journal of nuclear materials, 274 (1999), 273.
[21] V. Slugen, V. Magula, Nuclear engineering and design 186/3, (1998), 323.
[22] R. Ilola, V. Nadutov, M. Valo, H. Hanninen, Journal of nuclear Materials 302 (2002) 185-192
[23] E. De Grave, In: Report 96/REP/EDG/10, RUG Gent 1996.
[24] K. Varga et al., Journal of nuclear materials 348, (2006), 181-190.
[25] A. Szabo, K. Varga, Z. Nemeth, K. Rado, D. Oravetz, K.E. Mako, Z. Homonnay, E. Kuzmann, P. Tilky, J. Schunk, G. Patek, Corrosion Science, 48 (9), (2006) 2727-2749.
[26] M. Prazska, J. Rezbarik, M. Solcanyi, R. Trtilek, Czechoslovak Journal of Physics, 53, (2003) A687-A697.
[27] J. Kučera V. Veselý, T. Žák, Mössbauer Spectra Convolution Fit for Windows 98/2K, (2004), ver. 4.161.
[28] R. M. Cornell, U. Schwertmann, In: The Iron Oxides, (1996), ISBN 3-527-28576-8.
[29] J. Lipka, M. Miglierini, Journal of Electrical Engineering 45, (1994), 15-20.

[30] S. Morup, H. Topsoe, J. Lipka, Journal de Physique, 37, (1976), 287-289.
[31] V. G. Kritsky, Water Chemistry and Corrosion of Nuclear Power Plant Structural Materials, (1999), ISBN 0-89448-565-2.
[32] J. Korecki et al., Thin Solid Films 412, (2002) 14-23.

Permissions

The contributors of this book come from diverse backgrounds, making this book a truly international effort. This book will bring forth new frontiers with its revolutionizing research information and detailed analysis of the nascent developments around the world.

We would like to thank Dr. Soon Heung Chang, for lending his expertise to make the book truly unique. He has played a crucial role in the development of this book. Without his invaluable contribution this book wouldn't have been possible. He has made vital efforts to compile up to date information on the varied aspects of this subject to make this book a valuable addition to the collection of many professionals and students.

This book was conceptualized with the vision of imparting up-to-date information and advanced data in this field. To ensure the same, a matchless editorial board was set up. Every individual on the board went through rigorous rounds of assessment to prove their worth. After which they invested a large part of their time researching and compiling the most relevant data for our readers. Conferences and sessions were held from time to time between the editorial board and the contributing authors to present the data in the most comprehensible form. The editorial team has worked tirelessly to provide valuable and valid information to help people across the globe.

Every chapter published in this book has been scrutinized by our experts. Their significance has been extensively debated. The topics covered herein carry significant findings which will fuel the growth of the discipline. They may even be implemented as practical applications or may be referred to as a beginning point for another development. Chapters in this book were first published by InTech; hereby published with permission under the Creative Commons Attribution License or equivalent.

The editorial board has been involved in producing this book since its inception. They have spent rigorous hours researching and exploring the diverse topics which have resulted in the successful publishing of this book. They have passed on their knowledge of decades through this book. To expedite this challenging task, the publisher supported the team at every step. A small team of assistant editors was also appointed to further simplify the editing procedure and attain best results for the readers.

Our editorial team has been hand-picked from every corner of the world. Their multi-ethnicity adds dynamic inputs to the discussions which result in innovative outcomes. These outcomes are then further discussed with the researchers and contributors who give their valuable feedback and opinion regarding the same. The feedback is then collaborated with the researches and they are edited in a comprehensive manner to aid the understanding of the subject.

Apart from the editorial board, the designing team has also invested a significant amount of their time in understanding the subject and creating the most relevant covers. They scrutinized every image to scout for the most suitable representation of the subject and create an appropriate cover for the book.

The publishing team has been involved in this book since its early stages. They were actively engaged in every process, be it collecting the data, connecting with the contributors or procuring relevant information. The team has been an ardent support to the editorial, designing and production team. Their endless efforts to recruit the best for this project, has resulted in the accomplishment of this book. They are a veteran in the field of academics and their pool of knowledge is as vast as their experience in printing. Their expertise and guidance has proved useful at every step. Their uncompromising quality standards have made this book an exceptional effort. Their encouragement from time to time has been an inspiration for everyone.

The publisher and the editorial board hope that this book will prove to be a valuable piece of knowledge for researchers, students, practitioners and scholars across the globe.

List of Contributors

Branko Kontić
Jožef Stefan Institute, Slovenia

Lu Guangyao, Ren Junsheng, Huang Wenyou, Xiang Wenyuan, Zhang Chengang and Lv Yonghong
China Guangdong Nuclear Power Holding Co. Ltd., China

Robertas Alzbutas
Lithuanian Energy Institute, Lithuania
Kaunas University of Technology, Lithuania

Andrea Maioli
Westinghouse Electric Company, USA

Egidijus Norvaisa
Lithuanian Energy Institute, Lithuania

Eugenijus Uspuras and Algirdas Kaliatka
Lithuanian Energy Institute, Lithuania

Daniela Hossu, Ioana Făgărăşan, Andrei Hossu and Sergiu Stelian Iliescu
University Politehnica of Bucharest, Faculty of Control and Computers, Romania

Do-Young Ko
Central Research Institute, Korea Hydro & Nuclear Power Co. Ltd., Republic of Korea

Zaffar Muhammad Khan, Shahab Khushnood , Muhammad Akram Javaid, Luqman Ahmad Nizam, Khawaja Sajid Bashir and Syed Zahid Hussain
University of Engineering & Technology, Taxila, Pakistan

Muhammad Afzaal Malik and Zafarullah Koreshi
Air University, Islamabad Pakistan

Mahmood Anwer Khan
College of Electrical & Mechanical Engineering NUST, Rawalpindi, Pakistan

Arshad Hussain Qureshi
University of Engineering & Technology, Lahore, Pakistan

Aluísio Sousa Reis, Júnior, Eliane S. C. Temba, Geraldo F. Kastner and Roberto P. G. Monteiro
Centro de Desenvolvimento da Tecnologia Nuclear – (CDTN), Brazil

Rashydov Namik, Kliuchnikov Olexander, Seniuk Olga, Gorovyy Leontiy, Zhidkov Alexander, Ribalka Valeriy, Berezhna Valentyna, Bilko Nadiya, Sakada Volodimir, Bilko Denis, Borbuliak Irina, Kovalev Vasiliy, Krul Mikola and Georgy Petelin
Institute Cell Biology & Genetic Engineering of NAS of Ukraine, Ukraine
Institute for Safety Problems of Nuclear Power Plants NAS of Ukraine, Ukraine
National University "Kyievo-Mogiljanskaja Academy", Ukraine

Fulvio Mascari and Giuseppe Vella
Department of Energy, The University of Palermo, Italy

Brian G. Woods
Department of Nuclear Engineering and Radiation Health Physics, Oregon State University, USA

Kent Welter
NuScale Power Inc., USA

Francesco D'Auria
San Piero a Grado - Nuclear Research Group (SPGNRG), University of Pisa, Italy

Jürgen Rudolph, Steffen Bergholz, Benedikt Heinz and Benoit Jouan
AREVA NP GmbH, Erlangen, Germany

V. Slugen, J. Lipka, J. Dekan, J. Degmova and I. Toth
Institute of Nuclear and Physical Engineerining, Slovak University of Technology Bratislava, Bratislava, Slovakia